Systematics of Ornamental Crops

NIPA® GENX ELECTRONIC RESOURCES & SOLUTIONS P. LTD.
New Delhi-110 034

About the Authors

Dr. Manjunath S. Patil has completed his bachelor degree in Horticulture with distinction from the Mahatma Phule Krishi Vidyapeeth, Rahuri in 2002. He completed M.Sc. in Horticulture obtaining first rank with gold medal from the University of Agricultural Sciences, Dharwad in 2004. He completed his Ph.D. in Horticulture in 2008 with first class from the University of Agricultural Sciences, Bangalore. For his Ph.D. dissertation he worked on Breeding and development of molecular markers linked to bacterial blight resistance in anthurium (*Anthurium andreanum* Lind.). He was awarded with University merit scholarship for Master's and Devki Devi Ahuja award Doctoral degree programmes. Initially he worked as Senior Research Fellow at the Indian Institute of Horticultural Research, Hessaraghatta, Bengaluru in a CSIR funded project on Development of molecular markers for bacterial blight resistance. He started his carrier as Assistant Professor of Horticulture at College of Agriculture, Pune, MPKV, Rahuri, Maharashtra, from 2009 to 2017. Currently he is working as Assistant professor of Horticulture at ZARS, Solapur in the same university. During the period he taught courses like Plant Propagation and Nursery Management, Commercial Floriculture, Ornamental Horticulture, Landscape Gardening, Breeding of flower crops, Commercial floriculture, Value addition in flowers, Breeding of fruit and plantation crops, for Under Graduate and Post Graduate students in Horticulture. He also guided 15 students as Chairman or member of Advisory Committee for post graduate degree programmes. So far he has published more than 35 research articles apart from co-authoring a text books on Strategies for *In vitro* conservation of Gladiolus published by Lambert Academic Press, Germany, edited a chapter on Biodiversity of Anthurium, Daya Publishing House, and Essence of Horticulture and Flower Breeding and Genetics, New India Publishing Agency, New Delhi.

Dr. Anil. R. Karale was born in 1954 at Satara District of Maharashtra State in India. He completed his bachelor degree in Agriculture with first class from the Mahatma Phule Krishi Vidyapeeth, Rahuri in 1975. He completed Master Degree in Horticulture obtaining first class in 1977 and doctoral degree in Horticulture in 1997 from the same university. He acted as Co-course Director, CAS trainings at Rahuri. He was superannuated as Associate Dean, College of Agriculture, Pune, Mahatma Phule Krishi Vidyapeeth, Rahuri, Maharashtra, in 2014. During this period he taught courses like Breeding of fruit crops, Plant Propagation and Nursery Management, Commercial Floriculture, protected floriculture, etc. for Under Graduate and Post Graduate and Ph.D. students in Horticulture. He was associated with the development of pomegranate varieties like G-137, P-26, Mridula, Phule

Arakata, Bhagwa. So far he has handled more than 3 research projects on various aspects of Horticulture and published more than 60 research articles. He acted as PI of ICAR *Ad-hoc* Project Improvement of Tamarind by selection and Niche Area of Excellence, High-Tech Floriculture Project. He also guided 24 students as Chairman or member of Advisory Committee for post graduate and Ph.D. degree programmes. So far he has published more than 140 popular articles, apart from he had written 8 books, 44 book chapters, 18 radio talks, etc.

Dr. B. Jayanna Naik was a subject expert in plant biotechnology. He was born in 1984 in Bommaganipalli village, Bramhasamudram Mandal, Anantapuramu District, Andhra Pradesh, India. He completed his M.Sc. in the Department of Biotechnology from Sri Krishnadevaraya University in the year of 2007. He obtained his Ph. D from the Acharya Nagarjuna University, Guntur in the Department of Biotechnology in the year 2016. He did his post-doctoral research at the National Institute of Horticultural and Herbal Science, Jeju, South Korea. Later he was joined as an Assistant Professor in the Department of Molecular Biology, Central University of Andhra Pradesh, Ananthapuramu, Andhra Pradesh, India.

Dr. Venkatesh S.C. is working as Associate Professor of Crop Improvement and Biotechnology and presently working at Regional Horticulture Research and Extention Centre, Bengaluru, University of Horticultural Sciences Bagalkot. Having 27 years of research experience in the field of Molecular Biology, Biotechnology and Molecular breeding and teaching experience of 15 years. Taught courses in the field of Biotechnology, Biochemistry, Crop Physiology, Breeding for UG and PG students. Guided 2 M.Sc. students of Molecular Biology & Biotechnology department and one M.Sc. student of Plantation Spices, Medicinal & Aromatic crops department as a major advisor. Served as a member for more than 15 M.Sc. students. Published more than 25 research articles in the peer reviewed International and National Journals. Visited John Innes Centre Norwich, UK and Tanzania. Undergone several international trainings. Part of the Ph.D. work was conducted at John Innes Centre Norwich, UK.

Systematics of Ornamental Crops

M.S. Patil
Assistant Professor of Horticulture
Zonal Agriculture Research Station, Solapur
Mahatma Phule Krishi Vidyapeeth
Rahuri, Maharashtra (India)

A.R. Karale
Ex. Associate Dean
College of Horticulture, Pune
Mahatma Phule Krishi Vidyapeeth
Rahuri, Maharashtra (India)

B. Jayanna Naik
Assistant Professor
Department of Molecular Biology
Central University of Andhra Pradesh
Ananthapuramu, Andhra Pradesh (India)

Venkatesha S.C.
Associate Professor
Plant Biotechnology
Department of Biotechnology and Crop Improvement
College of Horticulture
Mysuru University of Horticultural Sciences
Bagalkote, Karnataka (India)

NIPA® GENX ELECTRONIC RESOURCES & SOLUTIONS P. LTD.
New Delhi-110 034

NIPA® GENX ELECTRONIC RESOURCES & SOLUTIONS P. LTD.
101,103, Vikas Surya Plaza, CU Block
L.S.C.Market, Pitam Pura, New Delhi-110 034
Ph : +91 11 4386 0225, 9717133558, 9540816132
E-mail: newindiapublishingagency@gmail.com
Website: www.niparesources.com

Print ISBN: 978-93-58877-37-3

ebook ISBN: 978-93-58878-65-3

Composed and Designed by NIPA®.

भा.कृ.अनु.प
ICAR
डॉ. जे. दिनकर अडिग
निदेशक
Dr. J. Dinakara Adiga
Director

ಭಾ.ಕೃ.ಸಂ.ಪ - ಗೇರು ಸಂಶೋಧನಾ ನಿರ್ದೇಶನಾಲಯ
भा.कृ.अनु.प - काजू अनुसंधान निदेशालय
ICAR-DIRECTORATE OF CASHEW RESEARCH
(Department of Agricultural Research and Education,
Ministry of Agriculture and Farmers' Welfare, Govt. of India)
P. O. Darbe, Puttur - 574 202, Dakshina Kannada, Karnataka
(An ISO 9001:2008 Certified Institution)

काअनि
DCR

Foreword

People have an ancient and strong bond to ornamental crops, which are known to have a positive effect on the human mind. Ornamental crops are plants that are grown primarily for their aesthetic value and are used to enhance the beauty of gardens, landscapes, indoor spaces, and various other settings. Ornamental crops can include flowering plants, foliage plants, trees, shrubs, and even certain types of grasses. These crops play a dynamic role in augmenting the beauty of our surroundings, reinforcing environmental sustainability, contributing to the economy, enriching cultural practices, promoting social well-being, and offering several health benefits. Their importance extends beyond mere decoration and embraces a wide range of positive impacts on individuals and society.

Systematics is a branch of biological science that studies the distinctive characteristics of species and how they are related to other species through time. Thus, it is the basis used to understand the evolution of life. Plant systematics deals with interrelation among plants and their evolutionary descent. It allows for the identification of unspecified species and their classification by assessing them to known species level. This systematics method can be used to analyze genetic constituents. It is used to scientifically name any species, which sustains in name consistently across the globe and averts confusion.

For thorough systematics of ornamental crops, knowledge, primarily those of taxonomy, morphology, nomenclature and classification is very essential. In a good reading material, it is essential to use information of the associated sciences with appropriate explanations and applications. In this context, there is need for assemblage of systematics of ornamental crops in the form of book. The present book titled "Systematics of Ornamental Crops" by Dr. M.S. Patil, Dr. A. R. Kalale, Dr. B. Jayanna Naik and Dr. Venkatesha S.C is a commendable effort. This book provides fundamental knowledge to the students, researchers and other stakeholders in the field of systematics of ornamental crops.

I congratulate the authors for bringing out this book.

Dated: 19.07.2024

Dr. J. Dinakara Adiga
Director

Preface

Ornamental plant systematics has lacked an appropriate text, especially at the undergraduate and postgraduate level. Currently, students search various voluminous reference books for basic and advanced studies. The aim of this book is to provide a general overview of ornamental plant taxonomy in an effective manner- a simplified but factual account.

There are about 2,45,000 species of flowering plants on earth and these have been grouped into 197-430 families, depending on the authority concerned. However, a basic knowledge of angiosperms can be acquired by a study of selected group of families. We have selected 33 families of flowering plants, prominent in India.

The subject matter is divided into 55 chapters which are by and large independent so that the teacher may choose the desired sequence of topics. Chapters 1-3 discuss the traditional aspects of plant taxonomy including the history and classification. Chapters 4-6 discuss the methods, techniques and principles of plant collection and plant nomenclature. Chapter 7, 8, 9 deal with chemotaxonomy, palynotaxonomy and serotaxonomy. Chapter 10, 11, 12, 13 and 14 expand on botanical libraries, biodiversity, botanical gardens, red data book and herbarium. Floral formulae and floral diagrams, from which it is generally possible to identify a family, are explained in Chapter 15. Description of important families is mentioned from Chapters 20 to 54.

This book is meant to meet the needs of students preparing for two year postgraduate degree program of Indian universities, we are also hopeful that it will benefit teachers and professionals, too. With this end in view, an attempt has been made to cover as far as possible the syllabus of the course. We are thankful to Dr. M.N. Jagtap, Head, Department of Botany, DBF, Dayanand College of Arts and Science, Solapur and Mr. A.B. Deshmukh, Librarian, College of Agriculture, Pune. The authors had tried to present the subject-matter in an easily understandable way without sacrificing the essential details and general principles and yet avoiding redundant matter and unnecessary complications. The book, as it stands, is expected to meet adequately the needs of the students for whom it is meant. We welcome comments, and suggestions from students and fellow teachers.

Authors

Contents

1

Introduction

Ornamental horticulture is the branch of horticulture concerned with cultivating plants of all kinds for show and to satisfy the eye rather than for food (Bailey, 1960). It includes floriculture, the culture of trees for shades, display and landscape gardening. Gill (1988) defined taxonomy as a study aimed at producing a system of classification of biological organisms, which best reflects the totality of their similarities and differences. According to him, there are three inter-related aspects of taxonomy namely: Identification, Nomenclature and Classification. Identification deals with the determination of a taxon as being identical or similar to another and already known organisms. Nomenclature on the other hand has to do with determination of the correct name for the organisms and is governed by International Code of Botanical Nomenclature. It permits only a single valid name for each plant. Classification is the placing of plants in groups or categories according to particular plan or system.

The objectives of taxonomy are to provide: a convenient method of identifying, naming and describing plants, a classification which as far as possible express the natural relationships of organisms, detect evolution at work, discovering its processes and interpreting its results, an inventory of plant resources-local, regional and continental and to provide an integrating and unifying role in the training of biology.

1.1 Taxonomy and Systematics

The science of the classification of organisms according to their resemblances and differences is called as taxonomy. According to them the word 'systematics' is often used synonymously with taxonomy, but sometimes interpreted more widely to include also the identification, practice of classification and nomenclature.

Jones and Luchsinger (1987) believe that although there is no agreement among the botanists for the distinctions between 'systematics and taxonomy', some botanists treat them as two separate branches. According to these botanists, systematics is the 'study of the diversity of plants and their identification, naming, classification and evolution' while taxonomy is 'restricted to the study of classification'.

Solbrig (1966) maintained that the two fields of enquiry (viz. taxonomy and systematics) have absolutely different concern. But Heywood (1967) and Morse (1974) treated systematic as a field covering the scientific study of the diversity and differentiation of organisms and the relationships that exist between them and consider taxonomy as a part of systematics. Stace (1980) also considered these two as separate disciplines in his book entitled 'Plant Taxonomy and Biosytematics'.

However, the terms taxonomy and systematics have been so loosely and interchangeably used in the past that to establish a proper delineation between the two is extremely difficult. In actual practice, and also in the present text, the two terms are used synonymously and deal with the study of classification, its principles, procedures and rules. Lam (1959) used them as synonyms.

The Greek meaning of taxonomy is "arrangement by rules" and of systematics is "to put together". The term taxonomy was first coined and used by the famous French botanists A.P. de Candolle in 1813 in his book 'Theorie Elementaire de la Botanique'.

1.2 Systematics and the Systematist

1.2.1 Systematics

Radford (1986) defined systematics as 'the study of phenotypic, generic and phylogenetic relationship among taxa', and mentioned that the science of systematics possesses following essential qualities.

From the fundamental point of view systematics is the study of the nature, causes, patterns and trends in variation among taxa.

From the structural point of view systematics is the study of the basic taxonomic components and evolution.

From the functional point of view systematics is the study of characters from many fields of evidence for establishing relationships among taxa.

From the developmental point of view systematics is the study of variation among taxa for the determination of character correlations and relationships.

From the theoretical point of view systematics is based on the idea that because of great diversity in the biological world there exist some 'discontinuous units' that can be identified, classified, described and named on the basis of evolution.

From the philosophical point of view systematics is an ever-evolving and unending field for understanding taxonomic and evolutionary processes, principles and concepts.

1.2.2 Systematist

The systematist is a student, researcher and scholar who studies, classifies, identifies, describes, names, observes, synthesizes or analyses the variations within the population, species and higher taxa.

From the functional standpoint, a taxonomist has the following activities or duties to perform:

1. As a classifier, a taxonomist determines the position and rank for new taxa.
2. As a identifier, a taxonomist distinguishes or identifies new taxa and establishes the diagnostic character for old taxa.
3. As a describer, taxonomist determines the circumscription for a specimen according to an established system of classification.
4. As a nomenclaturist, a taxonomist assigns names to new taxa and determines the correct name for old taxa according to the rules laid down by the latest International Code of Botanical Nomenclature.
5. As a coiner, a taxonomist selects the proper Greek or Latin prefixes or suffixes for scientific names.
6. As an observer, a taxonomist observes the characteristics of organisms, population and taxa for a better understanding of their taxonomy.
7. As an analyzer, a taxonomist decides which characters of the individual, population or taxa are to be analyzed.
8. As a synthesizer, a taxonomist collects information from various aspects (morphology, cytology, anatomy, etc.) of the organisms for the development of new systems of classification and identification.
9. As a theorizer, a taxonomist develops new concepts on the basis of recent findings and reinterprets old concepts.
10. As a historian, a taxonomist provides better understanding of the old classical concepts in the light of the new researchers in the field.
11. As a generator, a taxonomist generates effective methods for collection, analysis and presentation of data.
12. As a conserver, a taxonomist suggests new approaches for the preservation of threatened and endangered species.

1.2.3 Human Being as A Taxonomist

Every human being always behaves as a taxonomist. It is so because each one of us identifies, classifies, describes and names while taking decisions about the food we eat, the beverages we drink, the homes we purchase, the clothes we wear, the friends we like, the games we play, the religion we believe in, and the politics we exercise. While taking about sex, sickness, sin and sorrow we talk differently to children, old people, specialists and parents and thus behave as taxonomists. We describe that road is long, grass is green and ball is round, and thus work as a taxonomist.

1.3 Objectives, Goals and Aims of Plant Systematics

1.3.1 Objectives

- To prepare a scheme of classification that provides phonetic, natural or Phylogenetic relationships among plants.
- To establish a suitable method for identification, nomenclature and description of plant taxa.
- To provide an inventory of plant taxa that suits local, regional and continental needs.
- To create an understanding of the evolutionary processes.
- To train the students of plant sciences in regard to the diversity of organisms and their relationship with other biological branches.

1.3.2 Goals

- To acquire the fundamental values of plant systematics.
- To know about the basic concepts and principles of plant systematics.
- To be aware of the importance of taxonomic relationships in plant systematics.
- To develop the knowledge of applicability of plant systematic studies.

1.3.3 Aims for the Study of Regional Flora

- To know how to collect specimens.
- To know how to prepare specimens for future preservation.
- To know how a manual should be used.
- To know how to use identification keys.
- To recognize divisions, classes, families, genera and species.
- To know how the plants are described.

- To know how the diversity in species may be related with the regional habitat diversity.
- To become familiar with the basic taxonomic principles, and with at least one system of plant classification.

1.4 Hierarchical Stages of A Systematist

The following may be the hierarchical grades or stages of a systematist.

A natural scientist is a systematist who provides or gathers information about the relationships between phonetic, genetic, and ecologic variations among the organisms.

A systematist is a taxonomist who develops a profound knowledge of the specific aspects of systematics.

A taxonomist is a classifier who names, describes, classifies and identifies the organisms.

A classifier is a student who establishes the named, delimited and identified taxa for establishing the diversity among organisms.

A student of diversity among organism is a classifier, taxonomist, systematist and natural scientist who provides information to the society about the phenetic, genetic, and ecologic variations among the organisms for the upliftment of mankind.

1.5 Hierarchical Categories of Taxonomy

The International Code of Botanical Nomenclature (Voss *et al.*, 1983) has recognized the following categories or ranks of taxa along with their endings, as mentioned in Table 1.1

Table 1.1: Series of ranks and endings provided by the International Code of Botanical Nomenclature (1983)

Ranks of taxa	Endings of ranks above genus	Examples
Division	phyta	Pterophyta, Magnoliophyta
- Subdivision	phytina	Pterophytina
Class	opsida	Pteropsida, Magnoliopsida
- Subclass	opsidae	Pteropsidae
Order	ales	Rosales, Asterales
- Suborder	ineae	Rosineae
Family	aceae	Rosaceae, Asteraceae
- Subfamily	oideae	Rosoideae
- Tribe	eae	Roseae
- Subtribe	inae	Rosinae

Ranks of taxa	Endings of ranks above genus	Examples
Genus	us, a, um, on, es	*Rosa, Pinus*
- Subgenus	etc.	
- Section		
- Subsection		
- Series		
- Subseries		
Species		
- Subspecies		
- Variety		
- Subvariety		
- Form		
- Subform		

1.6 Basic Components of Taxonomy

Classification, identification, description and nomenclature are the four basic components of taxonomy.

Classification is the arrangement of botanical groups with definite circumscriptions by position and rank according to artificial criteria, phenetic similarities or Phylogenetic relationships.

Identification is the determination of similarities or dissimilarities between the two elements. Under identification we make a direct comparison of the characteristic features of a specimen with those present in the already existing keys for identification.

Description is the orderly recording of maximum possible characters of a taxon, individual plant, plant part, or object.

Nomenclature is a simple system under which the individual taxonomic groups of plants are scientifically named.

1.7 Future of Plant Taxonomy

The evidence gathered from written records indicates that man has been categorizing or classifying organisms for more than 2000 years. Incorporation of new researches, evidences, methods and techniques have developed the science of systematics to its present status. But a lot more is still to be done. Introduction of mathematical data, statistical details and the use of computers in the recent past have brought tremendous changes in the field of taxonomy. The use of electron microscopy in several cases has entirely changed the old concepts. Chromatography and spectrochemistry have also brought unexpected changes in several existing ideas.

In the future, several new aspects in the field of systematics, will be explored. Some of them are undermentioned.

1. According to Stace (1980) about 300,000 species of green plants, over 1,50,000 fungi and a few thousand bacteria are known to the biologists. But in future several thousand of new species of plants still await discovery, description and naming in different parts of the world.
2. Monographs of many plant groups are to be prepared.
3. Many groups, treated more than a century ago, badly need a revision in the light of new discoveries and techniques.
4. Explorations of several unknown floristic regions are highly essential.
5. Several old floras are to be revised.
6. Speciation understanding on several groups is yet to be done.
7. Many old basic fields and taxonomic principles need re-evaluation.
8. Biology of majority of the plant species has not been studied so far.
9. Several mathematical and statistical schemes for the determination of relationships have been developed. These schemes are yet to be used in many cases. More mathematical logic is to be used in taxonomic investigations.
10. Several 'anomalous groups' are still waiting for their proper placing in the systems of classification.
11. In future, taxonomists will have to use the expertise and knowledge of genetic diversity. This would help in the introduction of new crops and also in the improvement of old crops.
12. New and improved methods of studying nucleic acids will have to be used by the taxonomists in future for the basic understanding of the relationships among the groups of organisms.

References

Bailey, L.H. 1960. The Standard Encyclopedia of Horticulture. The Macmillian Company, New York. pp. 2407.

Gill, L.S. 1988. Taxonomy of flowering plants. Africana-Fep publishers Limited, 116-119.

Heywood, V.H. 1967. Plant taxonomy, Edward and Arnold Publishers, London.

Jones, S.B. and Luchsinger, A.E. 1987. Plant systematics (2nd ed.), McGraw-Hill, New York.

Lam, H.J. 1959. Taxonomy: General principles and angiosperms, '*Vistas in Botany*' **2:** 4-75.

Morse, L.E. 1974. Computer assisted storage and retrieval of the data of taxonomy and systematics, *Taxon*, **23**: 29-73.

Radford, A.E. 1986. Fundamentals of plant systematics, Harper and Row, New York.

Solbrig, O.T. 1966. Evolution and Systematics. Current Concepts in Biology Series-New York: The Macmillan Company.

Stace, C.A. 1980. Plant taxonomy and biosystematics, University Park Press, Baltimore.
Voss, E.G. 1983. International Code of Botanical Nomenclature, Bohn, Schelterma and Holkema, Utrecht, The Netherlands.

2

Origin of Ornamental Crops

The concept of Centers of Origin given by Russian scientist N.I. Vavilov (1935) is based on his studies of vast collection of plants at the Institute of Plant Industry, Leningrad. According to him, crop plants evolved from wild species in the areas showing great diversity and termed them as primary Centers of Origin; but in some areas, certain crop species show considerable diversity of forms although they did not originate their, such areas are known as secondary center of origin of these species. He identified 8 primary centers (Chinese, Indian, Inner Asiatic, Asia Minor, Mediterranean, Abyssinian, South Mexico, Central America and South America Andes Center) and three sub-centers (Indo-Malayan, Chilean and Brazilian-Paraguayan sub-center) of origin for crop plants. Basilevskaya (1960) first investigated the geographical origin of garden plants. Coats (1968) who first stressed out the ornamental herbs whereas Hui-Lin-Li (1974) wrote about the introduction of woody plants. According to Tay (2007) survey of thirty common flowers and their related species show a similar pattern of concentrated occurrence of species in certain geographic regions.

Exploration and exploitation of gene centers need to be proceed with the systematic cooperation of taxonomists, geneticists and plant pathologists. The 17 centers of origin identified for different flower and ornamental crops are given (Table 2.1)

Table 2.1: Origin of flower and ornamental crops (Collected from published sources)

Centre of origin	Flower crops
North American region	White acacia (2n=26), Anchusa (2n=18), Coreopsis (2n=18, 20), Evening primrose (*Oenothera* sp. 2n=14, 22, 28), Gaillardia (*Gaillardia pulchella*, 2n=36), Phlox (*Phlox drummondi*, 2n=28), Perennial lupin (*Lupinus polyphyllus*, 2n=48), Purple coneflower (*Echinacea purpurea*, 2n=11, 22), Cone flower (*Rudbeckia hybrida*, 2n=38), Tuberose (2n=60)
Meso-American and Caribbean region	Floss flower (*Ageratum houstonianum*, 2n=20), Begonia (*Begonia imperialis*, 2n=26, 27, 28), Cacti (Opuntia, Echinocereus, Mammilaria, 2n=22, 24), *Cosmos bipinnatus* (2n= 24, 48), *Dahlia variabilis*, 2n= 32,

Centre of origin	Flower crops
	64), Globe amaranth (*Gomphrena haageana*, 2n=32), Marigold (*Tagetes erecta, T. patula, T. tenuifolia*, 2n=24, 48), Orchids (*Odotonglossum, Oncidum, Vanilla*), *Zinnia elegans* (2n=24).
South western South American region including Chile & Peru west of Andes	*Amaranthus caudatus* (2n=32), *Anthurium* sp. (2n=30, 60, 90), *Bougainvillea* sp. (2n=34, 51), *Calendula officinalis* (2n=32), *Caladium hybridum* (2n=28), *Dieffenbachia* (2n=34), Gloxinia (2n=26), *Salvia splendens* (2n=12, 54), *Heliotropium arborescens* (2n=14, 20, 22, 26, 28), *Petunia hybrida* (2n=14, 56), *Philodendron* (2n= 32, 34, 36), Morning glory (*Ipomoea alba, I. purpurea, I. tricolor*, 2n=30), *Verbena* hybrids (2n=10, 15).
Northern south American region including Columbia, Venezuela, Guyana, Surinam, French Guyana, northern Brazil & part of northern Peru	*Cuphea* (2n=34), *Xanthosoma* (2n=26).
South-eastern South American region including southern Brazil, Bolivia, Paraguay, Uruguay and Argentina	*Salvia* (2n=14-60).
European region including Scandinavia	*Adonis vernalis,* Common daisy (*Bellis perennis*, 2n= 18), Horn violet (*Viola corunata,* 2n= 12, 14, sweet violet (*V. odorata*, 2n=54).
North African & Mediterranean region	African violet (*Saintpaulea ionantha*, 2n=30), *Sansevieria trifasciata* (2n=20), Corn flower (2n=20), Cyclamen (2n= 34, 51, 68), Daffodil (2n= 14, 28), Hyacinth (2n= 20), Iris (2n= 40, 80), Oleander (2n= 22), Pansy (2n= 22, 26), Peony (2n= 10, 20), Stock (2n= 14), Wind flower.
Balkan and Anatolia region including Central Asia (Afghanistan, Turkistan)	*Tulipa kaufinanniana, T. fosteriana, T. greigii* (2n= 24, 60), Water lily (2n= 28, 56).
East and central African region	Bamboo
South Africa	Belladonna lily (*Amaryllis belladonna*, 2n=22, 33), Bird of paradise (*Strelitzia reginae*, 2n=22), Cape primrose (*Sreptocarpus hybrida*, 2n= 30, 32), Crassula (2n=14), Gasteria (2n=14), Gazania (2n=10, 12, 14, 16), Geraniums (Pelargonium, 2n=16, 88), Giant summer hyacinth (*Galtonia candicans*, 2n=16), Gladiolus (2n=30, 120), *Haworthia* (2n=14), Protea (2n= 24).

Centre of origin	Flower crops
Far eastern region including China, Japan	China aster (*Callistephus chinensis*, 2n=18), China wisteria (*Wisteria sinensis*, 2n=14, 16), *Chrysanthemum* sp. (2n=18, 72), Day lily (*Hemerocallis*, 2n=22, 33), Royal lily (*Lilium regale*, 2n= 24), *Peony lactiflora* (2n=10, 20).
Southern China to eastern foot hills of Himalayan regions including northern Burma, Thailand & Vietnam	Aglaonema (2n=34), Balsam (*Impatiens balsamina* 2n=14,16,20), *Begonia rex* (2n=16, 156), *Codiaeum*, Cockscomb (*Celosia argentea*, 2n=36), Croton (2n=20), Globe amaranth (*Gomphrena globosa*), Orchids (2n=36, 38, 40, 42, 70), Indian rubber tree (*Ficus elastica*).
South Asian region	Orchids
South east Asian region	*Acacia auriculiformis* (2n=26), *Acacia catech*u (2n=26).
Australian region	*Acacia longiflora* (2n=16, 26, 52, 78), Acroclinium (2n=44, 45), *Brassaia actinophylla* (2n=11, 12), Brachycome (2n=22, 30). *Callistemon lanceolatus* (2n=34, 36), *Casuarina* (2n=18), Clianthus (2n=32), Eucalyptus (2n=22), *Helichrysum* (2n=10, 28).
Caucasian Western Asia	Autumn crocus (2n=38, 42), crested gentian (2n=40), *Pyrethrum coccineum* (2n=54), Pin-cushion flower (*Scabiosa*, 2n=18).
Macronesian including Canary Islands and Madeira	*Aloe sp.* (2n=14, 28), Campanula (*Azorina vidalii*, 2n=34).

These centers coincide to a significant extent with the regions of the Conservation International's Biodiversity Hot spots in the world. No one single country is self sufficient in all the germplasm of flowers. The centers of origin of ancient flowers from China and Turkey coincide with the Vavilov's food crop centers of origin. Domestication of food crops might also lead to the domestication of those flowers. Flowers were part of crop weed complex from the beginning and those with beautiful flowers would have been selected for domestication.

2.1 Ornamental Plants Native to India

The important flowers/ornamentals which are native to India cultivated in different parts of the world are orchids, rhododendron, musk rose, begonia, balsam, globe amaranth, Gloriosa lily, Foxtail lily (*Eremurus hilamaicus*), *Achyranthes* (2n=42, 84), *Alternanthera* (2n= 34), *Amranthus viridis* (2n=34), *Aphelandra* (2n=56), aster, barleria (2n=40), *Bauhinia*, begonia, crinum, gardenia (2n= 44), hydechium, *Ipomoea* (2n=30), iris (2n=48), ixora (2n=22), *Lilium* (2n=24, 36), honeysuckle (2n=18), lotus (2n=16), orchids, *Pandanus* (2n=60), rhododendrons (2n=26, 156), *Tabernaemontana coronaria* (2n=22, 23, 28, 33, 66), *Thunbergia* (2n=52, 66, 60, 64), *Tulipa clusiana* var. *stellata* (2n=36) and *T. aitchisonii* (2n=24), water lily (*Nymphaea* spp.). Other such ornamentals are listed in Table 2.2.

Table 2.2: Ornamental plants native to India

Annuals	Shrubs	Climbers	Trees
Balsam (2n=14, 16, 20), *Celosia argentea* (2n=72), Blue poppy (*Meconopsis)* (2n=82), *Gomphrena globosa* (2n=26), Foxtail lily *(Eremurus himalaicus*) (2n=14*)*, Lady's lace (*Pimpinella monoica*) (2n=20, 22), primula *(Primula denticulata, P. rosea)* (2n=16, 24).	*Clerodendron inerme, Crossandra infundibuliformis* (2n=30), Croton (2n=20), *Daedalacanthus nervosus,* Glory lily (2n=22, 88)*, Hamiltonia sauveolens, Holmskioldia sanguinea, Jasminum spp.* (2n= 26, 52), *Justicia spp.* (2n=30, 36), *Mussanda, Nyctanthus arbortristis* (2n=36), *Plumbago zeylanica* (2n=28), *Rosa moschata* (2n=14, 56), *Russelia juncea*	*Climatis paniculata, Clitorea ternatea* (2n=16)*, Ficus repens* (2n=26)*, Hiptage madablota* (2n= 58), *Mussaenda frondosa* (2n= 22)*, Osmanthus fragrance, Passiflora leschenaulti, Porana paniculata* (2n= 26).	Amaltas (2n=28), Kadamba (2n=44) *Bauhina variegata* (2n=28), *Butea monosperma* (2n=36+ 2B), *Cordia sebestena* (2n=18, 38), *Erythrina indica* var. *Parcellii* (2n=42), *Cassia nodosa* (2n=28) , *Erythrina variegata* var. *Orientalis* (2n= 42), *Ficus religiosa* (2n=26), *F. benghalensis* (2n= 26)*, Lagerstroemia speciosa,* Indian coral tree (*Erythrina blakei*), Karanj (*Pongamia pinnata*) (2n=20,22), *Michelia champaca* (2n=38), Pride of India (2n=48, 96)*, Tecomella undulate* (2n= 36), *Thespesia populnea* (2n=26)*,* Yellow Silk Cotton (*Cochlospermum gossypium*) (2n=12).

Globally, most of the ornamentals are confined to north-eastern Eurasia. There are about 8,000 species of ornamentals in world trade for commercial purposes. Amongst them, *Gladiolus, Iris,* Narcissus, *Rosa*, *Tulipa* are grown over large areas. Out of 120 perennial plants, peony, phlox, iris, lily, tulip, daffodil and 100 annual species, China aster, petunia and marigold showed greater diversity.

2.2 Evolution of Ornamental Plants

Ornamentals are believed to have been evolved from gene mutation, recombination, interspecific hybridization, polyploidy and natural selection. Interspecific hybridization and polyploidy are considered as the most important sources of evolution and domestication of flowering plants. In ornamental plant breeding, these phenomena can be observed in the breeding history of many ornamental crops viz. *Rosa*, *Chrysanthemum, Gladiolus, Alstroemeria, Lilium* and orchids (Dey and Bhattacharjee, 2011).

A) Through polyploidization

1. Single diploid parent: Sweet pea, *Primrose, Primula sinensis*, *Hyacinth*,
2. Several diploid parent: Narcissus
3. Diploid and polyploid parent: *Iris*, *Rosa* (2n=14)
4. Complex of polyploids: *Chrysanthemum, Dahlia*
5. Autotriploidy with reduced seed fertility: *Chrysanthemum frutescetts, Tulipa gesneriana*.
6. Autotetraploidy: *Primula sinensis*, *Tulipa clusiana*
7. Successive polyploidy: *Hyacinthus orientalis, Primula malacoides, Chrysanthemum maximum*
8. Crossing followed by segregation: Garden *Streptocarpus*,
9. Allotetraploid: *Tagetes patula*

B) Through Interspecific Hybridization

1. Annuals: Modern ageratum, modern geranium, triploid marigold, annual Marguerite carnation (2n=30, 60), hybrid antirrhinum (2n= 16), hybrid verbena.
2. Bulbous ornamentals: *Gladiolus victorialis, G. gandavensis, G. brenchleyensis, G. turicensis,* Homoglad hybrids, *Polianthes blissii, Polianthes bundrantii,* Rainbow tuberose, hybrid dahlia, long trumpet hybrids of amaryllis.

3. Ornamental shrubs: Bourbon rose, Polyantha rose, Damask rose, Noisette rose, Park rose, White rose, *Bougainvillea buttiana, B. spectoglabra, B. specto-perviana*
4. Other commercial flowers: *Gerbera cantabrigensis,* perpetual carnation, hybrid anthurium.

References

Coats, A.M. 1968. The origin of garden plants and the FSU contribution. www.mobot.org

Dey, L.C. and Bhattacharjee, S.K. 2011. *In*: Ornamental Crop Breeding. Aavishkar Publishers, Jaipur, pp. 24-26.

Hui-Lin-Li. 1974. The origin of garden plants and the FSU contribution. www.mobot.org.

Tay, D. 2007, In: Flower Breeding and Genetics, (Ed. Anderson N.O.), Springer Press, pp. 113-175.

Vavilov, N.I. 1935. Theoretical Basis for Plant Breeding, Vol. 1. Moscow. Origin and Geography of Cultivated Plants. *In*: D. Love (Transl.). The Phytogeographical Basis for Plant Breeding, Cambridge University Press, pp. 316-366.

3

History of Plant Taxonomy

3.1. History and Development of Plant Classification

The botanical term "Angiosperm", from the Ancient Greek 'angeíon' (receptacle, vessel) and seed, was coined in the form Angiospermae by Paul Hermann in 1690, as the name of that one of his primary divisions of the plant kingdom.

This included flowering plants possessing seeds enclosed in capsules, distinguished from his *Gymnospermae*, or flowering plants with achenial or schizocarpic fruits, the whole fruit or each of its pieces being here regarded as a seed and naked.

Its use with any approach to its modern scope only became possible after 1827, when Robert Brown established the existence of truly naked ovules in the *Cycadeae* and *Coniferae*, and applied to them the name gymnosperms.

From that time onwards, so long as these Gymnosperms were, as was usual, reckoned as dicotyledonous flowering plants, the term Angiosperm was used antithetically by botanical writers, with varying scope, as a group-name for other dicotyledonous plants.

In 1851, Hofmeister discovered the changes occurring in the embryo-sac of flowering plants, and determined the correct relationships of these to the Cryptogamia.

This fixed the position of Gymnosperms as a class distinct from Dicotyledons, and the term Angiosperm then gradually came to be accepted as the suitable designation for the whole of the flowering plants other than Gymnosperms, including the classes of Dicotyledons and Monocotyledons. This is the sense in which the term is used today.

The Wettstein system and the Engler system use the name *Angiospermae*, at the assigned rank of subdivision.

The Reveal system treated flowering plants as sub-division Magnoliophytina, but later split it to *Magnoliopsida*, *Liliopsida* and *Rosopsida*.

The Takhtajan system and Cronquist system treat this group at the rank of division, leading to the name *Magnoliophyta* (from the family name *Magnoliaceae*).

The Dahlgren system and Thorne system (1992) treat this group at the rank of class, leading to the name Magnoliopsida. The APG system of 1998, and the later 2003 and 2009 revisions, treat the flowering plants as a clade called 'angiosperms' without a formal botanical name. However, a formal classification was published alongside the 2009 revision in which the flowering plants form the sub-class Magnoliidae.

The internal classification of this group has undergone considerable revision.

3.1.1 Earlier than Man Could Read and Write

Not much a known today about the botanical knowledge of our preliterate ancestors. But it is clear that they knew by experience the plants that were edible and others which were not. The preliterate mankind also possessed some kinds of linguistic mechanisms for correct distinctions among different kinds of plants.

3.1.2 Age of Theophrastus, Secundus, Dioscorides and Parasara

The advent of writing and printing changed the entire picture of the ancient preliterate mankind. Theophrastus (370-285 B.C.), the 'grandfather of the modern botany', was the greatest botanical writer of the distant past. He was a student of Plato and Aristotle, and studied botany under their philosophic guidance at Athens. He classified the plants into four groups: herbs, sub-herbs, shrubs and trees. He also distinguished between the non-flowering plants (crytogams) and flowering plants (phanerogams). He suggested that calyx and corolla are the modified leaves. He described nearly 500 plants in details, and certain names (e.g. *Asparagus*, *Daucas*, and *Narcissus*) are used even today in the same sense. The details of his works are available to the world in the form of books entitled 'Enquiry into plants' and 'The causes of Plants'.

Caius Plinius Secundus (23-79 A.D.) known to the botanical world as Pliny the Elder, was a lawyer and also served in the Roman army till his death caused by the heart failure. Voluminous and important works of Pliny are available in the form of 37 volumes of Natural History was among the first books to be printed by the movable type in the late 15th century. The word 'stamen' in its modern sense was first used by Pliny.

Pedanios Dioscorides (62-128 A.D.) was a contemporary of Pliny the Elder. He was actually a Physician in the Roman army and belonged to a Roman province, Cilicia. His monumental work is compiled in the form of 'Materia Medica'. It described the botany, mainly the medicinal aspects, of about 600 species of plants. Meteria Medica was considered as a highly esteemed source book for those practicing medicine in those days. The importance and fame of *Materia Medica* of Dioscorides may be guessed from the fact that the Emperor Flavius Anicius Olybrius (500 A.D.) presented a beautifully illustrated copy of the book to his daughter (Princess Juliana Anicia) as a precious gift at the time of her marriage. Juliana's copy *Materia Medica* remained for several years in Constantinople (Istanbul) and was later transferred to Vienna, where it still exists under the name of Codex Juliana. Several plant names (e.g. *Aloe, Aristolochia, Anemone, Phaseolus*) as suggested by Dioscorides are still in use in the present botanical literature.

The book titled 'Vrikshayurveda', written by an Indian, Parasara, is one of the earliest Indian works describing plants in a scientific manner. Though it was written before the beginning of the Christian era, the scientifically described plant classification and distribution in this book led the famous systematist Albert E. Radford (1986) of USA to state that Parasara had some kind of hand lens or microscope. There exist several other references which show that the early civilizations of India, Egypt and China did have a definite knowledge of the plant taxonomy.

3.1.3 Taxonomy in Middle Ages

Little is known about the development of taxonomy during the early one thousand years of the Christian era. Then came the Middle Ages or Medieval Ages (about A.D. 1100-1500). Albert Magnus (A.D. 1200-1280) has been the most famous plant taxonomy worker of this period. Commonly called 'Doctor Universalis' or 'Aristotle of the Middle Ages' by his contemporaries and historians, Magnus employed a scheme of classification of plants that recognized monocots and dicots, and separated nonvascular plants from vascular plants. Worth mentioning are the names of two Muslim scholars of twelfth century. *Ibn-Sina* who authored Canon of medicine, and *Ibn-al-Awwam* of Spain who described over 600 species of plants and interpreted accurately the sexuality in plants.

3.1.4 Herbalists

After the Medieval Ages (A.D. 1500) the history of plant taxonomy was influenced tremendously by two things, the invention of printing and the development of the science of navigation. The printing technology lowered the cost of books and increased literacy. During the early years of printing the medically oriented books on plants became quite popular. Printed forms of ancient texts had many superfluous and irrelevant writings and this actually prompted several interested persons to write and publish their own botanical medical books. These books were called herbals and their authors were called herbalists.

The science of navigation prompted sailors to go on long voyages. This resulted in the exploration of several new areas of the world, and, in turn, increased man's practical knowledge of plant taxonomy.

At the advent of 16th century the first herbals published were '*Gart der Gesundheil*' and '*Hortus Sanitalis*'. These herbals have crude illustrations of plants and were published without an attribution of authorship. However, the 16th century is considered as the 'time of great herbalists'. The best known among the herbalists belong to Germany. Among them were Otto Brunfels (1464-1534) known for his herbal Herbarium Vivae Eicones, Jerome Bock (1489-1554) for his herbal Neu Kreuterbuch, and Leonhart Fuchs (1501-1566) for his herbal De Historia Stirpium. All these herbalists are considered as the 'German Fathers of Botany' and their herbals exhibit some excellent illustrations and detailed taxonomic descriptions of several available plants. However, they did not emphasize on any system of classification of plants.

3.1.5 Taxonomy During Seventeenth Century

Andrea Caesalpino (1519-1603), an Italian, was the 1st scientist who worked for achieving a rational scheme of classification of plants. Another name of repute among the taxonomists of the 17th century is that of Kaspar Bauhin (1560-1624), an Italian botanist. He compiled all the different kinds of plants known to the science till then in a register called 'Pinax Theatri Botanici'. His register which had an account of different names used by various workers for each plant is considered an authoritative discussion of synonymy in systematic botany. He, first, established the distinction between the concept of genus and species and also initiated the use of the binomial nomenclature, in some cases though not consistently.

John Ray (1627-1705), an English taxonomist, is another great contributor of the 17th century. His works are published in two books entitled '*Methodus Plantarum Nova* (1682) and *Historia Plantarum* (1686). He also suggested a scheme of classification of plants.

Joseph Pitton de Tournefort (1656-1708) is another significant name in the field of taxonomy, mainly for his publication titled '*Institutiones Rei Herbariae*'. He arranged over 9,000 kinds of plants in about 700 genera grouping them in classes.

3.1.6 Period of Linnaeus

The 18th century belongs clearly to Carl Linnaeus (1707-1778) from the point of view of history of taxonomy. He is the creator of the modern system of nomenclature.

A son of a Swedish clergyman, Linnaeus was educated at the universities of Lund and Uppasala, and obtained the degree of M.D. in Netherlands. After practicing medicine for a few years, he became a professor of natural history at the University of Uppasala, where he spent the rest of his life.

Linnaeus is considered as the 1st taxonomist who showed that the reproductive features were of paramount importance in taxonomy. He was the originator of the sexual system of classification in which he recognized 24 classes, mainly on the basis of number, length, union and certain other characters of stamens. He was the 1st to use consistently the '*Binomial System of Nomenclature*', in which each organism is represented first by its genetic name followed by the name of its species. This scheme of nomenclature provided by Linnaeus is used throughout the world till today. The plant taxonomy of findings of Linnaeus were published in May 1753 in the form of his work '*Species Plantarum*'.

The name of Linnaeus is commemorated today in the form of a well established scientific society. The Linnaean Society of London, and journals, such as Linnaea. Contribution of Linnaeus has been so great that it may well be stated that so far the 'plant taxonomy' would survive in the world, Linnaeus would be remembered,

3.1.7 Natural System Approach

Though Linnaeus (1753) was the 1st one to establish a system of classification of plants based on reproductive parts, several totally unrelated plants were classified together (e.g. *Prunus* was classified along with *Cactus* because of the same number of stamens) in his system of classification. This fact compelled the botanists after Linnaeus to realize that no single character is intrinsically or naturally more important than any other character. An approach to a natural system of classification first took seed in France, where Michel Adanson (1727-1806) emphasized the fact that in several cases natural characters are more useful than the others. This theory of Adanson was later recognized as phonetic taxonomy.'

The 1st scheme of classification based on natural characters was presented in 1789 by Antoine-Laurent de Jussieu (1748-1836), a great botanist of France. All the four members of A.L. de Jussieu's family (Antoine, Bernard,

Joseph and Antoine-Laurent) made notable contributions to the science of plant taxonomy. The plants resembling each other in a set of characters were grouped together in A.L. de Jussieu's scheme of classification, and therefore, it was purely natural in its approach. He presented his scheme of classification in his *Genera Plantarum Secundatum Ordines Naturales Disposita.*

Another family of botanists, contemporary of A.L. de Jussieu (1748-1836) was that of Augustin Pyramus de Candolle (1778-1841). A.P. de Candolle presented a new classification of plants in his book '*Theorie Elementaire*'. He followed the approach of the natural system in his scheme and put all alike plants together. An attempt to prepare an account of all available higher plants of the world was undertaken by A.P. de Candolle in his '*Prodromus Systematis Naturalis Regni Vegetabilis*' in 1816, and he continued with this unfinished project till his death in 1841. The same project was continued by his son, Alphonse (1806-1893) until 1873. After 1873, Alphonse and his son, Anne Casimir de Candolle (1836-1918) published the details of this project in the form of monographs, but this great project could never be completed.

The latest, the best and a highly recognized natural system of classification was proposed by George Bentham (1800-1884) and Joseph Dalton Hooker (1817-1911). They classified plants strictly on the basis of a natural scheme. They made thorough observations of the material from the herbaria, took very little help from the literature existing at that time, and presented their well known scheme of classification in their book titled '*Genera Plantarum*'. The Bentham and Hooker's system of classification is still supposed to be the best classification system, especially from the practical laboratory point of view.

3.1.8 The Phylogenetic Approach

Publication of Darwin's theory of organic evolution in 1859 in the form of his well-known book '*On the Origin of Species*' provided the real base for the development of modern systematics. Scientists started working on the line of thought that the life of plants as well as animals has continually changed on the earth over a period of time. Botanists started working on the concept of evolution regarding the development of a classification system of plants.

S. Endlicher (1805-1849) and A.W. Eichler (1839-1889), the two German botanists were the first to start working along this line of thought. They proposed the phylogenetic systems of the classification of plants, and their schemes were later modified and developed by A. Engler (1844-1930) and K. Prantl (1849-1893). Engler and Prantl's monumental work was published in several volumes in the form of '*Die Naturlichen Pflanzenfamilien*' between 1887 and 1915. Published in the 1915 volume of Pflanzenfamilien their scheme indicates that they started from simplest plants (e.g. Salicaceae in case of dicotyledons because the members of this family possess simplest floral structures) and ended with plants of complex floral structures (e.g. Compositae or Asteraceae in case of dicotyledons) because members of this family possess the height of floral complexity). Majority of the present day botanical institutions and publications follow Engler and Prantl's scheme of classification.

Another important work contemporary to Engler and Prantl, was that of C.E. Bessey (1845-1915). a professor of botany at the University of Nebraska. Bessey grouped the flowering plants on the basis of their evolutionary relationships. He categorized them on the basis of characters of primitiveness and advanceness. These guiding characters or principles were named as 'dicta' by Bessey. Richard von Wettstein (1862-1931), an Austrian botanist, and Hans Hallier (1868-1938). A German botanist, are two other plant taxonomists who suggested phylogenetic systems of classification.

John Hutchinson (1884-1972), a British botanist also suggested a widely recognized Phylogenetic system of plant classification which he published in two volumes of his well known book '*The families of flowering plants*'. First published in 1926, the 3rd edition of this great book appeared in 1973, shortly after his death on 2nd September 1972 at the age of 88. He classified the flowering plants on the basis of 24 general principles.

3.1.9 Some Recent Contributions

Four systems of plant classification have been published during the last few years, by Armen Takhtajan (1910-) of Leningrad (Russia), Arthur Cronquist (1919-) of New York (USA), Rolf Dahlgren (1919-) of Copenhagen (Denmark), and R.F. Thorne (1920-) of Claremont (USA).

Armen Takhtajan (1910-) of Botanical Institute of Academy of Sciences, Leningrad, proposed a system of classification of flowering plants in Russian in 1954. Its English version was 1st published in 1969 in '*Flowering Plants: Origin and Dispersal'*. A new version of his classification is published in Botanical Review of 1980. According to him the angiosperms are monophyletic and evolved from some very ancient group of gymnosperms. He divided the angiosperms (Division: Magnoliophyta) into two classes, i.e. Magnoliopsida (= dicotyledons) and Liliopsida (= monocotyledons).

Arthur Cronquist (1919-) of New York Botanical Garden proposed a comprehensive system of classification of angiosperms. First appeared in 1968 in 'The Evolution and Classification of Flowering Plants', the latest version of Cronquist classification is published in 1981 in '*An Integrated*

System of Classification of Flowering Plants'. Cronquist considered the seed ferns as the probable ancestors of angiosperms. He divided angiosperms into Magnoliopsida and Liliopsida. The Magnoliopsida, equivalent to dicotyledons consists of 64 orders, 318 families and about 165,000 species, whereas the Liliopsida consists of 19 orders, 65 families and about 50,000 species.

Rolf Dahlgren (1919-) of Botanical Museum of University of Copenhagen proposed a new scheme of classification of angiosperms. First published in 1975 in '*Botanische Notiser*', the revised and improved versions of his scheme appeared in 1980 in '*Botanical Journal of Linnean Society*', in 1981 in '*Phytochemisty and Angiosperm Phylogeny*' and in 1983 in Nordic Journal of Botany. He also considered angiosperms to be monophyletic, and divided the class Magnoliopsida (= Angiospermae) into two sub-classes viz. Magnoliidae (= Dicotyledoneae) and Liliidae (= Monocotyledoneae).

Robert F. Thorne (1920-) of Rancho Santa Ana Botanic Garden, Claremont, California, published a Phylogenetic system of classification of plants. First published in 1968 in *Aliso*, the revised and enlarged outlines of Thorne's scheme were published in 1981 in '*Phytochemistry and Phylogeny*' and in 1983 in '*Nordic Journal of Botany*'. Thorne has tried to establish the phylogenetic relationships among the higher taxa of flowering plants. He divided the class Annonopsida (= Angiospermae) into sub-class Annonidae (= Dicotyledoneae) and Liliidae (= Monocotyledoneae). Thorne divided Annonidae into 19 superorders, 41 orders, 56 sub-orders, 297 families, 350 sub-families, 9,640 genera and 1,73,370 species and Liliidae into 9 superorders, 12 orders, 17 sub-orders, 53 families, 102 sub-families, 2,615 genera and 52,120 species.

3.2 Taxonomy- Our Contemporary

The plant kingdom is very large and diversified. We are faced with an estimated total of about 3,00,000 or if fungi included, 4,50,000 plants. This large number includes many different kinds of plants of a vase range in size and structure. This entire living world of plants forms the subject matter of the plant taxonomy. The raw material of the taxonomic botany is the diversity of plant life in all its aspects. One is, therefore, faced with exceedingly diverse taxonomical activities and techniques. The algomogist collecting algae in the arctic region, the plant collector in the tropical forest, the botanist describing plants of a herbarium, the biochemist studying amino acids by chromatography, the cytogeneticist studying breeding systems, the cytologist finding the chromosome number, the electron microscopist going down to molecular details, the botanical artist depicting plants with his pen, brush and ink, all contribute to the growth and development of plant taxonomy.

Taxonomy is the only branch of botany that involves the interests and abilities of so many practitioners. From a plant collector to a computer programmer, from a biochemist to an electron microscopist, and from a mycologist to a taxonomist, everyone has room to show his talents in this branch. Thus we can say that taxonomy is truly our contemporary.

3.3 Chronological Development of Taxonomy in India

Indians of very old times had a satisfactory knowledge of plant taxonomy is evident from the studies of the earliest Indian civilization as well as from the treatises such as *Vrikshayurveda* of Parasara, *Charaka Samhita*, *Sushrut Samhita* and voluminous Indian classics such as *Vedas*, *Ramayana*, *Mahabharata*, etc. However, a complete chronology of the development of plant taxonomy in India is beyond the scope of this book. The major attempts of the botanical explorations of India during the past few centuries and the important published works are under mentioned in Table 3.1.

Table 3.1: Chronology of the major published taxonomic works of India.

Year A.D.	Researcher/ explorer	Major published/ botanical work
1530	Brunfels (1464-1534)	Wrote a book on medicinal plants in 3 volumes
1542	Fuchs (1501-1566)	Historia of plants
1551	Turner (1515-1568)	A New Herbal in 3 volumes in 1551, 1562 & 1568; regarded as father of botany in England at that time
1565	Garcia d'orta	Published Os Colquios in Goa; described common Indian medicinal plants
1578	C. Acosta	Published Tractado de las Drogas
1583	Caesalpino (1519-1603)	Wrote a book 'De Plants' describing 1,520 plants
1596	Kasper Bauhin (1560-1624)	Described 2,700 plants
1670	Heinrich van Rheede	Published Hortus Malabaricus
1728-1785	John Gerard Koenig	Formed society called 'The United Brothers' in order to promote and study Indian plants
1746-1793	Robert Kyd	Founded Botanical Gardens of Kolkata
1751-1815	William Roxburgh	Published Flora Indica and Plantae Coromandelianus; prepared 2382 coloured drawings of Indian plants; called Linnaeus of India
1817	Nathaniel Wallich	Published Plantae Asiaticae Rariores
1824	J.F. Watson	Published Flora of Kumaon
1833	J.F. Royal	Flora of Kashmir
1839	J. Grahm	Flora of Bombay
1840	J.W. Masters	Flora of Calcutta
1844	W. Munro	Flora of Agra

Year A.D.	Researcher/ explorer	Major published/ botanical work
1857	J. Long	Flora of Bengal
1859	W. Elliot	Flora of Andhra
1859	T. Anderson	Flora of Lucknow
1869	J.L. Stewart	Flora of Punjab
1872-1897	J.D. Hooker	Flora of British India published in the form of seven volumes
1887	George King	Started the publication of the journal The Annals of the Royal Botanic Garden
1890	George King	Established Botanical Survey of India
1901-1908	T. Cooke	Flora of Presidency of Bombay
1902	H. Collett	Flora Simlensis
1903	D. Prain	Bengal Plants
1903-1922	J.F. Duthie	Flora of Upper Gangetic Plains and the Adjacent Siwaliks and Sub-Himalayan Tract
1908	T.F. Bourdillon	The Forest Trees of Travancore
1910	H.H. Haines	A Frost Flora of Chota Nagpur
1915-1936	J.S. Gamble and C.E.C. Fisher	Flora of Presidency of Madras
1921	K. Rangacharier	A Handbook of South Indian Grasses
1921-1925	H.H. Haines	The Botany of Bihar and Orissa
1925	A.J. Eames and L.H. MacDaniels	Introduction to plant anatomy
1923	C.E. Parkinson	Forest Flora of Andaman Islands
1928	V. Kanjilal and Gupta	Forest Flora of the Chakrata, Dehra Dun and Saharanpur Forest Divisions
1929	P.V. Mayurnathan	The Flowering Plants of Madras City and its Immediate Neighborhood
1932	P.F. Fyson	The Flora of South Indian Hill Stations
1932-1940	V. Kanjilal	Flora of Assam
1935	C.J. Chamberlain	Gymnosperms: Structure and Evolution
1938	G.M. Smith	Cryptogamic botany
1940	N.L. Bor	Common Grasses of United Provinces
1949	R.L. Bahadur and S. Ghosh	Poisonous Plants of India
1953	S.P. Agahakar	Medicinal Plants
1953	A.W. Haupt	Plant Morphology
1960	B.P. Pal	Beautiful Climbers of India
1962	H. Santapau	The Flora of Saurastra
1962	K. Subramanian	Aquatic Angiosperms
1962	B. Bhadri, B. Singh & B.L. Desai	Water Plants
1963	J.K. Maheshwari	Flora of Delhi
1964	P. Maheshwari and U. Singh	Dictionary of Economic Plants in India
1964	G.S. Puri *et al.*	Flora of Rajasthan

Year A.D.	Researcher/ explorer	Major published/ botanical work
1964	A.C. Dutta	A class-book of botany
1965	M.S. Randhawa	Flowering Trees, Beautiful Trees and Gardens, Garden Through Ages, Beautiful Cities of India
1966	B.P. Pal	The Roses in India
1966	V.D. Vartak	Enumeration of Plants from Gomantak (Goa)
1967	B.P. Pal	Flowering Shrubs
1968	R.K. Gupta	Flora Nainitalensis
1968	R.N. Chopra and S.L. Nayyar	Glossary of Indian Medicinal Plants
1968	S.L. Jindal	Ornamental Bulbous Plants
1970	B.L. Desai	Seasonal Flowers
1973	S.V. Ramaswamy and B.A. Raji	Flora of Bangalore District
-	B.N. Ghosh	Beautiful Indian Orchids
1973	Santapau and Henry	Dictionary of Flowering Plants in India
1974	B.P. Pal and Vishnu Swarup	Bougainvilleas
-	K. Gopal Swami Iyengar	Complete Gardening in India
1975	M.A. Rau	High Altitude Flowering Plants of Western Himalayas
1975	M.B. Raizada	Supplement to the Duthie's Flora of Upper Gangetic Plains
1975	M.A. Rau	High Altitude Flowering Plants of Western Himalayas
1976	T.N. Srivastava	Flora Gorakhporensis
1976	C.J. Saldanah and D.H. Nicolson	Flora of Hasan District, Karnataka
1976	R. Rao	Flowering Plants of Travancore
1976	M.M. Bhandari	Flora of Rajasthan Desert
1977	M. Oomachan	The Flora of Bhopal
1978	M.B. Raizada and H.O. Saxena	Flora of Mussoorie
1978	N.C. Nair	Flora of Punjab Plains
1979	A.S. Rao	Orchids of India
1979	V.N. Naik	The Flora of Osmanabad
1979	B.L. Sapru *et al.*	Flora of Ladakh
1981	S.K. Jain	Medicinal Plants
1984	B.D. Naithani	Flora of Chamoli
1984	H.J. Chowdhary and B.M. Wadhwa	Flora of Himachal Pradesh
1984	O. Polunin and A. Stainton	Flowers of Himalayas
1986	M.K. Kaul	Weed Flora of Kashmir Valley
1986	CSIR publication	The Useful Plants of India

Year A.D.	Researcher/ explorer	Major published/ botanical work
1986	G.S. Randhava & A.K. Mukhopadhya	Floriculture in India
1986	K.L. Chadha & B. Choudhary	Ornamental Horticulture in India
1987	Som Deva and H.B. Naithani	The Orchid Flora o N.W. Himalayas
1988	R.K. Gupta	The Living Himalaya Plant Exploration and Phytogeography
1988	S.P. Vij (ed.)	Biology, Conservation and Culture of Orchids
1989	T.K. Bose	Commercial Flowers
1989	Vishnu Swaroop	Ornamental Horticulture, Garden Flowers, Indoor Gardening
1990	P.P. Trivedi	Beautiful shrubs

3.4 Some Indian Journals Related to Taxonomy

1. Annals of the Royal Botanic Garden, Kolkata.
2. Bulletin of the Botanical Society of Bengal.
3. Bulletin of the Botanical Survey of India.
4. Indian Forester
5. Indian Forest Records.
6. Indian Journal of Forestry
7. Journal of Bombay Natural History Society.
8. Journal of Economic and Taxonomic Botany.
9. The Records of the Botanical Survey of India.
10. The Journal of the Asiatic Society of Bengal.

3.5 A note on Botanical Survey of India

George King, the Chief Executive Officer of the Royal Botanic Gardens, established the Botanical Survey of India in 1890 with the main objective to coordinate the botanical work, being done in different parts of India at that time. Regional offices of BSI were established at Madras with M.A. Lawson as the Director for southern India, at Saharanpur with J.F. Duthie as the Director for northern India, and at Pune with T. Cooke as the Director for the western India. The publication of a journal 'The Records of the Botanical Survey of India' was also started by George King. The Industrial Section of the Indian Museum at Calcutta and the office of the Reporter of Economic products were brought under BSI in 1911. But the coming few decades were a period of inactivity for BSI mainly because of lack of funds and manpower. In 1939, the conditions deteriorated so that the post of Director at Calcutta remained suspended, and BSI was surviving only in the form of a Curator, Industrial Section, and a Systematic Assistant in the herbarium at Sibpur (Howrah).

The Government of independent India then came to the rescue of BSI. E.K. Janaki Ammal was appointed as the Officer-on-Special Duty in 1952 to reorganize BSI. The first unit was established in the form of The Central Botanical Laboratory at Allahabad with Janaki Ammal as the Director. Four Regional Circles were established with their headquarters at Dehradun of Northern Circle, Coimbatore of Southern Circle, Shillong of Eastern Circle, and at Pune of Western Circle. In 1962, the Central Botanical Laboratory was shifted from Allahabad to Calcutta, and at Allahabad a new regional station for Central India was established. In 1972, two new circles (Andaman and Nicobar Circle, and Arid Zone Circle) were established. Arunachal Pradesh Circle with its headquarter at Itanagar was created in 1977, and Sikkim-Himalaya Circle with its head office a Gangtok in 1979. Recently, the Deccan Circle at Hyderabad, and High Altitude Circle at Solan (Himachal Pradesh) were established in 1984.

BSI headquarter office at Kolkata coordinates the research and other activities of all units, mainly on the basis of the scientific policies of the Central Government. It also maintains links of BSI with other major research institutions of the country such as CSIR, ICAR, ICMR, etc. A senior scientist of BSI is also posted at the Royal Botanical Gardens, Kew to maintain a link between the two organizations as well as for many technical inquiries and clarifications regarding nomenclature, etc.

Some of the major publications of BSI include the Bulletin of Botanical Survey of India, The Records of the Botanical Survey of India, Annual Reports of BSI and Newsletters.

Reference

Cronquist, A. 1968. The Evolution and Classification of Flowering Plants. Thomas Nelson and Sons Ltd. London.

4

Systems of Classification

This deals with the description, identification and naming of plants, and their classification into different groups according to their resemblances and differences, mainly in their morphological characteristics. So far as angiosperms or higher 'flowering plants' are concerned, it has been estimated that over 199,000 species are already known to us, and that thousands more have yet to be discovered and recorded. The object of systematic botany or taxonomy is to describe, name and classify plants in such a manner that their relationship with regard to their descent from a common ancestry may be brought out. The ultimate object of classification is to arrange plants in such a way as to give us an idea of their phylogenetic relationships, i.e. the sequence of their origin and evolution from simpler, earlier and more primitive types to more complex, recent and advanced types during different periods. The earlier classification of plants were based on their economic uses, e.g. medicinal plants, cereals, fibre yielding plants, oil yielding plants, etc. or on gross structural resemblances, e.g. herbs, shrubs, trees, climbers, etc. These classifications were incomplete and fragmentary as plants that did not fit into such classifications or were of no economic value were usually ignored. Therefore, an ideal system of classification should not only indicate the actual genetic relationship but should also be reasonably convenient for practical purposes.

4.1 What is Classification?

Andrew Sugden (1984) defined the world 'classification' in *Longman Illustrated Dictionary of Botany* as 'the naming of species and their grouping into families, orders, divisions, etc.' Radford (1986) stated that 'classification is the arrangement of groups of plants with particular circumscriptions by rank and position according to artificial criteria, phonetic similarities, or phylogenetic relationships.'

In its simplest form, classification is the placement of plants, animals and objects into groups and categories for a clear understanding, proper study and effective organization.

4.2 Ranks of Plant Classification

Species, genus, family, order, class and division are the six main ranks of plant classification in an ascending order. Each rank has its sub-categories, i.e. towards the higher ranks, sub-form, form, sub-varieties, varieties, and sub-species are the sub-categories of species; sub-section, section, and sub-genus are the sub-categories of genus; sub-tribe, tribe, and sub-family are the sub-categories of family; sub-order is the sub-category of order; sub-class is the subcategory of class; and sub-division is the sub-category of division.

Several species are included within a genus, several genera within a family, several families within an order, several orders within a class, and several classes are included within a division.

4.3 Units of Classification

1. **Species:** A species is a group of individuals having a very close resemblance with one another, structurally and functionally. The individuals of a species interbreed freely and successfully in nature so that they may be taken as having descended from the same parent stock. They also breed true to type, i.e. give rise to progeny of the same kind. They also normally have the same number of chromosomes in their cells $2n$ in somatic cells and n in reproductive cells. Thus, all rose plants together constitute a species. Species, however, often show a wide range of variations, with many intermediate forms or gradations, serving as connecting links between two allied species.

   Under the influence of external conditions, certain individuals of a species may often show a marked degree of variation in form, size, shape, colour and other minor characteristics. Such plants are said to form varieties. A species may consist of one or more varieties, or none at all. Thus, we have different varieties of common garden pea, potato, mango, etc. Varieties however, are not permanent. They tend to revert to the species from which they were originally derived.

2. **Genus:** A genus a collection of species which bear a close resemblance to one another as far as morphological characteristics of the floral or reproductive parts are concerned. For example, banyan, peepul and fig are different species because they differ from one another in their vegetative characteristics such as the habit of the plant, shape, size and surface of leaf, etc. But these three species are allied because of the similarity in their reproductive organs, namely, the inflorescence, flower, fruit and seed. Therefore, banyan, *peepul* and fig come under the same genus (*Ficus*).

3. **Nomenclature:** The name of a plant has two parts. The first refers to the genus and the second to species. The system of designating every type of plant with a binomial, i.e. a name consisting of two parts, is known as binomial nomenclature. It was first established by Linnaeus and finally settled by the International Botanical Congress held in Amsterdam in 1935. Thus, banyan has been named *Ficus bengalensis*, fig- *Ficus carica* and Indian rubber plant *Ficus elastica*. The name of the author who first described a species is also written in an abbreviated form after the name of the species, e.g. *Mangifera indica* Linn. Here Linn refers to 'Line' or 'Linnaeus', who first described the plant.
4. **Family:** A family is a group of genera which have general structural resemblances mainly in their floral organs. Thus, in the genera, Hibiscus, Thespesia, Abutilon, Malva, etc., we find free lateral stipules, epicalyx, twisted aestivation of corolla, monadelphous stamens, unilocular anthers, axile placentation, etc. So all the above mentioned genera belong to same family, and that is Malvaceae.

Conventional Classification of Flowers and Ornamental Plants-

4.4 Classification According to Growth Habit

4.4.1 Herb

Herbs or herbaceous plants are those which have succulent leaves and stem. Herbaceous plants may be annuals, biennials or perennials.

4.4.1.1 Bushy herb

Periwinkle, marigold etc.

4.4.1.2 Non-bushy herb

4.4.1.2.1 Vine type

***Creeper:** *Ipomoea, Oxalis, Dracaena, Tecoma grandiflora* etc.

***Climber:** *Cuscuta, Clitoria, Passiflora, Gloriosa* etc.

4.4.1.2.2 Non-vine types: Lilies.

4.4.2 Shrub

Shrub is a small to medium sized bushy plants with several semi-woody stems. It is distinguished from a tree by its multiple stems and shorter height, usually less than 6 m tall. They branch profusely from the base of the stem to present a bushy appearance, without having a clear trunk.

4.4.2.1 Flowering shrub

They are grown for flowers. e.g. Rose, jasmines, oleanders, camellia, etc.

4.4.2.2 Foliage shrub

They are grown for their beautiful leaves. e.g. Croton, acalypha, aralia, etc.

4.4.2.3 Hedge

These are special type of shrubs planted in border, usually very closely for protection of garden e.g. Duranta, *Streblus*, *Justicia*, *Sesbania,* etc.

4.4.3 Tree

Tree is a perennial, tall plant with clear trunk, supporting branches and leaves.

4.4.3.1 Monocot: Palms

4.4.3.2 Dicot

4.4.3.2.1 Flowering trees

The trees which are known for their beautiful flowers and are suitable for garden are called flowering trees e.g. Magnolia, *Bauhinia*, kadam, ashoka, peacock, Indian medlar, etc.

4.4.3.2.2 Foliage trees

The trees which are well known and attractive for their charming leaves and branches. Here the flower is of secondary importance e.g. *Ficus*, *Albizzia*, *Pinus*, *Eucalyptus*, etc.

4.4.4 Aquatic Ornamental Plants

Aquatic plants are those plants adapted to living in aquatic environments. They are also referred as hydrophytes or macrophytes. Aquatic plants can only grow in water or in soil that is permanently saturated with water.

A.4.4.1 Floating plants

Water lettuce, water chestnut etc.

A.4.4.2 Rooted floating

Water lily, red lily, blue lily, lotus, royal water lily etc.

A.4.4.3 Bog plants

Sweet flag, *Alocasia*, etc.

4.5 Classification According to Conventional Method-

4.5.1 Seasonal flowers

The flowering plants which complete their life cycle within one growing season are called seasonal flower. They are of three groups:

4.5.1.1 Winter seasonal flower

They complete their life cycle within winter season e.g. Aster, *Cosmos*, *Dianthus* etc.

4.5.1.2 Summer seasonal flower

They complete their life cycle within summer season e.g. Balsam, Cock's comb, *Zinnia*, *Petunia*, etc.

4.5.1.3 Year-round flower

They can be grown during both in winter and summer season e.g. African daisy, butterfly pea, *Arctotis*, *Coreopsis* etc.

4.5.2 Bulbous plants

The ornamental and flowering plants, which are usually propagated by underground modified structures like bulb, corm, rhizome, tuberous root etc. are called bulbous plants. The main features of them are long narrow leaf, lack of aerial stem, development of long peduncle from the base of the plant etc. e.g. Tuberose, gladiolus, caladium, asparagus, zinger lily, dahlia, etc.

4.5.3 Palms

The plants under the family palmaceae are called palms. They have long stem without branching, having a bunch of leaves at the top. They are of three groups:

4.5.3.1 Tall palms

Royal palm, oil palm, areca palm, coconut palm etc.

4.5.3.2 Medium sized palms

Wild areca, cane palm etc.

4.5.3.3 Pot palm

Butterfly palm, lady palm etc.

4.5.4 Climbers

A group of plants with weak stems and climb up the support with the help of modified organs e.g. *Cuscuta, clitoria, passiflora, gloriosa, bougainvillia, lonicera, hiptage,* etc.

4.5.5 Conifers

The word comes from Latin word *conus* meaning cone and *ferre* meaning to bear, that means, one that bear a cone. An evergreen gymnosperm belonging to the order coniferales bearing cones and generally needle shaped or scale like leaves e.g. Thuja, aurocaria, pinus, tamarisk, etc.

4.5.6 Cactus

The word comes from Greek word '*kaktos*' first used by Theophrastus for spiny plant. The plants with succulent stem and leaf under the family cactaceae are called cactus e.g. Prickly pear, star cactus, orchid cactus, rose cactus, etc.

4.5.7 Orchid

The plants under the family orchidaceae are called orchid. They grow well in humid and partial shady place but not in stagnant water. They exhibit a wide range of diversity in form, size, colour and texture of flowers beyond the imagination of human mind e.g. *Dendrobium*, *Vanda*, *Cattleya*, etc.

4.5.8 Fern

The cryptogams plant under the family polipodiceae are generally termed as fern. Tropical humid mountains are the natural house of fern. They need high temperature, humid and shady place e.g. *Adiantum*, *Polypodium*, *Lygodium*, *Pteris*, *Osmunda*, etc.

4.5.9 Succulents

Succulent plants also known as fat plants, having some parts that are thickened and fleshy, usually to retain water in arid climates or soil conditions. The word "succulent" comes from the Latin word *sucus*, meaning juice, or sap. Succulent plants may store water in various structures, such as leaves and stems e.g. *Agave*, *Opuntia*, *Euphorbia*, *Portulaca*, *Sansevieria*, etc.

Miscellaneous

1. Cycads

Cycads are group of plants have a stout and woody trunk with a crown of large, hard, stiff and compound evergreen leaves. Cycads are resembled to palms in overall appearance. They possess a pachycauls. The word is derived from the Greek *pachy* meaning thick or stout, and Latin *caulis* meaning the stem e.g. *Cycas*, *Macrozamia*, *Zamia*, etc.

Classification Based on Botany (Table 4.1)

Family	Common name	Botanical name	Origin	2n=
Monocots				
Agavaceae	Tuberose	*Polianthes tuberosa*	Mexico	60 (2x, single); 50 (2x, double)
Amaryllidaceae	Amaryllis	*Amaryllis* spp. (syn. *Hippeastrum* spp.)	South America	22 (2x), 44 (4x)
	Daffodil	*Narcissus pseudonarcissus*	Spain	28 (2x)
	Day lily	*Hemerocallis fulva*	Asia	44 (4x)
Apocynaceae	Periwinkle	*Catharanthus roseus*	Madagascar	16 (2x), 32 (4x)
	Adenium	*Adenium obesum*	Africa	22 (2x)
Araceae	Anthurium	*Anthurium andreanum*	Columbia	30 (2x)
	Calla lily	*Zantedeschia aethiopica*	South Africa	32 (2x)
Asteraceae	Dahlia	*Dahlia variabilis*	Mexico	16 (2x), 32 (4x)
Cannaceae	Canna/ Indian shot	*Canna indica*	Tropical America	18
Heliconiaceae	Heliconia	*Heliconia psittacorum*	Central America	24 (2x)
Iridaceae	Gladiolus	*Gladiolus hybridus*	South Africa & Asia Minor	30 (2x), 60 (4x)
	Iris	*Iris* x *germanica*	India	24 (2x), 48 (4x)
Liliaceae	Lilium	*Lilium* spp.	North America & Europe	24(2x), 36(3x)
	Tulip	*Tulipa gesneriana*	Central Asia	24 (2x), 36 (3x), 48 (4x), 60 (5x)
Nelumbonaceae	Lotus	*Nelumbo lutea*; *N. nucifera*	America & India	16 (2x)
Nymphaeaceae	Water lily	*Nymphaea spp.*	India, Europe, South America	28(2x), 42, 56, 84 to 224(16x).

Family	Common name	Botanical name	Origin	2n=
Orchidaceae	Orchids	*Cattleya, Dendrobium,* etc.	India, South America & C. America	36 (2x), 38 (2x), 40 (2x), 42 (2x)
Strelitziaceae	Bird of paradise	*Strelitzia reginae*	South Africa	22 (2x)
Dicots				
Acanthaceae	Crossandra	*Crossandra undulaefolia*	India, Malaysia, Sri Lanka, East Indies	20 (2x), 30 (3x), 40 (4x)
Alstroemeriaceae	Alstroemeria	*Alstroemeria spp.*	South America	16 (2x)
Amaranthaceae	Gomphrena	*Gomphrena globosa*	India	40 (2x)
Asteraceae (Compositae)	China aster	*Callisteohus chinensis*	China	18 (2x)
	Chrysanthemum	*Chrysanthemum spp.*	China	18 (2x), 36 (4x), 54 (6x)
	Cosmos	*Cosmos sulphureus*	Mexico	24 (2x)
	Gaillardia	*Gaillardia pulchella*	South Ameica	36 (2x)
	Gerbera	*Gerbera jamesonii*	South Africa	50 (2x)
	Marigold	*Tagetes erecta, T. patula*	S. America	24 (2x), 48 (4x)
	Pot marigold	*Calendula officinalis*	India	18 (2x)
	Zinnia	*Zinnia elegans*	Mexico	22 (2x), 24 (2x)
	Brahma kamal	*Saussurea obvallata*	India, Bhutan	-
Balsaminaceae	Balsam	*Impatiens balsamina, I. wallerana*	India, North America	12 (2x), 16 (2x)
Begoniaceae	Begonia	-	-	-
Bignoniaceae	Incarvillea	-	-	-
Caryophyllaceae	Carnation	*Dianthus caryophyllus*	Mediterranean region	30 (2x)
Cruciferae (Brassicaceae)	Stock	*Matthiola incana*	S-W Europe & West Asia	14 (2x)
Geraniaceae	Rose scented geranium	*Pelargonium graveolens*	South Africa	18 (2x), 44 (4x), 66 (6x), 88 (8x)

Family	Common name	Botanical name	Origin	2n=
Labiatae/ Lamiaceae	Salvia	*Salvia splendens*	Brazil, Western Asia & Mediterranean region	16(2x), 32(4x)
Malvaceae	Hibiscus	*Hibiscus rosa-sinensis*	India, China	18 (2x)
	Hollyhock	*Althea rosea*	Central Asia	18 (2x), 27 (3x)
Nyctaginaceae	Bougainvillea	*Bougainvillea spp.*	Brazil	34 (2x), 51(3x), 68 (4x)
Oleaceae	Jasmine	*Jasminum spp.*	Indo- Malaya region	26(2x), 39 (3x), 52 (4x)
Plantaginaceae (Scrophulariaceae)	Snapdragon	*Antirrhinum majus*	Europe	16 (2x)
Plumbaginaceae	Statice	*Limonium sinuatum*	Europe	18 (2x), 27(3x), 36 (4x), 54 (6x), 72(8x)
Polemoneaceae	Phlox	*Phlox cuspidata; P. drummondii*	N. America	14
Primulaceae	Cyclamen			
Ranunculaceae	Anemone			
Rosaceae	Rose	*Rosa hybrida*	Europe & N. America	14 (2x), 56 (8x)
Solanaceae	Petunia	*Petunia hybrida*	S. America	14 (2x)
Verbenaceae	Verbena	*Verbena hybrida*	South & Central America	14 (2x)
Violaceae	Pansy	*Viola tricolor*	S. America &Himalaya	24 (8x)

4.6 Systems of Classification

The past taxonomic literature described three basic categories of systems of classification viz. artificial, natural or Phylogenetic systems. But Radford (1986) has described four types of the systems of classification viz. artificial, mechanical, natural and phylogenetic systems. The systems have developed in a step-wise manner, from the earliest to the present times. Four periods, depending on the criteria used in classification, may be recognized. 1st period: This extends for about 2,000 years from about 1730s B.C. to the end of the 17th century. During this period, the systems were based on the habit of plants. 2nd period: This extends from the thirties to the end of the 18th century. There is a definite change in the pattern of these systems. The habit has given place to sex organs (stamens and carpels) of plants, and Linneaus' sexual system of classification (1735) is specially remarkable. 3rd period: This extends from the beginning to the eighties of the 19th century. Several famous botanists belong to this period. The systems propounded during this period were based on the natural relationships of forms. Bentham and Hooker's natural system of classification (1862-83) is of outstanding merit. 4th period: This period extends from 1875 to the represent times. The systems of this period have been based on the phylogenetic relationships of plants.

1. Artificial classification

These systems use the habit and importance to man as the taxonomic characters. Some advocates of artificial systems of classifications were Theophrastus (370-285 B.C.), Secundus (23-79 A.D.), Dioscorides (62-128 A.D.), Magnus (1200-1280 A.D.), Brunfels (1464-1534), Bock (1489-1554) and Fuchs (1501-1566).

In the artificial system only one, or at most a few, characteristics are selected arbitrarily and the plants are arranged into groups according to such characteristics. As a result, closely related plants are often placed in different groups, while unrelated plants are often placed in the same group because of the presence or absence of a particular characteristic. This system enables us to determine readily the names of plants but does not indicate the natural relationship that exists among the individuals forming a group. Nevertheless, the artificial system is of very great advantage in view of the fact that by following this system one can get to the name of an unknown plant without much difficulty. In other words, the identification of an unknown plant is rendered much easier by this system.

2. Mechanical Classification

These systems used one or a few selected taxonomic characters to group taxa. Some mechanical classifications were given by Caesalpino (1519-1603),

Bauhin (1560-1624), Ray (1627-1705), Tournefort (1656-1708) and Linneaus (1707-1778).

3. Natural Classification

These systems of classifications used as many taxonomic characters as possible to group taxa. Some of the natural systems of classification were given by Adanson (1727-1806), A.L. de Jussieu (1748-1836) and his three family members (Antoine, Bernard and Joseph), A.P. de Candolle (1778-1841) and his son Alphonse (1806-1893) and Bentham (1800-1884) and Hooker (1817-1911).

In the natural system, all the important characteristics are taken into consideration and plants are classified according to related characteristics. Thus, according to the resemblances and differences, mostly in the important morphological characteristics, plants are first classified into a few big groups. These are further divided and sub-divided into smaller and smaller groups until the smallest division is reached and that is a species. All modern systems of classification are natural and supersede the artificial ones in that, on the one hand, they give us a true idea of natural relationships existing between different plants, based on a detailed description of their characteristics and also of the sequence of evolution from simpler to more complex types, and on the other hand they meet the need to identify unknown plants.

According to the natural system, the plant kingdom has been divided into two divisions, viz. cryptogams or 'flowerless' plants and phanerogams or 'flowering' plants. Phanerogams have again been divided into two sub-divisions, viz. gymnosperms or naked-seeded plants and angiosperms or closed-seeded plants. Angiosperms have been sub-divided into two classes, viz. dicotyledons and monocotyledons. The classes have been divided into *orders*, the orders into *families*, the families into *genera* and *species*, and sometimes, species into varieties. If a greater number of intermediate categories are required, the prefix *sub* is added to the particular term.

4. Phylogenetic Classifications

These systems of classification used as many taxonomic characters as possible in addition to the phylogenetic (evolutionary) interpretations. Some of the phylogenetic systems of classification were proposed by Eichler (1839-1889), Engler (1844-1930), a German botanist, and Prantl (1849-1893), Bessey (1845-1915), Wettstein (1862-1931), Hallier (1868-1938), Hutchinson (1964-1967), an English botanist, Takhtajan (1980), Cronquist (1981), Dahlgren (1981) and Thorne (1976), Tippo (1942), an American botanist, are phylogenetic.

4.7 Some Important Systems of Classification

4.7.1 John Ray (1627-1725)

An English biologist, John Ray treated over 18,000 species in the last edition of his book '*Methodus Plantarum Nova*' published in 1703. He developed a classification system on the basis of the form relationships by grouping together the plants that resembled one another. Ray was the first to divide herbs, shrubs, and trees into dicotyledons and monocotyledons on the basis of the presence of two or one cotyledons.

4.7.2 Carl Linnaeus (1707-1778)

Linnaeus, the father of taxonomy and creator of binomial system of nomenclature is considered as "the most prodigious systematist of all times" (Lawrence, 1951). In 1727, at the age of twenty, Linnaeus entered the University of Lund to study medicine. In 1728, he was transferred to the University of Uppasala, and published his first research paper on sexuality of plants in 1729. In 1737, he received M.D. degree from the University of Harderwijk of Netherlands. In Netherlands, he became the personal physician of a wealthy banker George Clifford. Clifford was interested in plants. Linnaeus remained for three years in Netherlands, travelled extensively through Europe, and collected plants. These three years were the most creative years of his life. In 1741, he returned to Sweden, became the Professor of botany and medicine at the University of Uppasala, and continued in the same position till his death on 10th January, 1778.

Linnaeus was so popular as a teacher that his classes on field trips were attended often by 200 or more students. He was buried in the Cathedral at Uppasala "unshaven, unwashed, unclad, enveloped with a sheet" as per his own wish and request.

Three universally known works of Linnaeus are the *Systema Naturae* published in 1735 which contains the outline of his system of classification. He revised in his *Genera Plantarum* (1737) and *Species Plantarum* (1753) published in two volumes. By 1760, his system became popular in Holland and Germany, and partly in England. Linnaeus classified plants according to the characteristics of their reproductive organs, viz. the stamens and carpels. Since these are regarded as the sexual organs of plants, Linnaeus's system is commonly called 'Sexual System'. According to it, plants are divided mainly into 24 classes, including 23 of phanerogams and one of cryptogams. Phanerogams are further sub-divided according to whether they are monoecious or dioecious. Further classification is based on the number of stamens. Plants with bisexual flowers are classified according to whether the stamens are united with the carpels, or free from them. The next consideration is whether the stamens are free or united. Then the number of stamens, their length, ultimately, the number of carpels are taken into account.

Plant collections of Linnaeus were sold by his wife in 1783 to J.E. Smith, a British botanist, who was one of the founders of Linnean Society of London. The entire plant collection is now housed in the office of the Linnean Society in Burlington House, Piccadilly, London.

The Linnaeus's system of classification is an artificial sexual system. However, Radford (1986) described it as a mechanical system of classification. Linnaeus divided plants into 24 classes, mainly on the bases of number, union and length of stamens. The classes were divided into orders. On the basis of the number of stamens in each flower Linnaeus placed all algae, fungi, mosses and ferns under one class Cryptogamia.

4.7.3 Antonie-Laurent de Jussieu (1748-1836)

A natural system of classification was proposed by A.L. de Jussieu in 1789 in his *Genera Plantarum Secundus Ordines Naturales Disposita*. He recognized one hundred orders of the plants which are now called *families*. He divided the plants into three main groups, i.e. Acotyledons, monocotyledons, and dicotyledons. He mainly emphasized on the number of cotyledons and their present of absence, number of petals and their presence or absence, and position of stamens. This system firmly established the philosophy of the natural system among the botanical community. All the hundred orders (now called families) were arranged by A.L. de Jussieu into 15 classes.

4.7.4 Augustin Pyramus de Candolle (1778-1841)

One more natural system of classification was proposed by A.P. de Candolle in 1813 in his book "*Theorie Elementaire de la Botanique*". He divided plants into two major groups i.e. cellulares (non-vascular plants), and vasculares (vascular plants). In his another book, '*Prodromus*', he described all the species of the vascular plants known to the world then. In this work he described 58,000 species of dicotyledons belonging to 161 families. The system was easy and simple, and surpassed all other systems. But its major drawback was the inclusion of vascular cryptogams among the monocots.

4.7.5 *George Bentham (1800-1884) and Joseph Dalton Hooker (1817-1911)*

Bentham, a self-trained British botanist and Hooker, Director of the Royal Botanic Gardens, Kew (England) were contemporaries of Darwin, elaborated a natural system of classification of 'flowering plants', giving a more or less equal status to all three groups. The most accepted natural system of classification. Their system is based on (virtually an extension of) de Jussieu's system (1789) and de Candolle's system (1819). It was published in *Genera Plantarum*, in which dicotyledons are discussed first, then gymnosperms and finally, monocotyledons. They divided dicotyledons into 3 sub-classes and 14 series.

Bentham and Hooker's system of classification is still used and followed in several herbaria of the world. In most of the Indian herbaria too, the plants are arranged according to this system of classification. It is supposed to be the best system for the students, to identify plants in the laboratories. This is so because Bentham and Hooker prepared the generic descriptions of the plants from their own observations and not by copying from the available literature. Large genera are divided into sections and sub-sections, and the description of genera is complete and accurate.

The series were divided into cohorts (equivalent to orders), and cohorts into orders (equivalent to families). There are altogether 202 orders of angiosperms-165 orders of dicotyledons and 37 orders of monocotyledons. They started with the family Ranunculaceae (with polypetalous corolla and an indefinite number of free stamens and carpels), and ended with Labiatae (with gamopetalous corolla and a definite number of stamens and carpels), leaving out incomplete and an anomalous series. They divided monocotyledons into 7 series, and series directly into orders. They started with orders having epigynous flowers, e.g. Orchidaceae and Scitamineae, passed through orders with petaloid hypogynous flowers, e.g. Liliaceae, and then through orders with flowers which have lost their petaloid character, e.g. Palmae and Araceae. Finally, they came to Gramineae and Cyperaceae with simple construction of flowers and presence of glumes.

4.7.6 *August Wilhelm Eichler (1839-1889)*

Eichler (1883), a German botanist, put forth a widely accepted system of classification of plants. This system may be considered as a transitional phylogenetic system.

4.7.7 Adolf Engler (1844-1930) and Kral Prantl (1849-1893)

Engler and Prantl's system of classification has also been described as a 'transitional phylogenetic system' by Jones and Luchsinger (1987). However, Radford (1986) and most of the other taxonomists describe it as a phylogenetic system.

Engler's system is based on Eichler's system (1883), the major categories which Engler and his associates have accepted. Adolf Engler (1844-1930), Professor of Botany at the University of Berlin, 1st proposed his system in 1886 as a guide to the botanical garden at Breslau. The system was published in an expanded and elaborated form in *Die Naturlichen Pflanzenfamilien* (1887-1909) in 23 volumes covering the whole range of the plant kingdom, under the editorship of Engler and Prantl. Engler and Gilg's *Syllabus de Pflanzenfamilien*, published in 1892, is comprehensive work on the systematic classification of flowering plants and cryptogams, and is a very useful publication. The most noteworthy features of the Engler and Prantl's system of classification are that they (i) placed monocots before dicots, (ii) considered orchids to be more evolved than grasses, (iii) considered apetalous and catkin-bearing dicots primitive to the dicots bearing petals and simple unisexual flowers.

According to Engler's system, the plant kingdom is divided into 13 divisions, of which the seed bearing plants (spermatophyta) form the last division named 'Embryophyta Siphonogama'. This has been further divided into two sub-divisions- Gymnospermae and Angiospermae. The latter are further divided into two classes- Monocotyledoneae and Dicotyledoneae. Monocotyledoneae has been directly divided into 11 orders with 45 families, while Dicotyledoneae comprises two sub-classes- (i) Archichlamydeae, containing 30 orders with 190 families, representing lower dicotyledons, and (ii) Sympetalae (or Metachlamydeae), containing 10 orders with 53 families, representing the higher dicotyledons. The orders have been further sub-divided into sub-orders, families and genera. There are altogether 288 families. Monocotyledoneae begins with Typhaceae and ends with Orchidaceae, while Dicotyledoneae begins with Casurinaceae and ends with Compositae.

The principle involved in this system is tracing the increasing complexity of flowers, particularly their accessory whorls, viz. achlamydeous flower (no perianth), haplochlamydeous flowers (one whorl of perianth) and diplochlamydeous flowers (two whorl of perianth). Those with no perianth, or with only one whorl or with polypetalous corolla form the sub-class Archichlamydeae. The gamopetalous condition with 1 or 2 whorls of perianth comes under the sub-class Sympetalae. The latter represents more advanced groups of dicotyledons. Progress within the sub-class is indicated through hypogyny, perigyny and epigyny, and the transition from a variable number of stamens and carpels to a definite number. Among dicotyledons, Engler starts with the Casuarinaceae family on the assumption that woody plants with unisexual, apetalous flowers borne in catkins are the most primitive.

Merits

1) The whole plant kingdom has been classified with necessary sketches, records of numbers of species, and keys of identification of all known genera of plants.
2) It is a natural system based on relationships and is compatible with evolutionary principles.
3) The group have been built up step by step to form a generally progressional morphological series.
4) Many groups have been phylogenetically arranged.
5) Apetalae (monochlamydeae) have been abolished, being considered an artificial group, and amalgamated with polypetalae into a new group, i.e. sub-class Archichlamydeae.
6) Compositae, and Orchidaceae are considered, very reasonably, the highest families of dicotyledons and monocotyledons, respectively, in view of their highly evolved characteristics. Because of its merits, Engler's system was accepted by American and British botanists.

Demerits

1) Monocotyledons gain precedence over dicotyledons. The higher dicotyledons and monocotyledons are believed to have developed from some primitive dicotyledonous stock.
2) Casurinaceae is considered most primitive among dicotyledons and placed at the beginning of his system. This view is no longer acceptable.
3) Amentiferae, i.e. the catkin-bearing families with unisexual flowers, e.g. Fagaceae, Salicaceae, Betulaceae, and considered primitive among dicotyledons and placed early in his system. This group is now considered to be advanced rather than primitive. Ranales, like Ranunculaceae, and Nymphaeaceae, can never be derived from them.
4) The derivation of dicotyledons from gymnosperms with unisexual strobili is another weak point of his system.
5) The system, as a whole, cannot be regarded as phylogenetic.

6) The position is Helobiae i.e. Potamogetonaceae, Alismaceae, Hydrocharaceae, between Pandanales and Glumiflorae is very unsatisfactory. Both Pandanales and Glumiflorae are advanced groups.
7) Similarly, Araceae, Lemnaceae and Typhaceae are supposed to be derived from Liliaceae, and therefore, cannot precede it.

4.5.8 Charles Edwin Bessey (1945-1915)

A system of classification that tries to reflect evolution is called phylogenetic system. Bessey, a student of Asa Gray, and a professor at the University of Nebraska, contributed a pure phylogenetic system of plant classification in 1894 after being highly impressed by the evolutionary ideas of Darwin. It could be published in the final form in 1915, shortly after his death. Bessey closely followed Bentham and Hooker's system of classification. His system was based on certain guiding principles of primitive and advanced characters called 'dicta'. He considered angiosperms monophyletic and derived from a cycadeoid ancestor having bisexual strobili.

According to Bessey (1915), Ranales were the primitive angiosperms. One branch of Ranales developed into monocots, and the other into dicots. This system is often called Ranalian concept of evolution. On being diagramed, this system took the form of a cactus plant, and came to be called Bessey's cactus.

4.5.9 John Hutchinson (1884-1972)

John Hutchinson, formerly Director of Royal Botanic Garden at Kew, England, is a leadin g exponent of a phylogenetic system of classification. He classified only the angiosperms in his famous work *The Families of flowering plants*, Volume 1 (dicotyledons), which appeared in 1926 and Volume II (monocotyledons) in 1934. The system has been revised in his British Flowering Plants, published in 1948, and in the second edition of *The Families of flowering plants*, published in 1959. His system differs from all the previous ones in several fundamental respects. It is, however, closer to Bentham and Hooker's or Bessey than to Engler's. The main features of his systems are as follows-

It is based on a logical interpretation of the theory that parts of an angiospermic flower are modified leaves. He held the view that the angiosperms, having strobilus-like, hypogynous, bisexual and polycarpellary flowers, originated in an extinct Mesozoic group of gymnosperms- the Bennettitales. According to Hutchinson, two primitive orders of dicotyledons arose independently from this ancestral stock through an imaginary group of proangiosperms, and developed simultaneously on two distinct parallel lines. Thus, he considered the origin of angiosperms as monophyletic, and supported Hallier and Arber and Parkin. Accordingly, he recognized two distinct proceeding on two phyletic lines- woody and herbaceous. On one line evolved the fundamentally and predominantly woody orders, which he called Lignosae, and on the other line evolved the orders, fundamentally and predominantly herbaceous, which he called Herbaceae. The woody line started from the primitive order of Ranales (family Magnoliaceae), passed through Annonaceae, Lauraceae, Dilleniaceae, Rosaceae, etc., and ended in families like Apocynaceae, Rubiaceae, Verbanaceae, etc. The herbaceous line started from the primitive order of Ranales (family Ranunculaceae), passed through Nymphaeaceae, Crudiferae, Caryophyllaceae, etc., and ended in families like Umbelliferae, Compositae, Solanaceae, Scrophulariaceae, Labiatae, etc. Hutchinson recognized several climax orders in the phylogeny of the dicotyledons and rearranged them in proper sequences. Orders having both woody and herbaceous forms were considered polyphyletic, i.e. members originating from different ancestors, as found in Apetalae (without perianth), Urticales, Umbellales, Euphorbiales, etc.

Primitive and Advanced Characteristics

These characteristics, which formed the basis of his classification, are described in the following list.

1. The polypetalous condition is more primitive than the gamopetalous condition.
2. Unisexual flower are more advanced than bisexual flowers.
3. Apetalous forms have originated from petaliferous stock, and represent reduction and advance.
4. Free parts, on the whole, are considered more primitive than connate or adnate parts. Thus numerous free stamens preceded few or connate stamens. Similarly, the apocarpous pistil is more primitive than the syncarpous pistil.
5. Spiral arrangement of parts is more primitive than cyclic, etc.

Hutchinson revised the scheme of classification (1969) is as follows. He has divided 'flowering' plants into two phyla: phylum I- Gymnospermae (not elaborated by him) and phylum II- Angiospermae. The latter are divided into two sub-phyla: sub-phylum I- Dicotyledons and sub-phylum II- Monocotyledons. Dicotyledons are divided into two big divisions: division I- Lignosae (with 54 order and 246 families) and division II- Herbaceae (with 28 orders and 96 families). Monocotyledons are divided into three divisions: division I- Calyciferae (with 12 orders and 29 families), division II- Corolloferae (with

14 orders and 34 families) and division III- Glumiflorae (with 3 orders and 6 families). Thus, there are altogether 411 families of Angiospermae- 342 of dicotyledons and 69 of monocotyledons. In his classification, Hutchinson placed the Gymnospermae first, then the Dicotyledons, and finally the Monocotyledons.

Merits

1) Most taxonomists believe that this system has given us a much better idea of phylogenetic conception and has stimulated rethinking on the subject.
2) The primitive or basic orders are Magnoliales, representing arborescent families, and Ranales, representing herbaceous families, giving rise to woody and herbaceous forms, respectively, on parallel lines.
3) Bisexual and polypetalous flowers precede unisexual and gamopetalous flowers.
4) Amentiferae (catkin bearing families with unisexual, apetalous flowers), e.g. Fagaceae-oak, Betulaceae- birch, Juglandaceae- walnut and Salicaceae- willow and poplar, is regarded as advanced (and not primitive, as by Engler) and thus has been transferred (including Urticales) to a new phyletic position close to Rosales and Leguminosae. Their apparent simplicity represents reduction and specialization (and not primitiveness).
5) Casuarinaceae has been assigned an advanced position and placed at the top of the Amentiferae.
6) Several large orders have been split up into distinct small families, thus simplifying matters e.g. Rosales, Perietales and Malvales.
7) Many families have been raised to the rank of orders, e.g. Leguminosae (now to an oder), Saxifragraceae to Saxifragales and Podostemaceae to Podostemales.
8) Several orders and families have been reshuffled, e.g. Cactales placed close to Cucurbitales and Passiflorales.
9) The origin of monocotyledons has been traced to dicotyledons at an early stage of evolution, the point of origin being the Ranales.
10) Helobiae (of Engler) have been split up into separate orders and the order of Butomales is considered to be the starting point of monocotyledons.
11) The genera of Liliceae and Amaryllidaceae have been reshuffled on the basis of inflorescence characteristics (and not on ovary position).

12) Several orders have been rearranged, ending in Cyperales and Graminales.

Demerits

1) Many doubt the wisdom of Hutchinson's rigid bifurcation of dicotyledons into Lignosae (woody type) and Herbaceae (herbaceous types).
2) Many hold different views regarding the relationship between various orders and families.
3) The monophyletic view of the origin of angiosperms is not universally accepted.
4) The monophyletic origin of monocotyledons from the Ranales, as put forward by Hutchinson, clashes with the views of Lotsy (1911) and Halloer (1912), who put forward polyphyletic (diplhyletic) views.
5) Urticales, Umbellales, Euphorbiales, etc. are said to originate from different ancestors.

4.8 Comparison of Systems of Classification

Table 4.2 provides points of comparison between three main systems of classification proposed by Bentham and Hooker, Engler and Prantl, and Hutchinson.

Table 4.2 Comparison of systems of classification

Sl. No.	Bentham and Hooker	Engler and Prantl	Hutchinson
1	It is a natural system of classification	It is a transitional phylogenetic system	It is a phylogeneic system
2	It is based on de Candolle's system of classification	It is based on Eichler's system of classification	It is based on Bessey's system of classification
3	Fixity of species is the main idea behind this system	Darwinian theory of descent is the main idea behind this system	Considerable knowledge of phylogeny is the basis of this classification
4	Dicots are divided into (i) Polypetalae, (ii) Gamopetalae, (iii) Monochlamydae	Dicots are divided into (i) Archichlamydae, (ii) Metachlamydae	Dicots are divided into (i) Lignosae (ii) Herbaceae
5	Gymnosperms are placed in between dicots and monocots	Gymnosperms are placed before dicots	Gymnosperms are placed before dicots.
6	Monocots are treated after dicots	Monocots are treated before dicots	Monocots are treated after dicots
7	Orchidaceae and some other advanced families are regarded as primitive and treated in the beginning	Orchidaceae are treated as advanced	Orchidaceae are treated as advanced
8	Families with free petals are placed in Polypetalae, with fused petals in Gamopetalae, and without petals in Monochlamydeae	Polypetalae and Monochlamydeae are placed together under a single group Archichlamydeae. Families with fused petals are placed under Metachlamydeae	Emphasis is on resemblances in place of differences, and so the Gamopetalae and Monochlamydeae are distributed according to their relationship in Polypetalae.
9	Monocotyledonous families are arranged in 7 series	Monocotyledonous families are arranged in 11 series	Monocotyledonous families are arranged in 29 orders
10	All flowering plants placed in 200 families	280 families of flowering plants have been recognized	Number of angiospermic families is 411
11	Gramineae is the last family of monocots	Orchidaceae is the last family of monocots	Gramineae is the last family of monocots

4.9 Current Systems of Classification

Armen Takhtajan (1980) of Russia, Arthur Cronquist (1981) of USA, Rolf Dahlgren (1983) of Denmark, and Robert F. Thorne (1983) of USA published systems of plant classification during recent years. These workers used the recent data and information of palaeobotany, biochemical systematic, and the ultrastructural details unveiled by the electron microscope, along with the information of the traditional sources such as morphology and anatomy of plants.

4.9.1 Armen Takhtajan (1980)

The latest version of Takhtajan's classification is published in 1980 in Botanical Review. According to him, the angiosperms are monophyletic and developed from some very ancient group of gymnosperms. He is an adherent of the Besseyan philosophy of taxonomy.

Some of the criteria, used by Takhtajan in determining the relative degree of advancement of flowering plants are undermentioned: (i) Woody plants are primitive compared to herbaceous plants; (ii) Deciduous woody plants are evolved from evergreen plants; (iii) Parallel venation is most advanced; (iv) Most primitive leaf arrangement is alternate; (v) Stomata with subsidiary cells are primitive while those lacking subsidiary cells are advanced; (vi) Unilacunar nodes are derived from trilacunar nodes; (vii) Xylem fibres evolved from tracheids to libriformfibres, through fibretracheids, (viii) Cymose inflorescence is primitive while racemose is derived; (ix) Flowers with an indefinite or a variable number of their floral parts are primitive; (x) Pollen grains with their exine lacking any external sculpturing are primitive while those having various types of sculptures are advanced; (xi) Apocarpous gynoecium is the characteristic of primitive taxa; (xii) Unitegmic ovules developed from bitegmic ovules; (xiii) Basic type of ovule is anatropous type, all others are derived ones; (xiv) Basic and most primitive type of female gametophyte is 8-nucleate Polygonum-type; (xv) Primitive condition is porogamy, and the derived conditions are misogamy and chalazogamy; (xvi) Most primitive and basic type of fruit is a many-seeded follicle which develops from a multicarpellary apocarpous gynoecium.

TakhtajandevidedMagnoliophyta (= Angiospermae) into two classes- Magnoliopsida (dicots) and Liliopsida (monocots). He further divided Magnoliopsida into 7 subclasses, 20 superorders and 71 orders, and Liliopsida into 3 subclasses, 8 superorders and 21 orders.

Merits of the System of Takhtajan

1. Dicots are discussed prior to monocots.
2. Dicots begin with the Magnoliales which are universally considered to be the most primitive living angiosperms.
3. In this system the families are small homogeneous units made up of closely related genera.
4. Division of dicots into two traditional groups of Engler and Prantl, i.e. Archichlamydae and Metachlamydae, has been abolished in this system.
5. Alismatales which are considered to be the most primitive living monocots these days are the starting point of monocots in this system.

Demerits of the System of Takhtajan

One strong demerit is the extremely narrowly defined taxa in this system that has resulted in the unwarranted splitting of the related groups.

4.9.2 Arthur Cronquist (1981)

On the basis of a wide range of taxonomic characters of Phylogenetic importance, Cronquist (1981) presented a phylogenetic classification. With the help of synoptic keys and detailed charts, he grouped the angiospermic families into orders and subclasses on a worldwide basis. Pteridosperms (seed ferns) were considered by him as the probable ancestors of angiosperms.

Class Magnoliopsida (= dicots) in this system has been divided into six subclasses i.e. Magnoliidae, Hamamelidae, Caryophyllidae, Dilleniidae, Rosidae, and Asteridae. Of them, the Magnoliidae were thought to be the basal complex, and all the remaining five subclasses were derived from it separately. Asteridae formed the most advanced group of dicots.

Class Liliopsida(= monocots) in this system has been divided into five subclasses i.e. Alismatidae, Arecidae, Commelinidae, Zingiberidae and Liliidae. It is proposed by Cronquist (1981) that Liliopsida have arisen from aquatic ancestors.

Cronquist (1981) recognizes a total of 2 classes, 11 subclasses, 83 orders and 383 families and about 2,19,300 species among the angiosperms.

4.9.3 Rolf M.T. Dahlgren (1983)

Rolf Dahlgren (1980, 1981, 1983), professor at the Botanical Museum of the University of Copenhagen, Denmark, proposed a new classification on the basis of characters drawn from several branches of botany including anatomy, embryology, and chemistry. He believes that "none of the extant groups of flowering plants is ancestral to any other present-day group. Thus, the Magnoliaceae-Ranunculaceae assemblage is not ancestral to other flowering plants but has simply retained a large number of primitive features" (Jones and Luchsinger, 1987). Dahlgren has been of the view that angiosperms evolved in one particular line of gymnosperms, and are thus monophyletic. Magnoliiflorae have been considered as most primitive dicotyledons while Alismatiflorae have been regarded as most primitive monocots.

Dicots (Magnoliidae) have been divided into 25 superorders and 85 orders while monocots (Liliidae) have been divided into 8 superorders and 25 orders.

4.9.4 Robert F. Thorne (1983)

On the basis of characters of comparative morphology, ultrastructure, embryology, palaeobotany, cytology, plant geography, pollen and seed morphology, and host-parasite relationships, Thorne (1983) published a pure phylogenetic system. He also believed in the monophyletic origin of angiosperms.

Thorne divided class Angiospermae (Annonopsida) into 2 sub-class, 28 superorders (with ending of *iflorae*), 54 orders, 73 sub-class, 350 families, 12,255 genera and 2,25,490 species.

Merits of Thorne's system

1. Annonales are universally accepted as the most primitive living angiosperms, and they form the starting point of this system.
2. Traditional division of dicots into Archichlamydeae and Metachlamydeae has been abolished in this system.
3. In this system, by and large, the closely related taxa are placed nearer to one another.
4. The families of "Amentiferae" (a group of unrelated families) have been distributed in different orders and the group Amentiferae has been abolished.
5. Inclusion of orders Malvales, Urticales, Rhamnales and Euphorbiales under one superorder Malviflorae is also a commendable aspect of this system.
6. Placing of the related orders Cornales and Dipsacles in one superorder Corniflorae is also the merit of this system.

Demerits of the Thorne's system

1. This system is not of much practical utility in the identification of plants.
2. The views of Thorne that angiosperms might have originated from some Pteridospermous members in early Cretaceous times is also not accepted by several taxonomists.

4.10 Angiosperm Phylogeny Group (APG) system

The Cronquist system, proposed by Arthur Cronquist in 1968 and published in its full form in 1981, is still widely used but is no longer believed to accurately reflect phylogeny. A consensus about how the flowering plants should be arranged has recently begun to emerge through the work of the Angiosperm Phylogeny Group (APG), which published an influential reclassification of the angiosperms in 1998.

Updates incorporating more recent research were published as APG II in 2003 and as APG III in 2009.

Traditionally, the flowering plants are divided into two groups, which in the Cronquist system are called Magnoliopsida (at the rank of class, formed from the family name *Magnoliacae*) and Liliopsida (at the rank of class, formed from the family name *Liliaceae*). Other descriptive names allowed by Article 16 of the ICBN include Dicotyledones or *Dicotyledoneae*, and Monocotyledones or *Monocotyledoneae*, which have a long history of use.

In English a member of either group may be called a dicotyledon (plural dicotyledons) and monocotyledon (plural monocotyledons), or abbreviated, as dicot (plural dicots) and monocot (plural monocots).

These names derive from the observation that the dicots most often have two cotyledons, or embryonic leaves, within each seed.

The monocots usually have only one, but the rule is not absolute either way. From a diagnostic point of view the number of cotyledons is neither a particularly handy nor reliable character.

Recent studies, as by the APG, show that the monocots form a monophyletic group (clade) but that the dicots do not (they are paraphyletic).

Nevertheless, the majority of dicot species do form a monophyletic group, called the eudicots or tricolpates. Of the remaining dicot species, most belong to a third major clade known as the *Magnoliidae*, containing about 9,000 species.

The rest include a paraphyletic grouping of primitive species known collectively as the basal angiosperms, plus the families Ceratophyllaceae and Chloranthaceae.

An important motivation for the ARG was what they viewed as deficiencies in prior angiosperm classifications, which were not based on monophyletic groups (i.e. groups consisting of all the descendants of a common ancestor). APG publications are increasingly influential, with a number of major herbaria changing the arrangement of their collections to match the latest APG system.

4.10.1 Angiosperm Classification and the APG

Until detailed genetic evidence became available, the classification of flowering plants (also known as angiosperms, *Angiospermae*, *Anthophyta* or *Magnoliophyta*) was based on their morphology (particularly that of the flower) and their biochemistry (what kinds of chemical compound they contained or produced).

Classification systems were typically produced by an individual botanist or by a small group. The result was a large number of such systems (see List of systems of plant taxonomy).

Different systems and their updates tended to be favoured in different countries; e.g. the Engler system in continental Europe; the Bentham & Hooker system in Britain (particularly influential because it was used by Kew); the Takhtajan system in the former Soviet Union and countries within its sphere of influence; and the Cronquist system in the United States.

4.10.2 Flowering Plant Diversity

The number of species of flowering plants is estimated to be in the range of 250,000 to 400,000. The number of families in APG (1998) was 462. In APG II (2003) it is not settled; at maximum it is 457, but within this number there are 55 optional segregates, so that the diversity of flowering plants is not evenly distributed. Nearly all species belong to the eudicot (75%), monocot (23%) and magnoliid (2%) clades. The remaining 5 clades contain a little over 250 species in total, i.e., less than 0.1% of flowering plant diversity, divided among 9 families.

The most diverse families of flowering plants, in their APG circumscriptions, in order of number of species, are:

- *Asteraceae* or *Compositae* (daisy family): 23,600 species;
- *Orchidaceae* (orchid family): 22,075 species;

- *Fabaceae* or *Leguminosae* (pea family): 19,400;
- *Rubiaceae* (madder family): 13,150;
- *Poaceae* or *Gramineae* (grass family): 10,035;
- *Lamiaceae* or *Labiatae* (mint family): 7,173;
- *Euphorbiaceae* (spurge family): 5,735;
- *Melastomataceae* (melastome family): 5,005;
- *Myrtaceae* (myrtle family): 4,620;
- *Apocynaceae* (dogbane family): 4,555.

In the list above (showing only the 10 largest families), the *Orchidaceae* and *Poaceae* are monocot families; the others are eudicot families.

In the late 1990s, an informal group of researchers from major institutions worldwide came together under the title of the 'Angiosperm Phylogeny Group' or APG.

Their intention was to provide a widely accepted and more stable point of reference for angiosperm classification. Their first attempt at a new system was published in 1998 (the APG system).

As of 2010, two revisions have been published, in 2003 (APG II) and in 2009 (APG III), each superseding the previous system. Eight researchers have been listed as authors to the three papers, and a further 33 as contributors.

However, the APG publications are increasingly regarded as an authoritative point of reference. Some examples:

- A significant number of major herbaria, including Kew, are changing the order of their collections in accordance with APG.
- The influential World Checklist of Selected Plant Families (also from Kew) is being updated to the APG III system.
- In the USA, a recent photographic survey of the plants of the USA and Canada is organized according to the APG II system.
- In the UK, the latest edition of the standard flora of the British Isles (by Stace) is based on the APG III system. The previous edition was based on the Cronquist system.

4.10.3 APG-I system

The initial 1998 paper by the APG made angiosperms the first large group of organisms to be systematically re-classified primarily on the basis of genetic characteristics. The paper explains the authors' view that there is a need for a classification system for angiosperms at the level of families, orders and above, but that existing classifications are "outdated".

The main reason why existing systems are rejected is because they are not phylogenetic, i.e. are not based on strictly monophyletic groups (i.e. groups which consist of all descendants of a common ancestor). An ordinal classification of flowering plant families is proposed as a "reference tool of broad utility". The broad approach adopted to defining the limits of orders resulted in the recognition of 40 orders, compared to, for example, 232 in Takhtajan's 1997 classification.

Other features of the proposed classification include

- Formal, scientific names are not used above the level of order, named clades being used instead.
- Thus eudicots and monocots are not given a formal rank on the grounds that "it is not yet clearvat which level they should be recognized".
- A substantial number of taxa whose classification had traditionally been uncertain are givenvplaces, although there still remain 25 families of "uncertain position".
- Alternative classifications are provided for some groups, in which a number of families can either be regarded as separate or can be merged into a single larger family. For example, the *Fumariaceae* can either be treated as a separate family or as part of *Papaveraceae*.

A major outcome of the classification is the disappearance of the traditional division of the flowering plants into two groups, monocots and dicots. The monocots are recognized as a clade, but the dicots are not, with a number of former dicots being placed in separate groups basal to both monocots and the remaining dicots, the eudicots or "true dicots".

APG I was superseded in 2003 by a revision, the APG II system, and then in 2009 by a further revision, the APG III system.

4.10.4 APG-II System

The second paper published by the APG presents an update to the original classification of 1998. The authors say that changes have been proposed only when there is "substantial new evidence" which supports them. The proposed classification continues the tradition of seeking broad circumscriptions of taxa, for example trying to place small families containing only one genus in a larger group. APG II continues and indeed extends the use of alternative "bracketed" taxa allowing the choice of either a large family or a number of smaller ones.

For example, the large *Asparagaceae* family includes 7 "bracketed" families which can either be considered as part of the *Asparagaceae* or as separate families.

Some of the main changes in APG II are:

- New orders are proposed, particularly to accommodate the 'basal clades' left as families in the first system.
- Many of the previously unplaced families are now located within the system.
- Several major families are re-structured.

It was a revision of the first APG system, published in 1998, and was superseded in 2009 by a further revision, the APG III system. APG II was published as:

- Angiosperm Phylogeny Group (2003). "An update of the Angiosperm Phylogeny Group classification for the orders and families of flowering plants: APG II". Botanical Journal of the Linnean Society 141(4): 399-436.
- The APG II system recognized 45 orders, five more than the APG system.
- The new orders were *Austrobaileyales*, *Canellales*, *Gunnerales*, *Celastrales*, and *Crossosomatales*, all of which were families unplaced as to order, although contained in supraordinal clades, in the APG system.
- APG II recognized 457 families, five fewer than the APG system. Thirty-nine of the APG II families were not placed in any order, but 36 of the 39 were placed in a supra-ordinal clade within the angiosperms. Fifty-five of the families came to be known as "bracketed families".

4.10.5 APG-III System

It is the third version of a modern, mostly molecular-based, system of plant taxonomy that was published in 2009 by the Angiosperm Phylogeny Group, 6½ years after its predecessor. In October 2009, members of the Linnean Society proposed an accompanying formal Phylogenetic classification of all land plants, compatible with the APG III classification.

In the APG-III the broad outline of the system remains unchanged, but the number of previously unplaced families and genera is significantly reduced. This requires the recognition of both new orders and new families compared to the previous classification. The number of orders goes up from 45 to 59; only 10 families are not placed in an order and only two of these (Apodanthaceae and Cynomoriaceae) are left entirely outside the classification (Chase *et al.*, 2009).

A major change is that the paper discontinues the use of 'bracketed' families in favour of larger, more inclusive families. As a result, the APG III system contains only 415 families, rather than the 457 of APG II. For example, the agave family (Agavaceae) and the hyacinth family (Hyacinthaceae) are no longer regarded as distinct from the broader asparagus family (Asparagaceae).

The APG III system recognized all of the 45 orders of the previous system, as well as 14 new ones. The order Ceratophyllales was erroneously marked as a new order, but it had been recognized in both of the previous APG systems. The newly recognized orders were: Amborellales, Nymphaeales, Chloranthales, Petrosaviales, Trochodendrales, Buxales, Vitales, Zygophyllales, Picramniales, Huerteales, Berberidopsidales, Escalloniales, Bruniales, and Paracryphiales.

The designation of alternative "bracketed families" was abandoned in APG III, because its inclusion in the previous system had been unpopular. APG III recognized 415 families, 42 fewer than in the previous system. Forty-four of the 55 "bracketed families" were discontinued, and 18 other families were discontinued as well.

About 20 families were accepted in the APG III system which had not been in the previous system, and a few families were moved to a different position. The newly recognized families are: Petermanniaceae, Schoepfiaceae, Limeaceae, Lophiocarpaceae, Montiaceae, Talinaceae, Anacampserotaceae, Centroplacaceae, Calophyllaceae, Guamatelaceae, Gerrardinaceae, Dipentodontaceae, Capparidaceae, Cleomaceae, Cytinaceae, Mitrastemonaceae, Metteniusaceae, Linderniaceae, Thomandersiaceae, and Quintiniaceae.

The number of families not placed in any order was reduced from 39 to 10. *Apodanthaceae* and *Cynomoriaceae* were placed among the angiosperms, incertae sedis, that is, not in any group within the angiosperms. Eight other families were placed incertae sedis in various supra-ordinal groups within the angiosperms. The unplaced families were: Apodanthaceae, Cynomoriaceae, Dasypogonaceae, Sabiaceae, Dilleniaceae, Icacinaceae, Metteniusaceae, Oncothecaceae, Vahliaceae, and Boraginaceae.

Three genera (*Gumillea*, *Nicobariodendron*, and *Petenaea*) were placed within the angiosperms incertae sedis. *Gumillea* had been unplaced in APG II. *Nicobariodendron* and *Petenaea* were newly added to the list.

Taxa of uncertain position

- *Apodanthaceae* Takht. (three genera)
- *Cynomoriaceae* Endl. ex Lindl.
- *Gumillea* Ruiz & Pav.
- *Petenaea* Lundell (probably in *Malvales*)
- *Nicobariodendron* (probably in *Celastraceae*).

Sub-families replacing discontinued families

- The number of sub-families replacing APG II bracketed families is reported in table 4.3.

Table 4.3. A number of subfamilies have been proposed to replace some of the families which were optional (i.e. bracketed) in APG II, but have been discontinued in APG III. These are shown in the table below.

Sub-families replacing APG II bracketed families	
APG II bracketed family	**APG III family: sub-family**
Agapanthaceae	Amaryllidaceae: Agapanthoideae
Agavaceae	Asparagaceae: Agavoideae
Alliaceae	Amaryllidaceae: Allioideae
Amaryllidaceae	Amaryllidaceae: Amaryllidoideae
Aphyllanthaceae	Asparagaceae: Aphyllanthoideae
Asparagaceae	Asparagaceae: Asparagoideae
Asphodelaceae	Xanthorrhoeaceae: Asphodeloideae
Hemerocallidaceae	Xanthorrhoeaceae: Hemerocallidoideae
Hesperocallidaceae	Asparagaceae: Agavoideae
Hyacinthaceae	Asparagaceae: Scilloideae
Laxmanniaceae	Asparagaceae: Lomandroideae
Ruscaceae	Asparagaceae: Nolinoideae
Themidaceae	Asparagaceae: Brodiaeoideae
Xanthorrhoeaceae	Xanthorrhoeaceae: Xanthorrhoeoideae

4.11 Features Used in Description of Angiospermic Plants

1. Habitat: Natural abode of the plant.
2. Habit: Herb (erect, prostrate, decumbent, diffuse, trailing, twining or climbing), shrub (erect, straggling, twining or climbing) or tree or any other peculiarity in the habit.
3. Root: Nature of root; any special form.

4. Stem: Kind of stem- herbaceous or woody; cylindrical or angular, hairy or smooth; jointed or not; hollow or solid; erect, prostrate, twining or climbing; nature of modification, if any.
5. Leaf: Arrangement- whether alternate, opposite or whorled; stipulate or exstipulate; nature of the stipules, if present, simple or compound; nature of the compound leaf and the number of leaflets; shape and size; hairy or smooth; deciduous or persistent; venation; margin; apex; and petiole.
6. Inflorescence: Type of inflorescence.
7. Flower: Sessile or stalked; complete or incomplete; unisexual or bisexual; regular, zygomorphic or irregular; hypogynous, epigynous or perigynous; bracteate or ebracteate; nature of bracts and bracteoles; if present; shape, colour and size of the flower.
8. Calyx: Polysepalous or gamosepalous; number of sepals or lobes; superior or inferior; aestivation; shape, size and colour.
9. Corolla: Polypetalous or gamopetalous; number of petals or lobes; superior or inferior; aestivation; shape, size, colour and scent; corona or any special feature. (When there is not much difference between the calyx and the corolla, the term perianth should be used. It may e sepaloid or petaloid, polyphyllous orgamophyllous or free or epiphyllous).
10. Androecium: Number of stamens- definite (ten or less) or indefinite (more than ten); free or united; nature of cohesion- monadelphous, diadelphous, polyadelphous, syngenesious or synandrous; nature of adhesion- epipetalous or gynandrous, or any special feature; whether alternating with the petals (or corolla-lobes) or opposite them. Length of stamens- general length; inserted or exerted; didynamous or tetradynamous; position of stamens- hypogynous, perigynous or epigynous; attachment of the anther and its dehiscence; anther lobes or appendages, if any.
11. Gynoecium or pistil: Number of carpels; syncarpous or apocarpous; nature of style- long or short; stigmas- simple, lobed or branched; their number and nature- smooth or papillose; ovary- superior or inferior; number of lobes; number of chambers (loculi); nature of placentation; number and form of ovules in each locules of the ovary.
12. Seeds: Number of seeds in the fruit; shape and size; albuminous or exalbuminous; nature of endosperm, if present.

References

Andrew Sugden, 1984. Longman Illustrated Dictionary of Botany. University of Wisconsin, Madison. pp. 1-192.

Angiosperm Phylogeny Group. 2009. "An update of the Angiosperm Phylogeny Group classification for the orders and families of flowering plants: APG III", Botanical Journal of the Linnean Society, **161** (2): 105-121.

Bentham, G. and Hooker, J.D. 1862-1883. Genera Plantarum, 3 vols. London.

Candolle, A.P. de. 1813. Theorieelementaire de la botanique, Paris.

Chase, M.W., Reveal, J.L. and Fay, M. F. 2009. Botanical Journal of the Linnean Society, 161 (2): 132-136.

Cronquist, A. 1981. An integrated system of classification of flowering plants. Columbia University Press, New York.

Dahlgren, R. 1981. Botanical Journal of Linnaean Society, **80:** 91-124.

Dahlgren, R. 1983. Nordic Journal of Botany, **3:** 119-149.

Eichler, A.W. 1875-1883. Blutthendiagrammeconstruirt und erlautert, 2 vols. Leipzig.

Hutchinson, J. 1964-1967. The genera of flowering plants, 2 vols. Clarendon, Oxford.

Hutchinson, J. 1969. Evolution and phylogeny of flowering plants, Academic press, London.

Jones, Jr. S.B. and Luchsinger, A.E. 1987: Plant Systematics and Evolution, McGraw-Hill International Editions, New Delhi, India.

Jussieu, A.L. de. 1789. Genera Plantarum secundumordinesnaturalesdisposita, Paris.

Linnaeus, C. 1735. Systema Naturae, Lugduni Batavorum.

Linnaeus, C. 1737. Genera Plantarum, Lugduni Batavorum.

Linnaeus, C. 1753. Species Plantarum, 2 vols. Stockholm.

Radford, A.E. 1986. Fundamentals of plant systematics, Harper and Row, New York.

Takhtajan, A. 1980. Botanical Review, 46: 225-359.

Thorne, R.F. 1976. Evolutionary Biology, **9:** 35-106.

Thorne, R.F. 1983. Nordic Journal of Botany, **3:** 85-117.

5

Classification of Ornamental Plant Nomenclature

5.1 Fundamentals of Plant Nomenclature

(A) Plant Nomenclature

5.1.1 Definitions

Assignment of definite names to plants is called nomenclature. In the present botanical world, the nomenclature involves the principles governed by rules formulated and adopted by International Botanical Congresses. The rules developed by IBC are listed formally in a code called International Code of Botanical Nomenclature (Voss *et al.*, 1983), abbreviated as ICBN. The major goal of ICBN is to provide one correct name for each taxon.

Taxa (singular, taxon) are the taxonomic groups of any rank. The ascending hieraracy of taxa include species, genus, family, order, class and division.

5.1.2 Who is a Nomenclaturist?

A nomenclaturist is a taxonomist who assigns names to new taxa, determines the right names for old taxa according to the rules of International Code of Botanical Nomenclature, and finalizes the right name for a specimen according to an established system of classification.

5.1.3 Why is Nomenclature Needed?

Imagine for a few moments about the state of affairs if there are no names for all the things we see, make or handle. How strange and chaotic the life would be in such a condition? The entire business affairs of the world would stop, and practically there will be no 'give and take' of knowledge. It would also be impossible for all of us to communicate our feelings to others. Practically, the life would stop. Therefore, nomenclature for everything is needed.

5.2 Common names and Scientific Names

Vernacular or common names are made up of words from the native language of the country or the region. They may vary in different countries as well as in different regions of the same country. They may be Portuguese words in Portugal; Spanish in Spain; English in England, USA and Australia; Hindi, Bangla, Oriya or Tamil, etc. in Uttar Pradesh, West Bengal, Odisha and Tamilnadu, respectively. 'Papita', 'Kela', 'Gajar', and 'Tamatar', of India and known in USA as papaya, banana, carrot and tomato and by other names in Spain. There is therefore no international uniformity in common names.

On the other hand, scientific names which are based mainly on Latin language have international uniformity. By the name *Carica papaya*, all botanists of the world would understand that it is papaya.

Jones and Luchsinger (1987) have stated that common names present five major problems mentioned in the following table:

Sl. No.	Common names	Scientific names
1	These are not universal. They vary in different languages.	These are universal and are recognized throughout the world.
2	They do not provide information indicating generic and family relationships	They provide information regarding these relationships
3.	A well-known plant may have hundreds of common names	A well-known plant has only one scientific name
4	Sometimes, two or more plants have the same common name	Two or more plants always have different scientific names
5	Many species do not have any common names	All known plants have a scientific name

The scientific name of a plant consists of two separate words. The first word designates the genus of the organism and the second word designates the species. Details of this two-name system were 1st given by Linnaeus (1753) in his famous *binomial system of nomenclature*.

5.3 Binomial Nomenclature

5.3.1 What is Binomial Nomenclature?

In the earlier days plant names were long and descriptive e.g. in the herbal of Clusius (1583), a species of willow is named *Salix pumila angustifolia altera*. Then, in 1623, Gasper Bauhin (1560-1624) devised a plant of adopting two names for each plant in his *Pinax Theatri Botanici*. But it was C. Linnaeus, to whom the actual credit goes for devising and methodically employing the binomial system of nomenclature. Linnaeus employed his system in the first edition of his *Species Plantarum* in 1753. According to this system, the scientific name of a plant consists of two Latin or Latinized words: (1) The first is the name of the genus, i.e. generic name or generic epithet and (2) The second is the name of the species, i.e. specific epithet. For example, the botanical name of poppy is *Papaver somniferum*. The first name (*Papaver*) designates the genus of the plant and the second word (*somniferum*) designates the species of this genus.

5.3.2 Generic name

It is always a noun, and written with a capitalized initial letter and the remainder small. The generic name is also always singular in number. It may have following types of origin:

1. Several generic names may be in honour of the names of well-known person, e.g. *Theophrasta* in honour of Theophrastus and *Candollea* in hononur of A.P. de Candolle.
2. They may be descriptive with reference to some common characteristics of the included species, e.g. *Cercocarpus* (coiled fruit), *Xanthoxylum* (yellow wood), etc.
3. They may be of poetic or mythological origin, e.g. *Theobroma* (god's food).
4. They may also be the aboriginal name of the plants, e.g. *Betula* and *Quercus* which were the old Greek names for the Birch and Oak.

5.3.3 Specific Epithet

According to the 1983 recommendations of the International Code of Botanical Nomenclature, all specific epithets be written with a small initial letter. But if the specific epithet is derived from common names, or from former generic names, or from a person's name, the initial letter of the epithet may be a capital letter. In the typed or handwritten matter, both the specific epithet and generic names should be underlined. They should be printed in *italics* and *boldface*. The authority, written after the species epithet, is never underlined.

The specific epithet is often as adjective. According to article 23 of ICBN, a specific epithet (i) may be a name in honour of a person, or (ii) may be derived from a geographical location, or (iii) may originate from an old common name, or (iv) may be derived from some characteristics of the plant, or (v) it may also be named arbitrarily.

5.3.4 Authority

The name of the species is incomplete if it is not followed by full or abbreviated name(s) of the author(s). For example *Pyrus malus* is incomplete. The complete name is *Pyrus malus* L. where L. is abbreviated for Linnaeus. Citation of the full or abbreviated form of the author is necessary because this will verify the date and time of the first valid publication of the name of a particular taxon.

5.4 International Code of Botanical Nomenclature (ICBN)

5.4.1 History of ICBN

Linnaeus in 1737 and again in 1751 proposed the elementary rules of naming plants in his Philosophia Botanica. Then in 1813, A.P. de Candolle set forth a detailed set of rules regarding plant nomenclature in his 'Theorie elementaire de la botanique'. The same rules of Linnaeus, A.P.de Candolle and his son Alphonse de Candolle were later evolved into our present International Code of Botanical Nomenclature (ICBN).

Alphonse de Candolle convened the 1st International Botanical Congress in 1867 in Paris. It was attended by the botanists of several countries. They adopted a set of rules of plant nomenclature, most of which were proposed by A. de Candolle. These excellent rules of plant nomenclature are known as de Candolle rules or Paris Code of 1867. Subsequent meetings of the International Botanical Congress were held in 1892 (Rochester code), 1905 (Vienna code), 1907 (American code) and 1910, but a general agreement, regarding the internationally acceptable rules of plant nomenclature, was reached in the meeting of the IBC at Cambridge in 1930. Lawrence (1951) has discussed the detailed history of the code and may be quoted that in 1930 at the Cambridge Congress "for the first time in botanical history, a code of nomenclature came into being that was international in function as well as in name". This code is called the International Code of Botanical Nomenclature. Scientists in the International Botanical Congresses suggest the modifications or amendments which are incorporated in the ICBN on a regular basis.

The latest International Code of Botanical Nomenclature, 1983 was adopted by the 13th International Botanical Congress, Sydney in August 1981, and the Chairman of the editorial committee was E.G. Voss. ICBN is divided into three parts, i.e. Principles, Rules and Recommendations.

5.4.2 Principles of ICBN

The philosophical basis of the code is formed by the following six principles:

1. "Botanical nomenclature is independent of zoological nomenclature.
2. The application of names of taxonomic groups is determined by means of nomenclature types.
3. The nomenclature of taxonomic group is based upon priority of publication.
4. Each taxonomic group with a particular circumscription, position and rank can bear only one correct name, the earliest that is in accordance with the rules, except in specific cases.
5. Scientific names of taxonomic groups are treated as Latin regardless of their derivation.
6. The rules of nomenclature are retroactive unless expressly limited".

5.4.3 Rules and Recommendation of ICBN

According to the latest available ICBN (1983) the detailed provisions of the code "are divided into Rules, set out in Articles, and Recommendations". As mentioned under point No. 4 of the Preamble of the 1983 ICBN "The object of the Rules is to the nomenclature of the past into order and to provide for that of the future; names contrary to a rule cannot be maintained".

As mentioned under point to 5 of the Preamble of 1983 ICBN "The Recommendations deal with subsidiary points, their object being to bring about greater uniformity and clearness, especially in future nomenclature; names contrary to recommendation cannot, on that account, be rejected, but they are not examples to be followed".

The rules and recommendations of ICBN apply to all organisms treated as plants (including fungi but excluding bacteria), whether fossil or non-fossil. Nomenclature of bacteria is governed by a separate code called International Code of Nomenclature of Bacteria (ICNB).

It is stated under point No. 10 of ICBN (1983) that the latest "edition of the code supersedes all previous editions".

5.5 Some Important Rules of Nomenclature

1. Ranks and endings of taxa: In an accepted system of classification, each individual plant is treated as belonging to a number of taxa of different ranks. Generally the species is considered as the basic unit of classification. Other main ranks in the flowering plants in an ascending

order are genus, family, order, sub-class, and class. However, recent ICBN (Voss *et al*., 1983) has mentioned 22 different ranks and some standardized grammatical endings (suffixes) for the ranks from division down to the level of genus. All these ranks are mentioned under Article No. 1.5

2. Principles of priority: One plant might have described under different botanical names by various plant nomenclaturists in different parts of the world. But according to the "principles of priority" "each taxon is to be known by its earliest name". For example, *Cleome gynandra* Linn. Was first described and named by Linnaeus in 1753. Then he himself changed its name as *Cleome pentaphylla* Linn. In 1824 de Candolle recognized three separate genera (*Cleome*, *Polanisia*, and *Gynandropsis*) and named the Linnaeus genus as *Gynandropsis pentaphylla* (Linn.) DC. Iltis in 1960 merged *Gynandropsis* and *Cleome* into one genus, i.e. *Cleome*. So according to the "Principle of priority", the oldest name (*Cleome gynandra* Linn.) is the correct valid name.

   Because of Principle of Priority, names of several plants have now been changed.

3. Type method: The type method is a legal device to provide the correct name for a taxon. A type specimen is a herbarium sheet (or rarely a drawing or a photograph) of a specimen which was used by the author to provide its authentic description. According to Article 9 of ICBN the type of a genus is a species (e.g. the type of genus *Vernonia* is *V. noveboracensis* (L.) Michx.), and the type of of a family is a genus (e.g. *Aster* is the type genus of Asteraceae). Several kinds of types designated by ICBN are undermentioned:

   a) Holotype: It is a specimen used by the author in the original publication as the nomenclatural type.

   b) Isotype: It is a duplicate specimen of the holotype, i.e. from the same collection, with the same locality, date and number as the holotype.

   c) Lectotype: It is a specimen selected by a competent worker from the original material studied by the author of the species, when no holotype was designated or when the holotype has been destroyed or lost.

   d) Neotype: It is a specimen selected to serve as a substitute for the holotype when all material, on which the name of the taxon was based, is missing.

   e) Nomenclatural type: It is that element with which the name of a taxon is permanently associated.

f) Syntype: It is one of the two or more specimens cited by an author of a species when no holotype was designated; or a syntype is any one of the two or more specimens originally designated as types.

g) Paratype or co-types: It is a specimen other than isotype or holotype. If two or more specimens have been cited as types by the author, the remaining cited specimens are called 'co-types or paratypes'.

h) Topotype: It is a specimen collected from the same locality from where the holotype was collected.

4. Synonyms and related definitions: A name rejected due to misuse or difference in taxonomic judgement is called synonym.

   A specific or intraspecific name which has priority and is retained when transferred to a new taxon, is called a basionym.

   A case in which two or more identical names are based on different types, of which only one can be a legitimate name, is called a homonym.

   An illegitimate binomial in which the name of the genus and the name of the species is the same is called tautonym, e.g. *Armoracia armoracia* (L.) Britton.

   An automatically created legitimate tautonym for infraspecific or infrageneric taxa is called an autonym, e g. *Hypericum* sub-genus *Hypericum* section *Hypericum*.

5. Citation of author: Some of the rules related to author citation are undermentioned:

   i) Original author: The name of a taxon is complete and accurate only when it is followed by a full or abbreviated form of the author(s) who first validly published the concerned name, e.g. Liliaceae and *Lilium superbum* Linn.

   ii) Joint author: If two authors have jointly published the name of a taxon, the names of both the authors should be cited, and linked by the words *et* or *&* e.g. *Illicium griffithi* Hook and Thoms (or Hook et Thomas).

   iii) Rank alteration: When a taxon of a lower rank is upgraded in a higher rank but retains its name, the author's name who published it first should be cited in the bracket. It is to be followed by the name of the author who made the alteration e.g. Allioni raised the rank of variety *Medicago polymorpha* var. *orbicularis* L. to the species rank. Therefore, it becomes *Medicago orbicularis* (L.) All.

iv) Name proposal: When the name of a taxon is proposed but not validly published by one author, and is later on validly published by another, the word *ex* should be used as a connecting link between the name of the former author and the name of the subsequent author e.g. *Gossypium tomentosum* Nutt ex. Seem.

6. Names of cultivated plants: Wild plants brought under cultivation retain their original names.
7. Latin diagnosis: The diagnosis, details and description of new taxa published before 1st January, 1955 were accepted by ICBN as valid, irrespective of the language. After this date the description of any new taxa would be considered valid only of accompanied by Latin diagnosis.
8. Effective and valid publication: The publication of new names and description are effective and considered valid only when they are distributed in a printed form to the general public or to at least ten well-established botanical institutions.
9. Choice of names when the taxon rank is changed: When the rank of a taxon is changed (viz. a species becomes a genus or *vice versa*) the earliest legitimate name in its new rank is its correct name.
10. Choice of names when same-rank taxa are united: When two or more taxa of the same rank (viz. two or more genera, two or more species, etc.) are united into one, the oldest legitimate name of these taxa would be retained as the name of the united taxon.
11. Retention of names of divided taxa: When a genus or a species is divided into two or more genera or species, respectively, the original name of the genus or species must be retained.
12. Retention of names of taxa on transference: When a sub-division of a genus is transferred to another genus, or a species is transferred to another genus without the change of the rank, the original legitimate name must be retained. For example, *Hydrocotyle asiatica* L., on transference to genus *Centella*, must be named as *Centella asiatica* (L.) Urban.
13. Rejection of names: Names of taxa must be rejected in the following conditions:
    a) When the names are illegitimate (i.e. if it is a tautonym, later homonym, rejectable generic name, nomenclaturally superfluous, etc.).
    b) When the names give different meanings, and become a permanent source of confusion.

c) When the characters of the name of the taxon are derived from two or more different elements.

d) When the generic names coincide with a morphological term, or are unitary designations to species, or are words not intended as names.

e) When the specific names, are tautonyms, or are published without any consideration of binomial system, or are words not intended as names.

14. Names of different taxa: According to the International Code of Botanical Nomenclature (1983) the following should be the pattern of naming different taxa:

a) Genera and higher ranks should be monomials, e.g. *Rosa* L., Rosaceae Juss.

b) Species should be binomials, e.g. *Bougainvillea spectabilis* Willde.

c) Subspecies should be trinomials, e.g. *Hibiscus moscheutos* ssp. *palustris* (L.) Clausen.

d) Varieties should be quadrinomials e.g. *Lilium catesbaei* Walter ssp. *catesbaei* var. *longii* Fernald.

(B) Botanical Names

Botanical names are either Latin words or words that have been Latinized from Greek or some other language.

The name of a species is a binomial and consists of a generic name and a specific epithet. A specific epithet is the 2nd part of the binomial. However, one must be quite clear about the fact that it is wrong to use the specific epithet alone to designate a particular species; it must always be used with a generic name to form the binary combination for that species. The specific epithets are formed from nouns, adjectives, etc. and may join these words with a large number of prefixes and suffixes.

5.6 Common Prefixes Used in Specific Epithets

Some Latin prefixes of numbers are as follows:

1. uni-(L).: *uniflorus* (one flowered)
2. bi-(L).: *bifoliatus* (two leaved)
3. tri-(L).: *triangularis* (with 3 angles)
4. quadri-(L).: *quadrangularis* (with 4 angles)
5. quinque-(L).: *quinquefolius* (5 leaved)
6. sex-(L).: *sexangularis* (6 angled)

7. septem-(L).: *septemlobus* (7 lobed)
8. octo-(L).: *octoflorus* (8 flowered)
9. noveme-(L).: *novemneris* (with 9 nerves)
10. decum-(L.): *decemlobus* (with 10 lobes)

Some of the Greek prefixes of numbers include *mon*-(for one, e.g. monandra), *di*-(for 2), *tri* (for 3), *tetra*-(for 4), *penta*-(for 5), *hexa*-(for 6), *hepta*-(for 7), *octa*-(for 8), *ennea* (for 9) and *deca* (for 10 e.g. *decapetalus*).

Some of the other common prefixes of numbers are *amphi*-(Gr.) for two kinds; *diplo*-(Gr.) for double; *haplo*-(Gr.) for single; *multi*-(L.) for many; *poly*-(Gr.) for many; *a*- or *ab*-(L.) away from; *ecto*-(Gr.) for outside; *endo*-(Gr.) for inside; *inter*-(L.) for between; and intra-(L.) for within.

5.7 Common Suffixes used in Specific Epithets

Some of the common suffixes along with one example of each are under mentioned:

-aceus: *crustaceus*; -alis: *digitalis*; -aris: *angularis*; -arium: *aquarium*; -aticus: *aquaticus*; -estris: *campestris*; -eus: *roseus*; -ilis: *sextilis*; -osus: *foliosus*.

5.8 Plant Parts Used As Epithets

Almost all plant parts including root, stem, leaf, bud, flower, fruit and seeds, have been used as names of specific epithets. Some of the specific epithets, along with the related plant parts in parenthesis, are under mentioned.

receptaculum (for receptacle), sepalum (for sepals), carpellum (for carpel), loculus (for locules), stylus (for style), petalum (for petals), discus (for disc), ovarium (for ovary), ovulum (for ovule), pedicellus (for pedicel), pistillum (for pistil), fructus (for fruit), sperma (for seed), folium and phyllon (for leaf), rhiza (for root), caulos (for stem), etc.

5.9 Specific Epithets Linked with Colour

Some of the specific epithets linked with colour, along with the name of the color in the parenthesis, are under mentioned alphabetically:

albicans (whitish), *albus* (white), *atrovirens* (dark green), *aureus* (golden yellow), *calcareous* (chalky white), *candidus* (shining white), *croceus* (saffron yellow), *flavidus* (slightly yellow), *flavus* (pale yellow), *fulvus* (dull yellow), *glaucus* (grey-green), *luteolus* (pale yellow), *niger* (black), *niveus* (snow white), *purpureus* (purple), *roseus* (rosy), *violaceous* (violet), *virens* (green) and *viridus* (green).

5.10 Specific Epithets Linked with Geography

Some of the geography-linked specific epithets, along with their related geographical places in parenthesis, are under mentioned alphabetically:

africans (of Africa), *americanus* (of America), *arabicus* (of Arabia), *argentinus* (of Argentina), *asiaticus* (of Asia), *australiensis* (of Australia), *austriacus* (of Austria), *brasiliensis* (of Brazil), *canadensis* (of Canada), *chinensis* (of China), *cubensis* (of Cuba), *europaeus* (of Europe), *germanicus* (of Germany), *belveticus* (of Switzerland), *indicus* (of India), *italicus* (of Italy), *mexicanus* (of Mexico), *sinensis* (of China), *viginicus* (of Virginia) and *zeylanicus* (of Ceylon).

5.11 Specific Epithets Linked to size

Some of such epithets are mentioned below alphabetically:

altus (altitude or tall), *exaltatus* (very tall), *giganteus* (gigantic or very large), *grandis* (large), *humilis* (dwarf), *major* (greater), *minor* (less), *minutus* (very small), *procerus* (very tall), *pumilus* (dwarf), and *robustus* (stout or robust).

5.12 Specific Epithets Linked With Habit

Epithets, along with the name of the related habit in parenthesis, are under mentioned alphabetically:

arborescens (arborescent), *dichotomus* (dichotomous).*erectus* (erect), *furcatus* (forked), *prostratus* (prostrate), *ramosus* (branched), *repens* (creeping) and *stoloniferus* (stoloniferous).

5.13 Specific Epithets Linked with Habitats

Some of such epithets, along with the name of the related habitat in the parenthesis are listed below alphabetically:

amphibious (living on land and in water), aquaticus (living in water), *arvensis* (growing in ploughed field), *campestris* (of field), *hypogeus* (underground), *lacustris* (of ponds or lakes), *littoralis* (growing on seashores), *maritimus* (of sea), *montanus* (of mountains), *palustris* (of swamps), *rupesris* (on rocks), *sativus* (cultivated), *sylvaticus* (of woods), *sylvestris* (growing in woods) and *terrestris* (growing in dry soil).

References

Jones, Jr. S.B. and Luchsinger, A.E. 1987: Plant Systematics and Evolution, McGraw-Hill International Editions, New Delhi, India.

Lawrence, G.H.M. 1951. Taxonomy of Vascular Plants, Macmillan, New York.

Voss, E.G. (ed.) 1983. International Code of Botanical Nomenclature (ICBN), Regnum Vegtabile, Vol. 97. Bohn, Scheltema and Holkema, Utrecht.

6

Plant Collection and Specimen Preparation

Plant collection and field preparation of specimens are the fundamental aspects of study, training and research in plant systematics. Herbarium specimens are the permanent records of plant species of a particular place at a given time. Therefore, the plants should be carefully collected, selected, and the herbarium specimens should be properly prepared and preserved.

6.1 Which Type of Specimens Should Be Collected?

The points to be kept in mind during a plant collection are:

1. Collect entire, vigorously growing typical specimens.
2. Select such individuals that represent almost all phases of the natural population.
3. Avoid collecting insect-damaged specimens.
4. Collect underground parts (e.g. roots, bulbs, rhizomes, tubers, etc.) of herbaceous perennials.
5. Collect those specimens of flowering plants that contain flowers, fruits, and seeds, because keys are prepared mainly on the basis of these characters.
6. Specimens larger than the size of a single sheet, should be divided and pressed on a series of sheets.
7. Collect plants with the leaves intact as different kinds of foliage prove helpful in identification.
8. Collect the bark and wood samples of the woody plants.
9. Avoid collecting rare or uncommon plants. Never collect the only plant of a species at a locality.

6.2 Field Equipment and Supplies

Some of the commonly used field equipment and supplies needed for plant collection and preparation of herbarium specimens are:

1. Field press: It is made up of a pair of hardwood, metal or strong plywood frames of 12 by 18 inch (Fig. 6.1).
2. Driers or blotters: These are the sheets of heavy blotting papers or of other moisture absorbing material of 11 by 16 inch dimensions. Old newspapers are also used as driers.
3. Straps or ropes: Rope, sash cord, or a pair of strong web straps of about 4-5 feet length are used to tighten the press.
4. Corrugate ventilators: Sheets of corrugated cardboard of 12 by 18 inch are used in between the driers (= blotters) as ventilators when plants are dried by artificial heat. They provide space for the passage of air through the press to remove moisture.
5. Field notebook: An indispensable item of a plant collector is permanently bound, small field notebook with ruled pages. It is used to record the full data (such as date, place, locality, habitat, elevation, local name, collector number, etc.) of the collected plant. The data is recorded in the field notebook in the field itself.
6. Digging and clipping tools: Trowel, diggers, hammer, pruning shears, garden clippers, geological pick, and heavy sheath knife, etc. are all used for digging or clipping the plants.
7. String tags: These are made of water proof material, and are used for labeling plants that are not immediately pressed.
8. Vasculum: Vasculum is a container made of tin or aluminium sheet and contains a hinged light lid. Plants, which are not pressed in the field, are placed in a vasculum to preserve their freshness for sometime (Fig. 6.2).
9. Collecting bags: These are plastic bags, used as containers for fresh specimens.
10. Collection bottles: These are glass or plastic bottles with leak-proof screw caps. These are used for collecting small-sized material, to be preserved in liquid preservatives.
11. Liquid preservative: Formalin-acetic acid-alcohol (FAA) is the most common liquid preservative used for anatomy materials. 6:3:1 mixture of chloroform, 95% ethyl alcohol and glacial acetic acid is often used for cytology materials.
12. Hand lens: A 5X or 10X lens is used in the field for observation and identification.
13. Waxed paper: Sheets of waxed paper are used for pressing viscid or weak-looking plants.

14. Cardboard storage boxes: These are the boxes or containers used to store dry materials.
15. Maps: These are useful in the field for determining the localities for particular species.
16. Colour charts: These are used for determining the actual colour of the flower parts in the field itself.
17. Camera and film: These are used for taking photographs of important plants in the field.
18. Other field equipment and supplies: Altimeter, compass, pocket knife, soft lead pencils, insect repellent, portable plant drier, and seed envelopes are some other useful equipment for plant collection.

Fig. 6.1: Field press

Fig. 6.2: Vasculum

6.3 Organization of the Field Trees

Following should be the organization of the field press:

Hardwood or plywood press

1. Corrugate ventilator
2. Drier or blotter
3. 10 sheets of torn newspapers (each sheet will contain one specimen).

6.4 How to Press Plant Specimens?

Specimens, after being cut or dug, should be pressed as soon as possible. The specimen can be placed carefully or a pressing sheet, i.e. a newsprint sheet or a blotter taking care so that there is no folding or overlapping of parts. The extra leaves or branches are removed if necessary. Plants too large to fit in the 11 x 16 inch fold of newspaper or a blotter may be bent into a V, N or M figure. The specimens should not protrude from the fold of the paper.

Specimens should be arranged in such a way so that some upper and some lower surfaces of leaves are exposed. Flowers or inflorescences should be spread out thoroughly for a complete view. A few flowers can be cut longitudinally and pressed, if possible. Each sheet should contain a collector's number which refers to the notes in the collector's field notebook.

In the press, the specimens in the specimen paper are placed in between two driers or blotters. A corrugate ventilator is often inserted before the next specimen papers and driers are added. The usual sequence, as shown above under section X.3 (organization of the field press), is corrugate ventilator, drier, sheets of specimens in newspaper, drier, corrugate ventilator, drier and so on.

All the specimens are placed for pressing in such a way so that the entire bundle is ultimately of almost uniform thickness in the middle and sideways. Once the plant specimens are arranged, the press is tightly bound with ropes or straps to prevent wrinkling of the specimens. The press is now ready for drying.

6.5 Drying of Specimens

Specimens should be dried as rapidly as possible to get the best results. In the usual process of drying, the press containing the specimens is placed in the sun. After about 24 hours, the press is opened, and the specimens are placed in fresh blotters. Any last rearrangement of the plant parts may be made at this time. The press along with the specimens in fresh blotters, is now again bound tightly. The wet blotters are dried in the sun for reuse. For three-four days the wet blotters or driers are changed daily, until the specimens are completely dried. The press becomes loose when the plants are completely dried.

Artificial heat may also be used for the drying process. But the specimens are never to be dried in an oven. In humid regions, or the rainy season, the plants may be dried in a drier. A drier is made in the form of a wooden box 3 feet in length and 18 inches in breadth made up of 1/10 inch thick boards. Five light bulbs of 60 watts are fitted inside for producing heat. Small openings are made of the bottom of the box for the entry of air which gets heated and will dry the plants. In this way, a press containing 100 specimens can be dried in 8-12 hours.

An excellent drying chamber may also be made by using a wooden or metal box with an open top that will accommodate a press. The box is equipped with an electric heater with a fan.

6.6 Mounting of specimens

The process by which a dry specimen is attached to a mounting paper or a herbarium sheet, and a label is affixed at the lower right corner of the sheet is called mounting. The standard size of a herbarium sheet is 28.75 by 41.25 cm. Heavy, hand made card sheets are used as herbarium sheets.

Before mounting, the dry specimens are dipped in a saturated solution of mercuric chloride in ethyl alcohol to prevent infection of fungi, insects, etc. In place of mercuric chloride, lauryl pentachlorophenate is also used, because the former is a virulent poison.

A good quality glue or paste is applied to the back of the specimens for affixing them onto the herbarium sheets. Common animal glue is used in India for this purpose. Some of the good quality glues, used in developed countries, include Wilhold 128, Elmer's Glue-All, Swifts Z-5032, or Nicobon B. Strips of gummed paper are used as an additional aid to hold heavy and woody specimens. Loose parts (e.g. seeds, fruits, flowers, etc.) and dissected parts are placed in a paper packet and pasted on the same herbarium sheet. Use of cellulose tapes or stitching of the plant parts with thread should be avoided.

6.7 Labeling of Specimens

A label, usually of 6.5 by 10.5 cm (Jones and Luchsinger, 1987) dimension, is pasted on the lower right side of a herbarium sheet. It contains definite information about the specimen pasted on the herbarium sheet. Ideally, it contains a miniature essay on the plant and its habitat. The paper of the label should be of high rag quality. Following information should be incorporated in a label:

1. Heading, indicating the name of the institution or person, along with the name of the state and country.
2. Family name.
3. Scientific/ botanical name i.e. name of genus, species along with authority.
4. Locality, indicating the place of collection with its latitude and longitude, and distance from a well-known place.
5. Habitat, indicating vegetation type, moisture, soil, slope, etc.
6. Date of collection.
7. Field notes from the collector's field notebook.
8. Collector, mentioning his/ her name.

9. Collection number, which is a numerical series starting with 1 and continuing throughout the life time of the collector.
10. Local or vernacular name.
11. Additional information, such as colour of the flower and leaves, associated plants, bark, economic uses, etc.

6.8 Identification of Specimens

Finally, pressed, dried and labeled specimens are identified with the help of appropriate floras, manuals, and monographs. A duplicate specimen is to be retained before sending a specimen to the specialist for identification.

Reference

Jones, S.B. and Luchsinger, A.E. 1987. Plant systematics (2nd ed.), McGraw-Hill publication, New York.

7

Chemotaxonomy

7.1 What is Chemotaxonomy?

The approach of taxonomy in which chemical features of plants are used in developing classifications or in solving taxonomic problems is called chemotaxonomy, chemosystematics, chemical taxonomy, chemical plant taxonomy or plant chemotaxonomy. Some of the basic chemical features or evidentiary characters include alkaloids, flavonoids, carotenoids, polysaccharides, terpenoids, fatty acids, amino acids, aromatic compounds, etc.

The chemical evidences are useful in establishing relationships among taxa is exemplified (Cronquist, 1980) by the (i) Presence of aromatic plants in Juglandales, (ii) Production of betalains and not anthocyanins by the members of caryophyllales, (iii) Presence of highly aromatic compounds in the members of Lamiaceae, (iv) Presence of alkaloids in Solanaceae, and (v) Presence of tanniferous plants in Sapindaceae.

7.2 Purpose of Chemotaxonomy

Chemotaxonomy has been used at all levels of the hierarchy of classification starting from the rank of 'variety' up to the rank of 'division' in plants. Moreover, chemical evidences have been used in all the groups of the plant kingdom starting from simple organisms, such as fungi and bacteria, up to the most highly advanced and specialized groups of angiosperms. These evidences are used in the classification of plants with two main purposes, (i) To develop such taxonomic characters which may improve the existing systems of plant classification and (ii) To develop the present day knowledge of phylogeny or evolutionary relationships of plants.

7.3 Brief History

The history of the knowledge or use of chemical characters of plants and their utility in plant classification is very old. Socrates, the great Greek philosopher, was forced to take his own life by drinking a broth of poison hemlock, i.e. *Conium maculatum* of family Apiaceae. In his *Materia Medica*, Dioscorides in the 1st century A.D. described several aromatic mints. But some of the pioneer botanists, who employed chemical evidences in plant classification, are Nahemia Grew (1673), William Withering (1785), A.P. de Candolle (1804), Abbott (1886) and Geshoff (1891). However, the modern phase of chemotaxonomy began with the publication of the work of Abbott (1886) in *Botanical Gazzette*. McNair published a series of papers on the use of chemical evidences in solving taxonomical problems during 1917 and 1945.

The main research centre in the field of chemotaxonomy in the world, is *The Institut Fur Pharmakognosie of the University of Kiel, Germany*.

Some of the books providing details of chemotaxonomy are the Hawkes (1968), Swain (1973), Gibbs (1974), Bisby *et al.* (1980), Young and Seigler (1981), and Harborne and Turner (1984).

7.4 Chemical Characters and Their use In Taxonomy

No suitable classification of the chemical characters and their use in taxonomy is developed so far. Naik (1984) divided these characters into three categories, i.e. (i) Directly visible characters, e.g. starch grains, raphides, silica, gypsum, etc., (ii) Chemical test characters, e.g. phenolics, betalains, oils, fats, waxes, alkaloids, etc., and (iii) Proteins. On the other hand Jones and Luchsinger (1987) divided the natural chemical plant products useful in taxonomy, on the basis of their molecular weight, into two major groups, i.e. (i) Low molecular weight compounds with a molecular weight of 1000 or less, called micromolecules, e.g. amino acids, alkaloids, fatty acids, terpenoids, flavonoids, etc., and (ii) High molecular weight compounds with a molecular weight of over 1000 called macromolecules, e.g. proteins, DNA, RNA, complex polysaccharides, etc. Sometimes, the term 'semantides' is used for the information carrying proteins, such as DNA, RNA, etc.

In the absence of a suitable classification, a discussion of some taxonomically important chemical compounds, along with a few examples of their systematic value, is under mentioned.

7.4.1 Flavonoids

Most widely and effectively used compounds in chemotaxonomy are the flavonoids. These are the phenolic glycosides consisting of two benzene rings linked together through a heterocyclic pyrane ring. The reasons for their wide use in chemotaxonomy include their widespread distribution, chemical complexity, the relative physiological stability, quick and easy identification, and great structural variation.

Some of the common classes of flavonoids include anthocyanidins (e.g. cyanidin, delphinidin, etc.), flavones (e.g. apigenin, levtolin), biflavonyls (e.g. amentoflavone), flavonols (e.g. kaempferol, quercetin), flavonones (e.g. naringenin), isoflavones (e.g. orobol), isoflavonoids (e.g. ferreirin), chalcone (e.g. butein) and aurone (e.g. sulphuretin).

Some of the specific examples of the use of flavonoids in chemotaxonomy are under mentioned-

1. Alston and Turner (1963) documented hybridization between the species of *Baptisia* by using chromatographic patterns of flavonoids.
2. Dahlgren (1980) used flavonoids in clarifying certain relationship while suggesting his famous system of angiosperm classification.
3. Psilotum contains biflavonyls, and this brings it closer to lycopods than leptosporangiate ferns, as proposed by some workers.
4. Ellagic acid, a polyphenol, is confined to woody dicots and not found at all in monocots.
5. Analysis of leaf flavonoids of Liliaceae, Juncaceae, Cyperaceae and Gramineae suggest that all these families have arisen from Liliaceous ancestors.
6. Williams *et al.* (1983) identified six ornamental species of *Fuchsia* on the basis of their leaf flavonoids.
7. Belzer and Ownbey (1971) studied the flavonoid pattern of 5 species of *Tragopogon* of Compositae and their F_1 and F_2 hybrids, and noted that all the species and their population show different chromatographical results.
8. Isoflavone iridin is found only in the section Pogonivis of genus *Iris*.
9. South American species of *Eucryphia* of the family Eucryphiaceae may easily be distinguished on the basis of their flavonoid glycosides.
10. Species of *Malus* are classified on the basis of their flavonoids by Williams (1982).
11. Presence of five leaf flavonoids (tricin, luteolin, glycosyl flavones, bisulphate complex and 5-glucoside) in both Gramineae and Palmae, link them together.
12. Species of *Spirodella*, unidentified on morphological grounds, may be distinguished on the basis of their flavonoid chemistry.
13. In gerbera, phytochemical compound 'flavonoid' is used as chemotaxonomic marker to acertain inter and intraspecific variability.

14. In anthurium, after conducting analysis of floral fragrance compounds on five *Anthurium* species (*A. armeniense, A. fragrantissimum, A. lindenianum, A. orchranthum* and *A. roseospadix*) and three hybrids of *A. armeniense*, Kuanprasert *et al.* (1998) suggested that concentrations (%) and chemical profiles of 28 compounds can be considered as useful taxonomic markers to clarify phylogenetic relationships in *Anthurium* spp.
15. Phenolic compounds in leaves and bracts analyzed by thin layer chromatography reveals inter and intra varietal similarities among bougainvilleas. Bract pigments analyzed by spectro-photometer (200-700 nm) revealed intervarietal differences.
16. Flavonoids are the pigments responsible for flower colour in China aster. The flavonoid biosynthetic pathway is very well developed (Seitz *et al.*, 2007). All the enzymes and their corresponding genes have been studied in detail. As biochemical data become available, these genes were assigned specific functions. Enzyme assays with petal extract showed that chalcone synthase, the key enzyme anthocyanin biosynthesis, catalyses the formation of precursor of pelargonidin derivatives in China aster lines.
17. Flavonoids, carotenoids, chlorophylls, anthocyanins are phytochemical compounds considered as useful taxonomic markers in large flowered chrysanthemum cultivars to compare their inter and intraspecific variability in morphological characters and biochemical composition of pigments (Banerji *et al.*, 2012). Such analysis also provides important information for preparation of checklist of chrysanthemum, helpful in identification of varieties and for selection of desirable characters for intervarietal breeding programme.
18. Flavonoids were studied in the genus *Pelargonium* to compare their inter and intraspecific variability. This compound is considered as useful taxonomic marker to clarify phylogenetic relationships among species. HPLC procedures for cultivar identification on the basis of geranium florets have been developed (Asen and Griesbach, 1983).
19. Flavanol compounds are currently considered as useful taxonomic markers to clarify phylogenetic relationships in *C. roseus* (Daniel and Sabris, 1978) and all contained syringic acid.
20. Survey of phenolic compounds in miniature roses helps in establishing relationships of different cultivars and their probable origin. Different types of phenolic- phytochemical compounds present in leaves, flowers, stems, roots and barks were studied in 16 miniture rose cultivars to

compare their phylogenetic relationships, determination of hybrid complexes, identification of ploidy nature, detection of mutation, etc. The results of analysis together with cytological evidence given by other authors are taken as proof for showing relationships between cultivars and their probable origin. It is suggested that the qualitative and quantitative changes in phenolic compounds might have taken place during their evolution due to extensive hybridization and/or natural mutation (Datta and Singh, 1995, 1999).

7.4.2 Betalains

Betalains differ from flavonoids and other phenolic compounds owing to the presence of nitrogen in them. However, these are functionally equivalent to phenolics. Betalains are popularly known as nitrogenous anthocyanins. They are water soluble nitrogen containing compounds derived from tyrosine and are taxonomically restricted to caryophyllales. They comprise of red to violet betacyanins and yellow betaxanthins. Some of the aspects of taxonomic value of betalains are under mentioned:

1. All the ten families of the order Centrospermae (i.e. Aizoaceae, Amaranthaceae, Basellaceae, Caryophyllaceae, Chenopodiaceae, Didieraceae, Molluginaceae, Nyctaginaceae, Phytolaccaceae and Portulacaceae) produce betalains and are placed together in a single order. All these ten families are placed under Caryophyllales in modern schemes of classification.
2. Betalains are also reported in the family Cactaceae. This family, once placed elsewhere in the classification schemes, should be placed under the betalains-containing order, i.e. Caryophyllales or in the older order Centrospermae.
3. In bougainvillea, colour of bracts (betalains pigments i.e. betacyanin and betaxanthin) act as the chemotaxonomic marker for identification of species and hybrid groups. The basal species and the three hybrid groups have a definite pattern of bract pigments. Interspecific hybridization brings about an interaction of two disparate genomes by bringing them together in an entirely different genetical environment. This is the basis of different novel colour forms (Kocher and Ohri, 1977). Mutations seem to bring about a partial or complete loss of betacyanins and/or qualitative or quantitative increase in betaxanthin production. A particular expression of bract pigments is also influenced by external and internal factors like light and temperature by which the genes responsible for the synthesis of these pigments can be repressed or depressed. For example,

light has been found to enhance the synthesis of betacyanin and some bands of betacyanin disappeared when the newly formed bracts were put in darkness. The nature of some of the cultivars like 'Mary Palmer' can be explained on this basis (Kocher *et al.*, 1979).

7.4.3 Terpenoids

These are a biogenic group of diverse compounds with mevolonic acid as their precursor (Goodwin, 1970). Some regard them as components of essential oils. According to the number of isoprene units present in a terpenoid molecule, they may be of five categories: (i) hemiterpenes (C_5H_8), e.g. tiglic acid, (ii) monoterpenes ($C_{10}H_{16}$), e.g. myroene, (iii) sesquiterpenoids ($C_{15}H_{24}$), e.g. parnesol, (iv) diterpenoids ($C_{20}H_{32}$), e.g. phytol and, (v) triterpenoids ($C_{30}H_{48}$), e.g. squalene.

Terpenoids have been usd extensively in the chemotaxonomy of mints, apiaceae, citrus plants, and gymnosperms. Some of the their chemotaxonomic applications are mentioned below:

1. Origin of certain *Citrus* cultivars was determined by studying their rind and leaf terpenoid pattern.
2. Geographical races of *Pseudotsuga menziesii* were distinguished by studying the terpenoids of their cortical oleoresins.
3. In the taxonomy of Asteraceae and tribe Genisteae of Fabaceae, the petal carotenoids, which form a prominent group of terpenoids, prove very useful.
4. Mirov (1961) used the gum terpentines in the taxonomy of *Pinus*.
5. Some specific terpenoids are found in certain families, e.g. cucurbitacins in Cucurbitaceae, asperuloside in Rubiaceae and sesquiterpene lactones in Asteraceae.
6. Floral scents of several carnation cultivars were reported to be composed of benzenoids, terpenoids and fatty acid derivatives; among which benzenoids are the most important fragrance factors (Clery *et al.*, 1999). Several classical terpenoids biosynthesized in these plants would be valuable target for fragrance breeding.
7. Scent of flower in lotus is due to 1-4-Dimethoxy benzene and 1-8-cineole, terpinen-4-01 and linalool. Among these 1-4-Dimethoxy benzene increases daily during anthesis and it is present at higher concentration in *N. pentapetala* or *N. lutea* than in *N. nucifera*.

7.4.4 Steroids

True steroids possess two methyl groups and are mostly alcohols or esters. They are located in plant cutins and perform the function of water proofing. Their distribution proved helpful in the taxonomy of the genera of tribe Veratreae of family Liliaceae. These genera contain the steroid veratum according to Kupchan (1961).

7.4.5 Iridoid Compounds

Monoterpenoid cyclopentanoid lactones represent a separate class of compounds, of taxonomic significance, called 'Iricords'. Some of the aspects of their taxonomic value are mentioned below:

1. Because of the presence of an iridoid compound, aucubin, genus *Buddleia* was transferred from Loganiaceae to Buddleiaceae, near Scrophulariaceae.
2. The iridoid compound, arsperuloside, is the characteristic of Rubiaceae.
3. On the basis of the presence of iridoids, the interfamilial relationships were clarified between Cornales, Scrophulariales, Caprifoliales, and Rubiales.
4. Several iridoid-families (Rubiaceae, Scrophulariaceae, Cornaceae, Dipsacaceae and Caprifoliaceae) are considered the putative progenitor of Asteraceae.

7.4.6 Alkaloids

These are the basic nitrogen-containing organic compounds, usually with a heterocyclic ring. Alkaloids show significant effects in the central nervous system of animals, and are well-known for their medicinal, chemical, poisonous and systematic view points. Some of their taxonomic aspects are under mentioned:

1. Jones and Luchsinger (1987) stated that alkaloids are useful in taxonomic studies in *Veratrum* (Liliaceae), Papaver and *Argemone* of Papaveraceae, as well as in *Lycopodium*, *Lupinus* and Caryophyllales.
2. Morphine is produced only by *Papaver somniferum*, and strychnine is found only in *Strychnos*.
3. Because of the accumulation of isoquincline alkaloids containing protopine, in both Fumariaceae and Papaveraceae, the two are considered closely related.
4. Lupin alkaloids are the characteristic of the tribes Sophoreae, Genisteae, and Podalyrieae of the sub-family Lotoideae of Leguminosae.

5. Distribution patterns of alkaloids in two genera (*Genista* and *Adenocarpus*) of the tribe Genisteae of Fabaceae and the genus *Ammodendron* of the tribe Sophereae of the same family suggest the transfer of *Ammodendron* from Sophereae to the tribe Genisteae.
6. Several modern taxonomists, including Takhtajan, Cronquist, Thorne and Dahlgen, rearranged the families with apocarpous gynoecia in Magnoliidae on the basis of the distribution of benzyl isoquinoline alkaloids.

7.4.7 Glucosinolates

Glucosinolates is a group of about 85 thioglucosides which on hydrolysis form glucose and a corresponding isothiocyanates. They are also called mustard oil glucosides because they are confined to the taxa of the mustard family and some related families of Capparales.

Formerly, Cruciferae, Capparaceae, Fumariaceae and Papaveraceae were all included under one order Rhoeadales. But the recent findings support the view that glucosinolate-producing families (Cruciferae, Capparaceae) should be placed separately under Capparales, and the alkaloid-containing families (Fumariaceae and Papaveraceae) under Papaverales.

7.4.8 Amino acids

These are the building blocks of proteins and are universally distributed in plant tissues. Several workers have studied the taxonomic potential of amino acids. A few examples of their taxonomic significance are mentioned below:

1. *Lathyrus* species could be grouped under seven infrageneric groups on the basis of the association of amino acids with the seeds.
2. Lathyrine is found only in the genus *Lathyrus* while azetidine-2-carboxylic acid is found only in Agavaceae, Liliaceae and Amaryllidaceae.
3. Acetyl ornithine is identified as the main amino acid of Fumarioideae.
4. Species of the section Gummiferae of the genus *Acacia* can be recognized easily on the basis of the amino acid contents of their seeds.

7.4.9 Lipids and Waxes

Lipids are the esters of fatty acids with glycerol, mostly made up of carbon, hydrogen and oxygen. Plant waxes are esters of long-chain alcohols with long chain fatty acids. They may contain free alcohols, free fatty acids, aldehydes, ketones or hydrocarbons. Some of the taxonomic applications of lipids and waxes are under mentioned:

1. Restricted distribution of some fatty acids is characteristic of certain families, e.g. petroselinic acid of Apiaceae, erucic acid of Cruciferae, chaulmoogric acid of some taxa of Flacourtiaceae, and xymenyric acid of Santalaceae.
2. Proportions of fatty acids in lipids are constant in both genera and species in Palmae.
3. Malvalic acid exists in the members of Malvaceae while capric acid is characteristic of Lauraceae and Lythraceae.
4. Fatty acids of bacterial lipids are not found in other plants.

7.4.10 Nucleic acids

Availability of the necessary expertise and equipment, required for the nucleic acid technique, nowadays, to most of the taxonomists, helped to progress in this direction. Nowadays, various species within the genus were compared on the basis of their DNA content.

7.4.11 Proteins

Proteins are used in chemotaxonomy owing to their large complex molecules, universal distribution, presence in large quantity in some organisms, and also because of the knowledge of several simple and rapid methods of protein analysis, such as electrophoresis. As the proteins represent the direct products of the DNA code, their comparative analysis may be used as fundamental taxonomic characters. Protein analysis by electrophoresis and other methods may be used to determine whether differences actually among plants that are treated as distinct species.

7.4.12 Cyanogenic Compounds

The ability of some plants to release poisonous compounds, such as hydrocyanic acid, amygdalin, etc. after injury to their cells, is called 'cyanogenesis' and these poisonous compounds are called 'cyanogenic compounds'. Plant containing these compounds are called 'cyanogenic plants'. Over two thousand vascular plants have so far been established as cyanophoric. Taxa belonging to Araceae, Gramineae, Juncaceae, Juncaginaceae and Scheuchzeriaceae are commonly cyanogenic. Several dicotlyledonous families belonging to Asteridae, Rosidae, and Dilleniidae are also cyanogenic.

References

Alston, R.E. and Turner, B.L. 1963. *American Journal of Botany*, **50**: 159-173.

Asen, S. and Griesbach, R. 1983. *Journal of American Society of Horticultural Science*, **108**(5): 845-850.

Banerji, B.K., Batra, A. and Dwivedi, A.K. 2012. *Journal of Horticultural Sciences*, **7**(1): 51-55.

Belzer, N.F. and Ownbey, M. 1971. *American Journal of Botany*, **58**: 791-802.

Bisby, F.A., Vaughn, J.G. and Wright, C.A. 1980. (eds.) Chemosystematics: Principles and Practices, Academic Press, London.

Clery, R., Owen, N.E. and Chambers, F.S. 1999. *Journal of Essential Oil Research*, **11**: 359.

Cronquist, A. 1980. Chemistry in plant taxonomy: An assessment of where we stand, (In) F.A. Bisby *et al*., Chemosystematics: Principles and Practices, Academic Press, London.

Dahlgren, R.M.T. 1980. *Botanical Journal of Linnaean Society*, **80**(2): 91-124.

Daniel, M. and Sabris, S.N.1978. *Indian Journal of Experimental Biology*, **16**(4): 512-513.

Datta, S.K. and Singh, M.S. 1995. *The Indian Rose Annual*, **13**: 60-70.

Datta, S.K. and Singh, M.S. 1999. *Scientific Horticulture*, **6**: 157-164.

Gibbs, R.D. 1974. Chemotaxonomy of flowering plants, 4 Vols. McGill-Queens University Press, Montreal.

Goodwin, T.W. 1970. Academic Press, New York. Pp. 1-442.

Harborne, J.B. and Turner, B.L. 1984. Plant Chemosystematics, Academic Press, New York.

Hawkes, J.G. (ed.) 1968. Chemotaxonomy and Serotaxonomy, Academic Press, London.

Jones, S.B. and Luchsinger, A.E. 1987. Plant Systematics, McGraw-Hall, New York.

Kocher, V.K. and Ohri, D. 1977. *Pfluzucht*, **79**: 47-51

Kocher, V.K., Kocher, S. and Ohri, D. 1979. *Current Science*, **48**: 163-164.

Kuanprasert, N., Kuehnle, A.R. and Tang, C.S. 1998. *Phytochemistry*, **49**: 521-528.

Kupchan, M. 1961. *Journal of Pharmaceutical Sciences*, **50**: 273-287.

Mirov, N.T. 1961. Composition of gum turpentines of pines (Technical bulletin / USDA). Pp. 1-158.

Naik, V.N. 1984. Taxonomy of Angiosperms, Tata Mcgraw Hill company, New Delhi.

Seitz, C., Vitten, M., Steinbach, P., Hartl, S., Hirsche, J., Rathje, W., Treutter, D., Forkmann, G. 2007. *Phytochemistry*, **68**: 824-833.

Swain, T. (ed.) 1973. Chemistry in evolution and systematics, Butterworth, London.

Turner, B.L. and Harborne, J.B. 1967. *Phytochemistry*, **6**(6): 863-866.

Williams, A.H. 1982. *Botanical Journal of the Linnaean Society*, **84**(1): 31-39.

Williams, C.A., Fronczyk, J.H. and Harborne, J.B. 1983. *Phytochemistry*, **22**: 1953-1957.

Young, D.A. and Seigler, D.J. 1981. Phytochemistry and Angiosperm Phylogeny, Praeger, New York.

8

Palynotaxonomy

In flowering plants, each anther has four numbers of pollen sacs. Two anther lobes contain 4 pollen sacs in which pollen grains are produced. Pollen grains develop from the diploid microspore mother cells in pollen sacs of anthers. Thus, palynology alynology is the science of pollen and spore morphology and its applications. It is derived from the Greek word 'palynein' meaning 'to scatter'. Morphological and comparative studies in palynology may solve intricate taxonomic problems. Pollen characters are useful in solving several complicated problems of interrelationships between various taxa and assessment of their status in the classification (Zhigila *et al.*, 2018). Pollen grains are in the binucleate stage at the time of pollination.

Fig. 8.1: Dr. P.K.K. Nair (Courtesy, Garg, 2015)

Wodehouse (1935) and Erdtman (1943, 1952, 1954, 1963 & 1969) made the most remarkable contribution in the field of pollen morphology. The significance of pollen attributes in taxonomy has been realized during the last three decade Erdtman (1952); Wagenitz (1976); Stix (1960), and Nair (1974). Dr. Parmeshwaran Krishnan Kutty Nair, fondly referred to as P. K. K. Nair (Fig. 8.1) and considered as the father of Indian palynology. Palynological renaissance, which started in 1952 with the contributions of Erdtman (1952) and Faegri and Iversen (1950), grew with the scientific activities of Nair and took a concrete shape in 1964 when he conveyed a symposium at NBRI, Lucknow.

The angiosperm pollen is either binucleate or trinucleate according to the precocity of division of the generative nucleus. The binucleate condition is considered as more primitive than the trinucleate. In the Centrospermae, the pollen is uniformly trinucleate, the monocot (Liliaceae) is binucleate, the apetalous and polypetalous dicot are binucleate and gamopetalous members trinucleate.

8.1 Pollen Characters and Related Studies

Study of pollen and spores is called palynology. The shape and symmetry of a pollen grain, the architecture of its wall, exine stratification, sculpture and structure, and type, number, position, shape and structure of its aperture, are some of the basic characters, which prove useful at all taxonomic levels. Pollen grains can be used as component of plant introduction.

In plant systematics, pollen morphology is useful to identify and classify various species in their respective families and pollen characters were used in combination with other morphological characteristics to delimit different taxa. For instance, the appearance of pores on the inner surface of the acetolysed pollen has proved to have reliable taxonomic characteristics (Perveen, 2003). It is used to identify hybrid plants and to solve the taxonomic problems of some medicinal and other economically important plants. Palynology has also been very helpful in phylogenetics, evolution and allergy but there is scanty research on such studies. Luo *et al.* (2015) suggested that pollen morphology has become integral part of understanding the evolution of the whole angiosperm.

Palynological characters have been used in solving several taxonomic problems, including the repositioning of several disputed taxa, and interpretation of problems relating to the origin and evolution of different groups (Nair, 1980). Cronquist (1981) and several other workers have made the exclusive use of pollen characters in providing classification of angiosperms. Erdtman (1966) used the pollen characters in discussing and solving the taxonomic problems of 105 families. Pollen morphology has proved to be of great value in the classification of Asteraceae, Acanthaceae, Saxifragaceae, Asclepiadaceae, Caryophyllaceae, etc. Heywood (1967) has gone up to the extent of stating that exine details of pollen are such that they can be used in plant identification much in the way that fingerprints are used for identification of criminals.

8.2 Monosulcate and Tricolpate Pollen

Jones and Luchsinger (1987) mentioned that angiosperms contain two basic kinds of pollen grains: monosulcate and tricolpate. Monosulcate pollen grains are characteristic of primitive dicotyledons, several monocotyledons, pteridosperms and cycads. Such pollen grains are boat shaped in outline and possess one long germinal furrow and a germinal aperture. Tricolpate pollen grains are characteristic of advanced dicotyledons. Such pollen grains have three germinal apertures and are globose in shape.

8.3 Stenopalynous and Eurypalynous

Such a tax on, in which the type of the pollen is constant and characteristic, is called stenopalynous or unipalynous, e.g. Gramineae is unipalynous having monoporate aperture and somewhat smooth exine surface. On the other hand, such taxa, in which pollen types vary in size, aperture, exine stratification, etc., are called eurypalynous or multipalynous, e.g. Euphorbiaceae is multipalynous with grains of various aperture forms.

Cruciferae, Labiatae and Asclepiadaceae are unipalynous families while Acanthaceae, Rubiaceae, Asteraceae and Verbenaceae are highly multipalynous families.

8.4 NPC System

The classification of pollen is based on the number-position-character-analysis, called NPC system suggested by Erdtman *et al.* (1963). Palynological studies suggest that the taxa with the same general NPC-formula be grouped together, and those showing different NPC, separately. NPC-system helps in providing a three-dimensional classification, and also in the preparation of diagnostic keys below the family level. Cronquist (1981) also emphasized on the importance of the NPC system in solving several taxonomic problems. For example, the pollen grains of Parietales (as investigated by means of "palynological compass needle") are 3-tremes (N_3), zonotreme (P_4) and colporate (C_5). Therefore, the NPC-formula for Parietales is 345.

8.5 Some Other Pollen Characters of Systematic Value

1. Size and shape of pollen are important diagnostic characters. The shape may be circular, triangular, etc. in polar view, whereas it may be prolate, perprolate, spheroidal, sub-polar, etc. in equatorial view.
2. Apertures may be proximal (gymnosperms and monocotyledons), distal (pteridophytes), zonal (dicotyledons) or global (dicotyledons).
3. Exine sculpturing may be of depression type (e.g. reticulate, foveolate, striate, fossulate, areolate, scrobiculate, etc.) or excrescence type (e.g. granulose, verrucate, spinulose, tuberculate, spinose, gemmate, clavate, etc.).
4. Pollen associations, such as their arrangement in tetrads, presence of pollinia, etc. are characteristic of several angiospermic families.
5. Number of nuclei in the pollen grains at the time of their release is a valuable taxonomic and phylogenic character.

8.6 Few Examples of Role of Palynology in Taxonomy

Morphological features of pollen grains have importance in plant taxonomy and are considered as supplementary parameters in taxonomic treatments. Cytopalynology studies indicate that *Celosia* is stenopalynous with pantoporate pollen grains. Cytotypic differences in pollen size, exine ornamentations and exine thickness are observed at varietal and cultivar levels. At varietal level, *C. cristata* exhibits distinguishing exine features. Exine is thinner in variety *cristata* than that of variety *plumosa*. At the cultivar level, spectacular pollen variation is found in garden cockscomb *C. cristata* var. cristata.

Pollen morphology (number of spores and spines) can be used for species *Chrysanthemum morifolium* and *C. zawadskii* identification.

1. Sharma (1987) investigated the pollen morphology of 16 Indian species of *Cyperus*, and prepared a key to differentiate all of them on the basis of pollen characters. Of the 16 investigated species, the pollen grains of six (*C. iria*, *C. difformis*, *C. squarosus*, *C. triceps*, *C. flabelliformis*, and *C. pumilus*) are predominantly 2-colpate, four (*C. rotundus*, *C. laevigatus*, *C. alulatus*, and *C. bulbosus*) are predominantly 4-aperturate with 3 colpi and 1 pore, three (*C. compactus*, *C. kyllingia* and *C. globosus*) are predominantly pantoaperturate, and one (*C. digitatus*) is predominantly 1-porate. Individual species are also differentiated on the basis of pollen characters.
2. On the basis of apertural morphoforms of pollen, Nair and Saxena (1974) proposed a palynological classification. He divided the entire plant kingdom into (i) Amorphosporophyta, including thallophyta with no apertural markings, (ii) Trimorphosporophyta, including Archegoniatae with monolete, trilete, or alete forms and (iii) Polymorphosporophyta, including angiosperms with different forms of apertural types.
3. Meyer (1975) gymnosperms contain alveolar or granular ektexine and laminated endexine, whereas angiosperms contain columnar or granular ektexine and non-laminated endexine.
4. A massive exine and thin intine is present in angiosperm pollens. But in certain taxa among monocots (e.g. Musaceae, Amaryllidaceae, Cannaceae, Zingiberaceae, Costaceae and Heliconiaceae), the exine is highly reduced and the intine is well-developed.
5. Several angiospermic taxa have distinctive pollen types. Gramineae have smooth sulcate pollen, Malvaceae and Asteraceae have typically spinulose exine and Plumbaginaceae have verrucate exine. Pollens are binucleate in Magnoliidae and trinucleate in Caryophyllaceae.

6. The exine pattern is useful in recognizing different species of a genus. According to Nair and Saxena (1974) pollens are pilate in *Bauhinia acuminata*, striate in *B. krugii*, verrucate in *B. retusa*, reticulate in *B. racemosa*, spinulate in *B. malabarica*, and reticulate tuberculate in *B. purpurea*.
7. Pollen grains are associated in tetrads in several families of dicots (Annonaceae, Mimosaceae, Droseraceae, etc.) monocots (Cyperaceae and Junaceae). In Asclepiadaceae and Orchidaceae, they form pollinia.
8. According to Kooiman (1972), 2-nucleate pollen grains of Lamiaceae are tricolpate while 3-nucleate pollen grains are of 6-colpate.
9. Palynological evidences have supported the separation of Paeoniaceae from Ranunculaceae; Nelumbonaceae from Nymphaeaceae; Fumariaceae from Papaveraceae, and Bombaceae from Malvaceae.
10. Removal of Podophyllum from Berberidaceae and its inclusion in a separate family Podophyllaceae is effected on palynological grounds.
11. Pollen characters prove useful in distinguishing the genera *Salix* and *Populus* of Salicaceae, *Phytolacca* and *Rivinia* of Phytolaccaceae, and several genera of Betulaceae, Primulaceae and Acanthaceae.
12. According to Nair (1980), three well-defined evolutionary stocks among angiosperms (i.e. monocot stock, Magnolian dicot stock, and Ranalian dicot stock) form the palynological basis of the triphyletic theory of angiosperms (Fig. 8.2). He enunciated the triphyletic theory of evolution of angiosperms in 1969, as a vital clue to unravelling the mystery of origin and evolution of the angiosperms.
13. The genus *Krameria*, variously placed under Leguminosae, Polygalales, etc. was recommended to be placed in a separate monotypic family Krameriaceae by Simpson and Skvarla (1981) on palynological grounds.
14. Pollen morphology (number of spores and spines) can be used for species *C. morifolium* and *C. zawadskii* identification.
15. Palynological studies suggest that Thunbergioideae of Acanthaceae should be given a family status.
16. Pollen types are considered to be characteristics of species. Meo and Khan (2006) reported that pollen morphology can be used for the identification of *Chrysanthemum* species. Pollen grains of *C. morifolium* and of *C. zawadskii* var. latilobum are tricolporate, suboblate and echinate. Shim *et al.* (1989) suggested that difference in the number of spines and spores could be used to differentiate between the cultivars of *C. morifolium* and *C. zawadskii* could easily be distinguished from *C.*

morifolium by the frequency of connected spores and spines.

17. The pollen of *Caesalpinia* species are tricolporate and trizonícolporate. Tricolporate pollens are recorded in three species of *Caesalpinia* viz. *Caesalpinia coriaria*, *C. pulcherrima* var. flava and *C. pulcherrima* var. pulcherrima while trizonicolporate is in *C. bonduc* (Prakash and Singh, 2020). Shape of pollen grains of *C. bonduc* and *C. pulcherrima* var. flava were oblate-spheroidal type while in *C. pulcherrima* var. pulcherrima was prolate-spheroidal and *C.coriaria* sub-oblate.
18. Crassulaceae is the largest family under order Saxifragales which is considered one of the taxonomically difficult families. Pollen morphology of different species of Crassulaceae has been described by Nair (1962) and Faegri & Iversen (1964). Palynological investigation of 19 species representing nine genera of Crassulaceae were studied using both light (LM) and scanning electron microscopy was studied by Shehata *et al.* (2023). Generally, the pollen grains were found to be monad, isopolar, radially symmetric, prolate-spheroidal, subprolate or prolate in equatorial view, triangular, tetragonal, or pentagonal in polar view.

8.7 Principles of Palynology

The gametophytic phase was represented by pollen, when it starts developing, till pollination leading to fertilization and seed set. During this, pollen released from the anthers moves in various directions for pollination and forms the subject matter of palynological studies such as:

(i) Aeropalynology and pollen allergy: the airborne pollen serves as indices of vegetational composition of a region and also causes allergy and hay fever in human beings;

(ii) Mellittopalynology and bee botany: the honey-bee collected pollen is used to feed the developing larvae in the beehive. This pollen also imparts properties to the honey manufactured here;

(iii) Palaeopalynology: the airborne pollen falls on the ground and gets fossilized over a time period, which serves as an indicator of the past vegetation;

(iv) Forensic palynology: pollen trapped in the body parts of victims of crime and the crime suspect is helpful in linking crime with the crime scene;

(v) Copropalynology: pollen content of animal faeces provides evidence of their feeding habits. It is thus the gametophytic cycle of angiosperms which forms the principles of palynology.

8.8 Contribution of Dr. P.K.K. Nair

1. Being the Founder of the Palynological Laboratory at NBRI and of the Palynological Society of India with activities involving organization of seminars, symposia, national conferences, presenting awards and above all, publication of the *Journal of Palynology*, amply testify Nair's organizational activities.
2. India gained pride of place in the formation of the International Federation of Palynological Science, in 1968, Nair as its first Vice President.
3. Under Nair's leadership, concrete proposal for a National School of Palynological Science also emerged and he became the founder editor of journals such as *Advances in Pollen-Spore Research*, *Glimpses in Plant Research* and *New Botanist*, to promote the development palynological science in India.

References

Cronquist, A. 1981. An Integrated System of Classification of Flowering Plants. Columbia University Press, New York.

Erdtman, G. 1943. An introduction to pollen analysis. Waltham, Mass: Verdoorn.

Erdtman, G. 1952. Pollen morphology and plant taxonomy-angiosperms. Stockholm (Sweden): Almqvist and Wikells Press.

Erdtman, G. 1954. Pollen morphology and plant taxonomy. *Botanical Notification*, 65-81.

Erdtman, G., Praglowski, J. and Nilsson, S. 1963. An introduction to a Scandinavian pollen flora- Almqvist & Wiksell, Stockholm

Erdtman, G. 1966. Pollen morphology and plant taxonomy, Vol. 1. Angiosperms, Hafner, New York.

Erdtman, G. 1969. Handbook of Palynology- Morphology- Taxonomy-Ecology. Hafner publishing Co., Inc., New York. 486 pp.

Faegri, K. and Iversen, J. 1950. *Text-Book on Modern Pollen Analysis*, Einar Munksgaard, Copenhagen, Denmark.

Faegri, K. and Iversen, J. 1989. "*Textbook of Pollen Analysis*", 4th ed., K. Faegri, P.E. Kaland, K. Krzywinski (Eds.). Chichester. John Wiley and Sons.

Garg, A. 2015. Current Science, 109(12): 2315-2318.

Heywood, V.H. 1967. Plant Taxonomy, Edward Arnold Publishers. London.

Jones, S.B. Jr. and Luchsinger, A.E. 1987. Plant systematics (2nd ed.) McGraw Hill, Singapore.

Kooiman, P. 1972. *Acta Botanica Neerlandica*, **21**(4): 417-420.

Luo, Y., Lu, L., Wortley, A.H., Li, D., Wang, H. and Blackmore, S. 2015. *Annals of Missouri Botanical Garden*, **101**(2): 406-455.

Meo, A.A. and Khan, M.A. 2006. *Pakistan Journal of Botany*, **38**(1): 29.

Nair, P.K.K. 1962. Pollen grains of Indian plants. II. *Bulletin of Garden,* Lucknow, No. **60**, 6-9.

Nair, P. K. K. 1974. *Evolutionary Biology*, **7**: 127-137.

Nair, P.K.K. and Saxena, A.K. 1974. *Acta Botanica Indica*, **2**: 23-31.

Nair, P.K.K. 1980. Glimpses in plant research V. Modern trends in plant taxonomy, Vikas Publication, New Delhi.

Perveen, N. 2003. Palynological study of selected medicinal plant of family Euphorbiaceae. University of Arid Agriculture, Rawalpindi, Pakistan, pp. 23-25.

Prakash, O. and Singh, M. 2020. *Journal of Pharmacy and Biological Sciences*, **15**(5): 37-41.

Sharma, O.P. 1987. *Bangladesh Journal of Botany*, **16**(2): 141-150.

Shehata, A.A., Taia, W.K., Shiha, M.A., Eltelet, S.A.M. and Salim, M.A. 2023. *Egyptian Journal of Botany*, **63**(3): 1101-1111.

Shim, K.K., Suh, B.K., Han, B.K. and Park, S.H. 1989. *Journal of Korean Society of Horticulture Science*, **30**: 224-233.

Simpson, B.B. and Skvarla, J.J. 1981. *American Journal of Botany*, **68**(2): 277-294.

Stix, E. 1960. *Grana Palynologica*, **2**:41-104.

Wagenitz, G. 1976. *Plant Systematics and Evolution*, **125**: 29-46.

Wodehouse, R. P. 1935. Pollen grains. New York.

Zhigila, D.A., Sawa, F.B.J., Akawu, J.J. and Ajiya, C.B. 2018. *Asuu Journal of Science*, **5**(1 & 2): 1-20.

9

Serotaxonomy

The application or utility of serology in solving taxonomic problems is called serotaxonomy or systematic serology. The study of antigen-antibody reactions is called serology. It had its origin towards the end of 20th century with the discovery of serological reactions and development of the discipline of immunology. The substance capable of stimulating the formation of an antibody is called antigen. A highly specific protein molecule produced by plasma cells in the immune system is called antibody. The antibodies combine chemically with specific kinds of antigens. Proteins are the useful carriers of taxonomic information and are most widely used antigens in serotaxonomy.

Serology is Useful in The Following Ways

1. It expresses similarities and dissimilarities amongst different taxa, and these data are helpful in taxonomy. It determines the degree of similarity between species, genera, families, etc. by comparing the reactions of antigens from various plant taxa with antibodies raised against the antigen of a given taxon.
2. It helps in comparing non-morphological characteristics, the knowledge of which is useful in taxonomy.
3. Single proteins from different plant taxa are also compared by serology techniques.

9.1 Some other Related Terms

1. Serology: According to Boyden (1964) it is the branch of biology which deals with "the nature and interactions of antigenic material and antibodies". But according to Smith (1976) serology is the "study of origin and properties of antisera".
2. Agglutinogens: Antigens are also called agglutinogens.
3. Agglutinins: Antibodies that participate in agglutination process.
4. Agglutination: A type of antigen-antibody reaction that results in visible clumps of organisms or other material.

5. Precipitins: These are the antibodies "capable of combining with and reacting upon antigenic material because of certain determinant groups" (Fairbrothers, 1968). Precipitins are the antibodies causing precipitation.
6. Precipitin reaction: It is the *in vitro* reaction between solutions or suspensions of antigen and antibody, with the formation of a precipitate.
7. Reference reaction: It is the "reaction between an antiserum and the antigenic material used to stimulate its formation" (Fairbrothers, 1968).
8. Haptens: These are the substances which are unable to induce antibody formation by themselves, but are able to react specifically with antibodies.
9. Serological reactions: Antigen-antibody reactions *in vitro* are known as serological reactions.

9.2 General Features Or Serological Reactions

1. The reactions is specific, an antigen combining only with its homologous antibody and vice versa.
2. Entire molecules react and not the fragments.
3. Antigens or antibodies do not pass through denaturation during these reactions.
4. Combination occurs at the surface, and it is firm but reversible.
5. Both antigens and antibodies take part in the formation of precipitates or agglutinates.
6. Antigens and antibodies can combine in varying proportions.

9.3 Brief History of Serotaxonomy

Nuttal (1901) was the first biologist to compare the immunochemical specificity of serum proteins for systematic purposes. Other notable serologists of that time were Kowarski (1901), Bertarelli (1902) and Magnus (1908). It was Dunbar (1910) who showed that proteins from pollen, seeds and leaves of rice were serologically distinct. A school of serology was founded in 1914 by Gohlke at Koenigsberg in Germany, and in the later years, Germany became the centre of serological studies. Mez (1926) and Moritz (1934) of Germany and Boyden (1942) of America developed the technique of serology in the later years. Notable contributions on the application of serology to taxonomy are those of Rives (1923), Hawkes and Lester (1966), Smith (1968, 1969, 1972), Kloz (1971), Fairbrothers (1967, 1975, 1977, 1978, 1983) and Crawford and Giannasi (1982).

9.4 Process of Serotaxonomy

In serotaxonomical studies, the protein extract of plant or animal origin (i.e. antigen), is injected into the blood stream of a rabbit or an experimental animal, to form the antibodies. In response to a specific antigen a specific antibody is produced. The serum, called antiserum, is then made to react *in vitro* with the antigenic protein as well as with proteins of other taxa, the affinities of which are to be determined. The amount of the precipitation shows the degree of the protein homology. The whole process may be explained as under:

Suppose we are to ascertain the closeness of a taxon A with taxa B, C and D. The proteins extracted from A are injected into a rabbit, in which the antibodies are produced. The antibodies are then extracted from the blood of the rabbit in the form of an antiserum. When the antiserum is allowed to react with the original protein extract from A, a total coagulation takes place. When this antiserum is allowed to react with the protein extracts from the taxa B, C and D, the degree of coagulation varies. This degree of coagulation varies. The degree of coagulation is related directly to the closeness between the taxa being compared.

9.5 Examples of Serotaxonomic Importance

Out of the several cases of the implication of serological data in the classification of angiosperms, a few are mentioned below:

1. According to Fairbrothers (1983) serological data have been used in the classification of orders and the assignment of families in Apiales, Caryophyllales, Capparales, Fagales, Cornames, Magnoliales, Juglandales, Papaverales, Rubiales, Ranunculales, Scrophulariales, Typhales, Primulales, etc.
2. Fairbrothers and Johnson (1959) separated six species of *Bromus* in the basis of serological studies.
3. According to Jensen (1968), serological characteristics within Ranunculaceae, show a close similarity between *Aconitum-Delphinium*, *Actaea-Cimicifuga*, *Anemone-Clematis* and *Ranunculus-Myosurus* and suggest a common ancestry for *Aquilegia*, *Leptopyrum* and *Thalictrum*.
4. While studying the serological assessment of relationships within Solanaceae, Hawkes and Tucker (1967) observed a strong serological relationship between *Solanum*, *Nicotiana* and *Hyoscyamus*, *Datura* and *Salpiglossis*.

5. Hillebrand and Fairbrothers (1970) made a phytoserological systematic survey of Caprifoliaceae and suggested the monophyletic origin of the family. According to these workers there is a close relationship between Caprifoliaceae and Cornaceae on serological grounds.
6. On the basis of serological studies, Johnson and Fairbrothers (1965) suggested that the genera *Magnolia* and *Michelia* show the closest affinity within Magnoliaceae.
7. Simon (1971) showed a close relationship between Nymphaeaceae and Nelumbonaceae on the serological ground.
8. Serotaxonomic findings of Pickering and Fairbrothers (1967) in Apiaceae support the classification of the family into Apioideae, Saniculoideae and Hydrocotyloideae and also suggest that Apioideae is more close to Saniculoideae than to Hydrocotyloideae.
9. *Hydrastis* of Berberidaceae has more serological similarities with Ranunculaceae than Berberisaceae.
10. Hawkes and Lester (1968) showed a relationship amongst the tuber bearing taxa of *Solanum* by serological techniques.
11. Recently, Piechura and Fairbrothers (1983), on the basis of their serotaxonomic studies, supported the view that taxa of Oleoideae evolved from those of Jasminoideae.

References

Boyden, A. 1964. Perspectives in systematic serology. In C.A. Leone (ed.) Taxonomic Biochemistry and Serology, Ronald Press, new York, pp: 75-99.

Fairbrothers, D.E. 1977. Prespectives in plant serotaxonomy. *Annals of the Missouri Botanical Garden*, **64**: 147-160.

Fairbrothers, D.E. 1983. Evidence from nucleic acid and protein chemistry, in particular serology, in angiosperm classification. *Nordic Journal of Botany*, **3**: 35-41.

Fairbrothers, D.E. and Johnson, M.A. 1961. The precipitation reaction as indication of relationships in some grasses. Recent Advances in Botany, 1: 116-120.

Hawkes, J.G. and Lester, R.N. 1968. Immunological studies on the tuber-bearing solanums. III. variability within *S. bulbocastanum* and its hybrids with species in series Pinnatisecta. *Annals of Botany*, **32**: 165-186.-

Hawkes, J.G. and Tucker, W.G. 1967. Family Solanaceae, In: Hawkes, J. G. (ed.): Serological assessment of relationships in flowering plants. Chemotaxonomy and Serotaxonomy. Academic Press, New York.

Hillebrand, G.R. and Fairbrothers, D.E. 1970. Phytoserological systematic survey of the caprifoliaceae. *Brittonia*, **22**: 125-133.

Jensen, U. 1968. Serologische Beitragezursystematik der Ranunculaceae. Botanical Jahrb Systematics, **88**: 204-268.

Johnson, M.A. and Fairbrothers, D.E. 1965. Comparison and interpretation of serological data in the Magnoliaceae. *Botanical Gazette*, **126**(4): 2650-269.

Pickering, J.L. and Fairbrothers, D.E. 1967. A serological and disc electrophoretic investigation of *Magnolia* taxa. *Bulletin of the Torrey Botanical Club*, **94**(6): 468-479.

Piechura, J.E. and Fairbrothers, D.E. 1983. The use of protein-serological characters in the systematics of the family oleaceae. *American Journal of Botany*, **70**(5): 780-789.

Simon, J. P. 1971. Comparative serology of the order Nymphaeales. II. Relationships of Nymphaeaceae and Nelumbonaceae. *Aliso: A journal of systematics and Floristic Botany*, **7**(3): 5-10.

10

Botanical Library

A botanical and horticultural library is a library specializing in the preservation and collection of literature and materials about plants. Botanical libraries are like special libraries are specialized information resources tailored to the unique needs of a particular organization, profession, or subject area. A botanical library is a specialized resource center dedicated to the study of plants and plant science. It typically houses a vast collection of books, journals, and other materials related to botany, horticulture, plant biology, and related fields. The mission of many botanical and horticultural libraries is to make accessible and available to those who use it the information on this topic. Botanical and horticultural libraries can be found in arboretums, botanical gardens, research institutions, horticultural societies, conservatories, governmental offices, colleges, and universities as part of a larger university library.

The collections within these types of libraries cover such topics as flowers, grasses, trees, fungi, aquatic plants, weeds, algae, pests, and fertilizers. The typical collection is composed of books, periodicals, and photo collections. The types of materials depend on the types of patrons. A public botanical garden library may contain books for children, nursery catalogs, seed catalogs, popular how-to publications, movies, and various other media resources. At an academic library, there may be such archived materials, such as historical letters, autographs of famous botanists, plant lists, field notes, and reports of botanic expeditions. These libraries might include:

1. **Reference Books**: Detailed works on plant identification, taxonomy, and botanical nomenclature.
2. **Scientific Journals**: Publications featuring research articles, reviews, and papers on the latest developments in plant science.
3. **Herbarium Specimens**: Preserved plant samples used for research and education.
4. **Historical Texts**: Rare or antique books that provide insights into the history of botany and plant exploration.

5. **Field Guides**: Practical guides for identifying plants in various regions.
6. **E**-resources: Now a days in the age of modern IT technology e-resources are key component of any library and botanical library is not exception for this.

Botanical libraries often support researchers, students, and enthusiasts by providing access to a wealth of information and sometimes hosting lectures, workshops, and educational programs. They might be part of larger institutions like botanical gardens, universities, or research centers. a botanical library is an essential resource for advancing knowledge and research in botany and horticulture. By providing access to specialized materials, supporting research efforts, and fostering education and collaboration, botanical libraries play a key role in the development and application of plant science.

Libraries may also hold highly specialized collections of certain species. For example, The Orchid Library of Oakes Ames is devoted exclusively to orchids. The collection is composed of more than 6,000 journals and books devoted to the classification and identification of orchid species around the world.

10.1 Botanical Library and Systematics

A botanical library is a storehouse of botanical books, journals, periodicals, magazines, photographs, movies, newspapers, dissertations, etc. All printed materials, drawings, films, paintings, and other similar articles belonging to the plant sciences are also stored there in a proper system.

It is also a documentation centre of botanical activities, and stores the data related to all the branches of plant science. It may also be called as a service facility to the botanists as it provides the required literature.

A botanical library exists for the benefit of the entire society in general and for the persons interested in botany in particular. However, it is such a huge repository of the world's botanical literature that no single individual can utilize this system in it's entirely.

The important taxonomic publications, available in the taxonomy section of a botanical library include floras, monographs, literature indexes, and name indexes. The information regarding the plants of a given geographical region are available in a *flora*, while a monograph contains information on the systematics of a given taxonomic group of plants on a worldwide basis. The binomial system of nomenclature is used in both the floras and monographs. Various systems of indexing and abstracting the taxonomic literature are available in *Biological abstracts* and *Bibliography of Agriculture*. For the name indexes of plants, the world botanists rely on books such as Index Kewensis, Gray Herbarium Index, and J.C. Willis's A Dictionary of the Flowering Plants and Ferns.

10.2 Common Terms Come Across in A Botanical Library

1. Document: A written item such as a letter, a manuscript, notes, an article or a book is called a document.
2. Documentation: An introduction to the written documents, specimens, etc. used as an evidence or a source of information in taxonomic studies is called documentation.
3. Literature retrieval: It is the process of locating and identifying by referring to the books, manuals, articles, the exact information within the entire published literature that is of interest to the user.
4. Document retrieval: It is the process of locating and identifying both the published as well as unpublished documents such as specimens, manuscripts, illustrations, notes, films, etc.
5. Voucher: It is a specimen or written document used for the verification of information.
6. Citation: A quotation from a book or author is called a reference.
7. Reference: A source of information with a definite direction is called as reference.
8. Reference book: An authoritative book consulted for facts or the background information is called a reference book.
9. Bibliography: The list of books, research articles, or other writings, usually of one author or one subject or topic is called bibliography.

10.3 Use of Botanical Library

Radford (1986) mentioned that the botanical library is useful to the systematics as an information source, a documentation centre, a data storehouse, and also as a service facility.

1. From the botanical library, taxonomists procure information to verify facts about the available literature, procedure, citation, classification, and all other related aspects of his/ her project and in that way the library is used as an information source.
2. The library is used as a documentation centre as it is a repository of rare publications, first editions, original manuscripts, illustrations, photographs, etc.

3. The library is also used as a data storehouse for research because it stores all the published data related to the systematic projects.
4. The big libraries are used as a service facility for providing the technical know-how of their available resources. A reference service of these libraries helps in procuring the requisite material, photocopies, microfilms, necessary for the projects in systematics. Most of the big libraries also have an interlibrary loan service for supply of materials not available locally.

10.4 Botanical Library As A Training Centre

Several big libraries train the students for using the library in a proper and systematic manner, and thus function as training centres. The trained library staff trains the young students by (i) introducing them the references of common use, such as manuals, dictionaries, atlases, maps, encyclopedias, indexes, etc., (ii) familiarizing them with the literature of the current developments such as periodicals, journals, monographs, recent bibliographies, etc., (iii) informing them of the proper rules for making catalogues, bibliography, etc., and (iv) providing them the way of analyzing the literature for scientific writing, etc.

10.5 Classification Used in A Botanical Library

In the botanical libraries, two systems of classification are used i.e. the Dewey Decimal (DD) system and the Library of Congress (LC) system. The Dewey Decimal system is an old system of classification and is now being replaced by the LC system in majority of the libraries of the world.

10.6 Major Botanical Libraries of the World

A huge taxonomic literature is available in five great national libraries of the world, viz. Library of Congress in Washington, USA; Library of the Academy of Sciences in Leningrad, Russia; the National Library of Beijing, China; Library of the British Museum in London, England; and the Bibliotheque Nationale in Paris, France.

The other big American university libraries are at California (Berkeley), Cornell, Harvard, Iowa State, Minnesota, Illinois, Columbia, Wisconsin, Chicago, Michigan, Yale and Texas.

Several large botanical libraries are associated with the botanical gardens, museums, and botanical societies. Chief among them are the libraries associated with the British Museum, the New York Botanical Garden, Missouri Botanical Garden, Gray herbarium, Massachusetts Horticultural Society at Boston and Academy of Natural Sciences at Philadelphia.

Some major Indian botanical libraries are associated with The Botanical Survey of India, Kolkata; National Botanical Garden, Lucknow; Forest Research Institute, Dehradun; Indian Agricultural Research Institute, New Delhi; and Universities of Delhi, Chennai, Meerut, Allahabad, and Kolkata. A big botanical library was established at Meerut College, Meerut, and Uttar Pradesh by Prof. V. Puri during 1948 and 1967. Till 1970, is used to subscribe 166 research journals, now also contains over 7,000 books.

10.7 Future Information Systems

The botanical libraries have been the main information systems during the last 500 years, i.e. since the invention of the movable type. But the introduction of the high speed electronic computers during the last few decades has brought about a colossus change in the entire scene. Instead of the printed pages, a large amount of the botanical information is now stored in computers, the plants may be identified more easily and correctly. One may say it with confidence that the computers will be the main information tools in future. Detailed taxonomic data would be written on the pages of a computer memory in future, and these data might be made available to the entire world via telecommunication. Nowadays, the taxonomists able to get the latest information on a topic of his/her choice through computers. But this does not mean that botanical libraries are going to be replaced by computers in future.

Reference

Radford, A.E. 1986. Fundamentals of Plant Systematics, Harper and Row, New York.

11

Biodiversity

Biodiversity in flora and fauna is the wealth of any nation. Billions of genes with considerable value are habitated in the bioreserves designated. Genetic resources are the backbone of crop improvement in horticulture. Term 'genetic resources' was coined by Sir Otto Frankelin 1967 to highlight germplasm as natural resource for improvement of crop plants. The ConsortiumofInternational Agricultural Research Centers (CGIAR) established in 1971 by United Nations Development Programme (UNDP) is a consultative group to support a network of 16 international agricultural research centres. It controls world's largest collection of genetic resources of more than 6 lakhs of accessions of over 3,000 crops, forage and pasture species (both cultivated & wild). International Bureau of Plant Genetic Resources (IBPGR, 1974 now Biodiversity International) situated at Rome, Italy had become the International Plant Genetic Resources Institute (IPGRI) in October, 1991 for collection, evaluation & conservation of germplasm for future use. IPGRI began independent operation as one of the Centres of CGIAR in 1994. Today, Biodiversity International is the only global research-for-development organization that investigates the conservation and use of agricultural and forest bio-diversity. It is in the forefront of assisting member countries in identifying genetic resources for nutritional and food security without hampering the fragile ecosystem.

Biodiversity of given region is assessed in terms of species richness. It boosts ecosystem productivity where each species, no matter how small, all have an important role to play. Plant Generic Resource (PGR) is fundamental to crop improvement programme and the key to establish future food and nutritional security. The 1st plant genetic resources centre of the world was situated at Leningrad, Russia. Later in McNeely *et al.* (1990) categorized India as one of the 12 mega biodiversity countries in the world and paved the way for its placement on the world's conservation map. Myers (2000) identified global bio-diversity hotspots and North East parts of India are the 2rd largest and next only to Mediterranean basin among the 25 identified globally.

The Year 2010 was declared as 'International Years of Biodiversity' by United Nations Organization. Biodiversity conference (COP-15) was held recently in

Montreal, Canada. World agreement on biodiversity has been accepted by > 196 member countries including India. The World Nature Conservation Day is observed annually on 28th July. The occasion aims to create awareness about the need to preserve the environment & natural resources in order to keep the world healthy.

Indian region has enormous diversity for ornamental plants which are grown commercially and in home gardens. India is one of 17 mega biodiversity hot spot regions of the world. India comprises 12 per cent of total global plant wealth. India is rich in medicinal plants biodiversity having 2.4 per cent world's area and 8 per cent global diversity. India is endowed with large germplasm pool (72,000 germplasm accessions of cultivated, wild and related taxa in arid zone fruit crops, vegetables, spices, condiments, plantation and tuber crops) is scientifically conserved in various institutes.

National Biodiversity Board, New Delhi was enacted in 2003. National Gene Bank was established at NBPGR, New Delhi in 1983. NBPGR acts as a nodal agency for implementation of plant germplasm registration. NBPGR ranks 3rd in collection of valuable heritage in the world after the USA and China. At NBPGR, nearly 4.52 lakhs seed species including 2.7 lakhs Indian seed species have been conserved in the gene bank. IIHR, Bengaluru has developed the technologies for conservation of nuclear genetic diversity for many horticultural crops.

Concept of biodiversity 'Hotspots' was given by Norman Meyer in 1988. He first identified 10 hotspots based on the high degree of endemism and habitat loss, i.e. the regions of richest biodiversity and most endangered. There are 36 total biodiversity hot spots in the world. Total land space covered by hot spots on earth is about 2.3 per cent. India and adjoining countries to it comprises four biodiversity hot spots and total agro-biodiversity hot spots in India are 25. About 90% tree coverage from these four hot spots has been destroyed and bio-diversity is in danger zone. 'Valley of flowers' in Uttarakhand, Himachal Pradesh is a part of Nanda Devi Biosphere Reserve. It conserves 300 rare species of wild flowers was declared as World Heritage site by United Nations Educational, Scientific and Cultural Organization (UNESCO). '*Kaas* plateau', Dist. Satara of Maharashtra which is also known as 'Valley of flowers of Maharashtra' declared as World Heritage site by UNESCO in July, 2012 for flower species. It has more than 850 botanical species including regional and extinct species.

11.1 India: A Centre of Rich Biodiversity

By the virtue of physiognomy and climatic diversity, India harbours three of the 34 biodiversity 'hot spots', the Himalayas, the Indo-Burma region and the Western Ghats, with thousands of endemic flora (Conservation International, 2005). Eastern Himalaya, North East India are included in global biodiversity hotspots. This region is known as mega biodiversity centre and hot spot area in the world. It is treasure house of variety of economic, medicinal, aromatic and plants of botanical curiosity. Some of the crop species are also used for cosmetics, neutraceuticals, food products, beverages, etc. The floristically rich India has about 141 endemic genera belonging to over 47 families. Botanical survey of India's easter circle office was established in April, 1956 at Shillong. Aromatic plants also constitute an important part of the world's floristic diversity. These plants produce unique aroma from their various parts i.e. leaf, seed, fruits, bark and roots.

As per Botanical Survey of India, India has 46,214 plant species. Of these, about 17,500 (7000 species in north-east region alone) represent flowering plants (7 per cent of the world flora); 37 per cent of them are endemic. Of the endemic species (4950), the largest number of (about 2532) species are located in Himalayas followed by peninsular region (1788 species) and Andaman Nicobar Islands (185 species). About 33 per cent of plants that occur in India are endemic. Endemicity of Indian orchids is very high and nearly 29 per cent of recorded taxa are endemic.

India is rich centre of floristic and economic plant diversity (Arora *et al.*, 2006), particularly for horticultural species like fruits, vegetables, spices, ornamentals, medicinal plants, etc. both at species and intra-species level. Thus, the amount of species diversity available offers substantial genetic diversity to meet the future needs, particularly for the present scenario of climate change.

11.2 The National Biodiversity Act

Biodiversity encompasses the variety of all life on earth. India is one of the 17-mega diverse countries of the world. With only 2.5% of the land area, India already accounts for 7.8% of the global recorded species. India is also rich in traditional and indigenous knowledge, both coded and informal.

Convention on Biological Diversity (CBD) is the 1st international agreement to study bio-diversity in India. CBD provides the facility of prior informed consent or material transfer agreement. It was signed during Earth Summit at Rio de Janeiro, 1992. As per report of CBD, > 122 botanical gardens exist in India. Empress botanical garden (Pune camp area) has maintained largest collection of exotic and indigenous flora and fauna & rich collection of very rare medicinal plants. Recognizing the sovereign rights of States to use their own biological resources, the Convention expects the Parties to facilitate access to genetic resources by other Parties subject to national legislation and on mutually agreed upon terms (Article 3 and 15 of CBD). Article 8(j) of the Convention on Biological Diversity recognizes contributions of local and indigenous communities to the conservation and sustainable utilization of biological resources through traditional knowledge, practices and innovations and provides for equitable sharing of benefits with such people arising from the utilization of their knowledge, practices and innovations.

Biological diversity is fundamental to agriculture and food production. From the millions of genes that serve as building blocks to the thousands of plants and animals that inhabitant the earth, almost limitless combinations of organisms that make up natural ecosystems. Under the contextual reference of modern intensive agriculture demanding many farmers to adopt high yielding varieties of plants and animals, Biodiversity makes an essential contribution for feeding the millions of population of the world. When farming communities abandon diversity, varieties and breeds of animals extinct and the specialized useful traits may be lost. Agricultural diversity forms an important component for tackling unforeseen effects of climate change and is one of the essential key component for developing varieties and animal species with standing temperature extreme floods, droughts, outbreak of diseases reducing the crop productivity and directly causing untold hardship to farming community.

Biodiversity is a multi-disciplinary subject involving diverse activities and actions. The stakeholders in biological diversity include the Central Government, State Governments, institutions of local self-governmental organizations, industry, etc. One of the major challenges before India lies in adopting an instrument, which helps to realize the objectives of equitable sharing of benefits enshrined in the Convention on Biological Diversity.

Salient Features of the Biological Diversity Act - 2002:

- After an extensive and intensive consultation process involving the stakeholders, the Govt. of India has brought Biological Diversity Act, 2002.
- To regulate access to biological resources of the country equitable share in benefits arising out of the use of biological resources.

- To conserve and sustainable use of biological diversity.
- Setting up of National Biodiversity Authority (NBA), State Biodiversity Board (SBB) and Biodiversity Management Committee's. (BMC's).
- NBA and SBB are required to consult BMCs in decisions relating to bioresource / related knowledge within their jurisdiction.
- To respect and protect knowledge of local communities traditional knowledge related to biodiversity.
- To secure sharing of benefits with local people as conservers of biological resources and holders of knowledge and information relating to the use of biological resources.
- All foreign nationals / organizations require prior approval of NBA for obtaining biological resources and / or associated knowledge for use.
- Indian scientists / individuals require approval of NBA for transferring results of research to foreign nationals / organizations.
- Conservation and development of areas of importance from the standpoint of biological diversity by declaring them as biological diversity heritage sites.
- Protection and rehabilitation of threatened species.
- Involvement of institutions of State Government in the broad scheme of the implementation of the Biological Diversity Act through constitution of committees.
- Protect India's rich biodiversity and associated knowledge against their use by foreign individuals and organizations without sharing benefits arising out of such use and check.

Definitions

1. "benefit claimers" means the conservers of biological resources, their byproducts, creators and holders of knowledge and information relating to the use of such biological resources, innovations and practices associated with such use and application:
2. "biological diversity" means the variability among living organisms from all sources and the ecological complexes of which they are part and includes diversity within species or between species and of eco-systems;
3. "biological resources" means plants, animals and micro-organisms or parts thereof, their genetic material and by- products (excluding value added products) with actual or potential use or value but does no include human genetic material:

4. "bio-survey and bio-utilization" means survey or collection of species, sub-species, genes, components and extracts of biological resources for any purpose and includes characterization, inventorisation and bioassay:

The National Biodiversity Authority shall consist of the following Members, namely-

(a) a Chairperson, who shall be an eminent person having adequate knowledge and experience in the conservation and sustainable use of biological diversity and in matters relating to equitable sharing of benefits, to be appointed by the Central Government.

(b) three *ex officio* members to be appointed by the Central Government, one representing the Ministry with Tribal Affairs and two representing the Ministry dealing with Environment and Forests of whom one shall be the Additional Director General of Forests or the Director General of Forests.

(c) seven ex officio members to be appointed by the Central Government to represent respectively the Ministers of the Central Government dealing with --------

(i) Agricultural Research and Education

(ii) Biotechnology

(iii) Ocean Development

(iv) Agriculture and Cooperation

(v) Indian System of Medicine and Homeopathy

(vi) Science and Technology

(vii) Scientific and Industrial Research

(d) five non-official members to be appointed from amongst specialists and scientisthaving special knowledge of, or experience in, matters relating to conservation ofbiological diversity, sustainable use of biological resources and equitable sharing ofbenefits arising out of the use of biological resources, representatives of industry,conservers, creators and knowledge-holders of biological resources.

11.3 Convention on Biological Diversity

Biological Diversity means the variability among living organisms from all sources, including interalia, terrestrial, marine and other aquatic ecosystems and the ecological complexes of which they are part and this includes diversity within species, between species and of ecosystems. Biodiversity is defined as the variety and variability among living organisms and the ecological complexes in which they occur is measured at three levels *viz.*, genes, species and ecosystem.

The Convention on Biological Diversity (CBD) is a landmark in the environment and development field, as it takes for the first time a comprehensive rather than a sectoral approach to the conservation of Earth's biodiversity and sustainable use of biological resources. It was in the year 1984 that the needs to have in place a global convention on biological diversity started gaining momentum. In response, the United Nations Environment Programme (UNEP) in the year (1987) recognized the need to streamline international efforts to protect biodiversity. It therefore established an *ad hoc* working group to investigate "the desirability and possible form of an umbrella convention to rationalize current activities in the field. This group by 1988 concluded that a) the existing treaties were inadequate to address the issue of conservation and sustainable use and b) a new global treaty on biological diversity was urgently needed. Organizations such as the World Conservation Union (IUCN) and the Food and Agricultural Organization (FAO) contributed draft articles in addition to specific studies commissioned by the UNEP. The UNEP Secretariat prepared the first draft and the formal negotiating process was started in 1991. The Inter-governmental Negotiating Committee for a Convention on Biological Diversity (INC) was given the task of ensuring the adoption of the Convention. Conservation of biological diversity, sustainable use of its components and fair and equitable sharing of the benefits arising out of the utilization are objectives of UN CBD. On May 22, 1992 the nations of the world adopted the CBD in Nairobi and on June 5, 1992 the CBD was tabled at the UN Conference on Environment and Development in Rio de Janeiro where a record 150 countries signed the Convention.

The Convention on Biological Diversity (CBD) was negotiated and signed by nations at the UNCBD Earth Summit at Rio de Janeiro in Brazil in June 1992. The Convention came into force on 29thDecember, 1993. India became a party to the convention in 1994. At present, there are 191parties to this convention and the treaty has been ratified by 168 countries.

As the national response to CBD, the Biological Diversity Act, 2002 was enacted on December 11, 2002 and came into force since April 15, 2003 with the notification of the Biological Diversity Rules 2003. The implementation of the Act is coordinated by three functional bodies namely the National Biodiversity Authority (NBA), The State Biodiversity Board (SBB) and the Biodiversity Management Committee (BMC).

11.4 Biodiversity Management Committees (BMCs)

Institutions of local state government will be required to set up biodiversity management Committees in their respective areas for conservation, sustainable use, documentation of biodiversity and chronicling of knowledge relating to biodiversity.

NBA and SBBs are required to consult the concerned BMCs on matters related to use of biological resources and associated knowledge within their jurisdiction.

11.5 Biodiversity Heritage Sites (BHS)

Biodiversity heritage sites (BHS) should includes both wild and domesticated biodiversity and human cultural relations with such biodiversity. The question will arise, how biodiversity. Heritage sites are different from the concept of "Protected areas"? Protected areas are covered in The Wild Life Act.

11.6 National Biodiversity Authority (NBA)

All matters relating to requests for access by foreign individuals, institutions or companies, and all matters relating to transfer of results of research to any foreigner will be dealt with by the National Biodiversity Authority.

- Establishment of NBA.
- The head office of the NBA is established at Chennai.
- NBA consists of the following members.

Members of NBA

- A Chairperson who shall be an eminent person having adequate knowledge on conservation and sustainable use of biological diversity.
- Three ex-officio members appointed by the Central Government. One representing the Ministry dealing with Tribal affairs.Two representing the Ministry dealing with Environment and Forests of whom one shall be the Additional Director General of Forests.
- Seven ex-officio members appointed by the Central Government to represent respectively the Ministries of the Central Government dealing with
    - Agricultural Research and Education
    - Biotechnology
    - Ocean Development

- Agriculture and Cooperation
- Indian Systems of Medicine and Homeopathy
- Science and Technology
- Scientific and Industrial Research
- Five non-official members appointed amongst specialists and scientists, representatives of industry, conservers, creators and knowledge holders of biological resources

Functions and Powers of NBA

- Regulate activities, approve and advice the government of India on research, commercial, bio-survey and bio-utilization.
- Grant approval to Section 3,4 and 6.
- Certain persons not to undertake Biodiversity related activities without approval of National Biodiversity Authority (Section 3).
- Results of research not to be transferred to certain persons without approval of National Biodiversity Authority (Section 4).
- Application for IPR rights not to be made without approval of National Biodiversity Authority (Section 6).
- Perform such other functions as may be necessary to carry out the provisions of this act.

11.7 State Biodiversity Board (SBB)

All matters relating to access by Indians for commercial purposes will be under the purview of the State Biodiversity Boards (SBB). The Indian industry will be required to provide prior intimation to the concerned SBB about the use of biological resource. The State Board will have the power to restrict any such activity, which violates the objectives of conservation, sustainable use and equitable sharing of benefits.

- Establishment of State Biodiversity Board in every State.
- State Government may by notification in the Gazette can establish the SBB in their State name e.g Tamil Nadu Biodiversity Board.
- No State Biodiversity Board shall be constituted for a Union Territory and in relation to Union Territory, the National Biodiversity Authority shall exercise the powers and perform the functions of a SBB for the Union territory.

11.8 National Bureau of Plant Genetic Resources (NBPGR)

National Bureau of Plant genetic Resources, New Delhi has national responsibility to carry out plant quarantine for exchange of germplasm meant for research. NBPGR in collaboration with other institutes, has undertaken 2415 germplasm explorations and assembled > 250000 accessions of agri-horticultural crops and their wild relatives. Pest Risk Analysis (PRA) standards are being worked out for 48 crops of horticultural importance. The import of transgenic material is regulated by DBT and NBPGR is the authorized agency for carrying out the quarantine test. The National Gene Bank housed at NBPGR is primarily responsible for conservation of all kinds of HGR on long term basis. Nearly 40,000 accessions have been conserved in the National Gene Bank for long term storage in the form of seed, while around 87,000 accessions of perennial crops are conserved in the field gene bank at 10 regional stations of NBPGR and 59 National Active Germplasm Sites (NAGS) located in different agro-climatic and phyto-geographical regions of the country. Besides 2000 accessions of 158 vegetative propagated crops in vitro and 8981 accessions of 726 species have been cryopreserved. There is a national network for evaluation and conservation of HGR which includes NBPGR, horticultural institutes, national research centres, AICRPs of ICAR, Department of biotechnology, Department of Environment and Forests, State Agricultural Universities, State department of Agriculture/ Horticulture/ Forestry. The documentation of PGR becomes important at the national, regional, and global levels for effective conservation of rapidly disappearing genetic stocks for their immediate and future uses. The Agricultural Research and Information System (ARIS) cell at NBPGR which is mainly responsible for the data base activities at NBPGR has developed many on line databases related to PGR and plant varieties, which are in use by researchers in India.

11.9 Horticultural Genetic Resources (HGR)

Indian region has enormous diversity for ornamental plants. The important ornamentals native to India are orchids, rhododendrons, primulas, musk rose (*Rosa moschata*), begonias (*Begonia* spp.), lily (*Lilium bulbiferum*), foxtail lily (*Eremurus himalaicus*), champaka (*Michelia champaka*), lotus (*Nelumbo nucifera*), water lily (*Nymphaea* spp.) and wild tulip (*Tulipa stellata* and *T. aitchisonii*). Orchids like *Arachnis*, *Ascocentrum*, *Cymbidium*, *Dendrobium*, *Paphiopedilum*, *Phalaenopsis*, *Phaius*, *Renanthera*, *Rhynchostylis* and *Vanda* have exceptionally beautiful and long-lasting flowers of myriad shapes, sizes and colours (Sharma and Rana, 2000). The important native ornamental trees are *Bauhinia variegata*, *Cassia fistula*, *Butea frondosa*, *Lagerstroemia flosreginae*, *Pongamia glabra*, *Tecomella undulata*, *Saraca indica*, *Ficus benghalensis*, *F. religiosa* and *Anthocephalus cadamba* while *Jaminum* spp., *Hiptage madablata*, *Mussaenda frondosa*, *Clerodendron inerme* and *Passiflora leschenaultii* are important shrubs and climbers.

11.10 Role of Sacred Groves in the Perspective of Biodiversity Conservation

Conservation of plant biodiversity assumes greater importance when the world is facing unprecedented loss of biological diversity. Over thousands of years, local people have developed a variety of vegetation management practices and conservation of elements of biodiversity on the religious basis, plant and animal worship and observing taboos on harvesting and hunting of plants and animals are characteristics of many indigenous communities in India. Sacred groves are special sites or areas that have one or more attributes which distinguish them as somehow extraordinary, usually in a religious or spiritual sense.

In recent years, the significance of sacred groves (patches of natural vegetation dedicated to ancestral spirits or deities and preserved on the basis of religious beliefs) has assumed immense importance from the point of view of anthropological, taxonomical and ecological considerations (Hughes and Chandran, 1998 and Tiwari *et al*., 1999). Such grove may consist of a multi-species, multitier primary forest, a clump of trees belonging to one species, or even a single old tree, depending on the history of the vegetation and local culture of the region or area. These groves are protected by local communities usually through customary taboos and sanctions with traditional, cultural and ecological implications. Named differently in different parts of the India are *Law lyngdhoh* in Meghalaya, *Kovil kadu* in Kanyakumari, *Dev bhumi* in Uttarakhand, *Kavu* in Kerala, *Sarna* and *Deorai* in Madhya Pradesh, *Oran* in Rajasthan, *Jaherthan* and *Garamthan* in West Bengal, *Deovan* in Himachal, *Ummanglai* in Manipur, etc. these groves are mainly found in areas dominated by tribals and managed by local people for various reasons.

Indian sacred groves are sometimes associated with temples/ monasteries/ shrines or with burial grounds. The attitude of pre-British village communities towards forests is reflected in the statement of the British traveler, Francis Buchnan, made near Karwar in Uttar Kannada "The forests are the property of the Gods of the villages in which they are situated, and the trees ought not to be cut without having leave from the *Gauda* or headman of the villages... who is also priest (*pujari*) to the temple of the village God". In Uttarakhand, Yumnotri, Badrinath, Kedarnath (Garhwal), etc. are the sacred places more than 3200 m in altitude. Several other sacred groves are known such as Thatyur, Maanthaat,

Danda Ka Deorana from Dehradun, Sem Mukhim Naag from Tehri Garhwal, etc. In Kumaun region, Chitai Gul, Jageswar, Dhaula Devi, Anyar Bunga Aeri, Sidh Baba from Almora, Garanath, Binsar, Minar, Chamarkhan, Bari Goljyu, Gauri Udiyar in Bageshwar district, etc.

Sacred groves play an important role in the religious and socio-cultural life of the local people. Rituals and ceremonies are often carried out in the sacred groves to propitiate ancestral spirits and deities for enhancing agricultural yields and for well-being of animals and human beings. In India, sacred groves are scattered all over the country and do not enjoy protection via federal legislation. Some NGOs work with local villagers to protect such groves.

11.11 Biopiracy

To check biopiracy, the proposed legislation provides that access to biological resources and associated knowledge is subject to terms and conditions, which secure equitable sharing of benefits. Further, it would be required to obtain the approval of the National Biodiversity Authority before seeking and IPR based on biological material and associated knowledge obtained from India.

- Indian Industry needs prior intimation to SBB to obtain bioresource. SBB has right to restrict if found to violate conservation and sustainable use and benefit sharing.
- Provisions for notifying heritage sites by State Government in consultation with local body.
- Creation of National, State and Local Biodiversity Fund and its use for conservation of biodiversity.
- Prior approval is needed from NBA for IPRs in any invention in India or outside India on bioresource.

Exemptions Provided in the Act

The Biological Act, 2002 provides for the following exemptions:

- Exemption to local people and community of the area for free access to use biological resources within India.
- Exemption to growers and cultivators of biodiversity and to *Vaidyas* and *Hakims* to use biological resources.
- Exemption through notification of normally traded commodities from the purview of the Act.
- Exemption for collaborative research through government sponsored or

government approved institutions subject to overall policy guidelines and approval of the Central Government.

Application for Intellectual Property rights not to be made without the approval of NBA

- No person shall apply for any Intellectual Property Right, by whatever name called in or outside India for any invention based on any research or information on a biological resource obtained from India without obtaining the previous approval of the NBA before making such application. (6(1)).
- NBA while granting the approval under this section, impose benefit sharing fee or royalty or both or impose conditions including the sharing of financial benefits arising out of the commercial utilization of such rights. (6(2)).
- The provisions of this section shall not apply to any person making an application for any right under any law relating to protection of plant varieties enacted by parliament. (6(3)).
- Where any right is granted under law referred to in sub-section (3), the concerned authority granting such right shall endorse a copy of such document granting the right to the NBA. (6(4)).

References

Arora, R.K., Nayar, E.R. and Pandey, A. 2006. Indian centre of floristic and economic plant diversity: A review. In: *Hundred Years of Plant Genetic Resources Management in India*, (Eds.)

Anurudh K. Singh, K. Srinivasan, S. Saxena, and B.S. Dhillon. National Bureau of Plant Genetic Resources, New Delhi, India, pp. 1-28.

Hughes, J.D. and Chandran, M.D.S. 1998. Sacred groves around the earth: An overview. In: Conserving the sacred for biodiversity management, (Eds.) P.S. Ramakrishnana, K.G. Saxena and U.M. Chandrashekhara, pp: 69-86.

Sharma, B.D. and Rana, J.C. 2000. Survey, collection and scope of floriculture germplasm. In: Horticulture Technology, Vol. 1. Production, (Eds.) V.K. Sharma and K.C. Azad. Deep and Deep Publication, New Delhi, pp. 381-398.

Tiwari, B.K., Barik, S.K. and Tripathi, R.S. 1999. Sacred groves of Meghalaya: Biological and Cultural Diversity. Regional Centre National Afforestation and Eco-Development Board North-Eastern Hill University, Shillong.

12

Botanical Gardens

12.1 What are Botanical Gardens?

Botanical gardens are the institutions that maintain the living plant collections of different varieties of plants, including the ornamental and cultivated ones, wild, medicinal, of economic importance, of various geographical regions, of special interest, etc. They are of value not only to the botanists, horticulturists and foresters but also to the millions of tourists.

A big botanical garden contains plant species from several corners of the globe. It also includes green houses, a library, a herbarium, research laboratories, and several miscellaneous resources including photographs, paintings, illustrations, reprints, note books, and specimens of several types. It is, therefore, not merely a garden but a botanical institution.

Modern botanical gardens serve as centres for documentation, research, reference, data storage, education, conservation, and several other biological facilities to mankind.At present, there are over 600 botanical gardens in the world.

12.2 Forty Fivemajor Botanical Gardens of the World: A Chronological Directory

The gardens were maintained in all parts of the world since the early times also, as in evident from the old Indian classics such as Mahabharat, Ramayan and Vedas. For food, drugs and beautiful flowers, man had begun to cultivate plants in gardens long before the dawn of history. The Hanging Gardens of Babylon are considered among the wonders of the ancient world. Several kinds of fruit, vegetable and drug plants were cultivated during the Middle Ages. Romans maintained small gardens as sources of medicine and as aids to medical studies.

Undermentioned are the historical details of a few modern botanical garden of the world in 1 chronological order.

12.2.1 Botanical gardens established during 16th century

1. The credit of establishing the 1st botanical garden of the world goes to Luca Ghini (A.D. 1490-1556), an Italian Professor of Botany. He established a botanical garden at Pisa, Italy in 1543. He was its 1st director followed by Andrea Cesalpino (1554-1558). This garden does not exist today.
2. Otto Botanico, The botanical garden of University of Padua, Italy, was established by Professor Francis Bonafede on 29th June, 1545. A large herbarium and a library are attached to this garden. More than 600 portraits of botanists are displayed in this garden.
3. University Botanical Garden at Florence, Italy was established in 1545 by Luca Ghini.
4. Vatican Garden at Rome was established in 1556 by Michele Mercati, a student of Andrea Cesalpino.
5. Bologna Botanical Garden at Bologna, Italy was founded in 1567 and its 1st director was Ulisse Aldrovandi (1567-1605).
6. Leiden University Botanical Garden was established in 1587 at Leiden, Netherlands. Carolus Clusius, the Flemish-Austrian botanist, who was the director of this garden, was the 1st to establish the 1st greenhouse of the world in this garden in 1599.
7. Montpellier University Botanical Garden at Montpellier, is one of the earliest botanical gardens in France. It was founded in 1593.
8. Botanical Garden, Heidelberg, Germany was also established in 1593.

12.2.2 Botanical gardens established during 17th century

1. Botanical Garden of Strasbourg, France comes next in the chronological order. It was established in 1619.
2. Oxford University Botanic Garden, Oxford, England was established in 1621. This was the 1st botanical garden of Great Britain, and Earl of Danby was its initiator. In 1734, the 1st greenhouse of England was established in this garden.
3. Jardin des Plantes, a botanical garden of Paris, was established in 1640 under the name 'Jardin du Roy'. King Louis XIII of France ordered in 1626 to establish this royal botanical garden for the purpose of growing all kinds of medicinal herbs. It was actually an independent and scientific institution, founded by the king to promote the teaching of pharmaceutical botany. Its present herbarium and palaeobotanical

departments are among the world's best. It is actually the oldest and most important non-university botanical garden, existing today in the world.

4. Botanical garden of Groningen, Netherlands, was established in 1642.
5. Botanischer Garden, one of the great botanical gardens of the world, was established in 1646 in Berlin, Germany. A.W. Eichler, Adolf Engler and L. Diels were the directors of this garden.
6. Botanical garden of Uppasala, Sweden was founded in 1655 by O.J. Rudbeck. Linnaeus worked as the director of this garden from 1741 to 1777, and the original part of this garden is now maintained as a memorial to Linnaeus. With the efforts of Linnaeus, a collection of more 300 species of this garden grew to over 3000 species within seven years.
7. Royal Botanical Garden of Edinburg, Scotland was established in 1670. Its 1st director was J. Sutherland.
8. The Chelsea Physic Garden was founded in London, England in 1673, as the garden of the Society of Apothecaries.
9. Amsterdam Botanical Garden, Netherlands, was founded in 1682.
10. Botanical Gardens of Tokyo Imperial University were founded in 1684.

12.2.3 Botanical gardens established during 18th century

1. Botanic garden of University of Moscow, Russia was founded in 1707.
2. Druggist's Garden of St. Petersburg, Russia, was established in 1713. St. Petersburg is the Leningrad of today.The reorganized form of this garden is now V.L. Komarov Botanical Institute of the Academy of Sciences, Leningrad, Russia. The herbarium attached to this garden is one of the largest in the world, holding over 5 million specimens.
3. 1st botanical garden in America was established by John Bartram in 1731 near Philadelphia.
4. Botanical garden of Vienna, Austria, was founded in 1754.
5. JardinBotanico de Madrid, Spain, was established in 1755.
6. Royal botanical garden, Kew England owes its establishment in 1760 (Radford, 1986). However, Core (1955) stated that these gardens "were officially opened in 1841, with Sir William J. Hooker as the 1st director". This is the largest botanical garden of the world.
7. Botanic garden of Cambridge University, Cambridge, England was founded in 1762.

8. 1st economic gardens, for the cultivation of spices, etc., were established in the island of St. Vincent, British West Indies, in 1764.
9. *HortusBotanicus* of University of Budapest, was established in 1771.
10. Botanical Garden of Coimbra, Portugal was founded in 1773.
11. Indian botanical garden, Kolkata, formerly known as Royal botanic garden or Kolkata botanic garden, was founded in 1787 by Lt. Colonel Robert Kyd. It is “one of the great botanical gardens of the world” (Core, 1955).

12.2.4 Botanical gardens established during 19th century

1. The JardinBotanico de Rio de Janeiro, Brazil was established in 1808.
2. UniversitetsBotaniske Have is the oldest botanical garden of Oslo, Norway, founded in 1814.
3. 1st botanical gardens of Australia are the Botanic gardens of New South Wales, Sydney. These were founded in 1816.
4. Botanic Garden at Buitenzorg, Java, was founded in 1817. It is well known for the cultivation of coffee and rubber.
5. Conservatoire etJardinBotaniques de Geneva were also founded in 1817 in Geneva. A.P. de Candolle was its 1st director.
6. Botanical garden of Cape Town was founded in 1848.
7. Government Botanical Garden, Kingston, Jamaica was laid out in 1857.
8. Singapore Botanic Gardens were established in 1859.
9. Missouri Botanical Garden, at St. Louis, is the 1st great garden of USA. It was established by Henry Shaw in 1859, and named as ‘Shaw’s Gardens’.
10. Arnold Arboretum at Harvard University, USA was founded in 1872 by James Arnold. Its 1st director was C.S. Sargent (1872-1927).
11. New York Botanical Garden in New York city, was established in 1891. It is one of the greatest gardens of the world. Its 1st director was N.L. Britton.
12. Jardin Botanico Municipal of Buenos Aires, Argentina is the last well known botanical garden of the world established during 19th century. It was founded in 1892.

12.2.5 Botanical Gardens Established During 20th Century

Almost every university with a botany department, now has a botanical garden. Some of the major gardens established during the early part of the century are under mentioned.

1. Botanical garden, National Museum of Natural History, Beijing was founded in 1930.
2. The Jardin Botanique de Montreal of Canada was established in 1936.
3. Botanical garden, Forest Research Institute, Dehradun, India was established in 1934 under the leadership of C.E. Parkinson.
4. National Botanic Garden, Lucknow, India was established in 1946.

12.3 Major Botanical Gardens of India

12.3.1 Lalbagh or The Mysore State Botanical Garden, Bengaluru

This famous botanical garden is considered to be the best in Southern India for its layout, maintenance, scenic beauty and scientific interest. Because of roses and other red flowers it was named as "Lalbagh" by Hyder Ali in 1760. Many new species of flowering and fruit plants have been introduced by Tipu Sultan. In 1799, the Lalbagh was taken over by British and Major Waugh was its director during 1799-1819. He introduced a number of foreign exotic plants in this garden. Garden plants and disseminating foreign plants are maintained with very special care. Garden remained under the control of Dr. Wallich from 1891 to 1931. Lalbagh was handed over to Agri-Horticultural Society in 1839 and was converted to Government Botanical Garden in 1856. However, Dr. Cleghorn made it a real botanical garden in 1856. Vigorous and systematic plant introductions and expansion of Lalbagh garden was done by Mr. Cameroon in 1874 and collection of best Cannas was made. A tropical nursery was established in the garden in 1908. Rao Bahadur H.C. Jayaraja was the 1st Indian director of this garden. The garden is now a big centre of horticultural activities. It now has well-equipped laboratories for seed testing and soil testing, and also a grape orchard, tree nursery, fruit nursery, pot garden, economic garden and a herbal garden. It has become the guiding centre for research, extension and development activities for horticulture. Tree nursery, landscape design, herbal garden, floriculture nursery, pot garden, arboretum, an economic garden and seasonal herb gardens are spread over 240 acres.

12.3.2 Lloyd Botanical Garden, Darjeeling

It was started as a branch of the Royal Botanic Garden, Kolkata on a 40 acres land piece in Darjeeling. The land was donated by Mr. William Lloyd. Mr. A.G. Jeffery was the 1st curator of this garden. The characteristic flora of Sikkim, Himalayas has been established and sustained in this garden. Since 1910, this garden has become a major institution for the distribution of seeds, bulbs, and plants of temperate Himalayas to different parts of the world. It has a vast collection of plants from north-western parts of India, Nilgiris, Burma,

China and Japan and garden has become a paradise for the students of botany. It has separate sections of coniferous and indigenous plants. A rock garden, orchidarium, bulbous section, succulent section, seed sections, herbarium of over 30,000 specimens, and Rosary are its major attractions. Its coniferous section has 45 species including Australian *Callitris*. The other attractions of this garden are- Sir John Anderson Rock Garden, Orchidarium, students' garden, cacti and succulent section, aquatics, herbarium and seeds section.

12.3.3 National Botanical Research Institute (earlier National Botanical Garden), Lucknow

This famous Indian botanical institution, established in its new form in 1946, is now known as National Botanical Research Institute, Lucknow. It is popularly known as 'Sikander Bagh', a name given by Nawab Wajid Ali Shah after his beloved begum Sikander Mahal. Sikander Bagh, originally laid out by Nawab Saadat Ali Khan (1781-1814). In 1857, one more adjacent garden was acquired by the British Government and Government Horticultural Garden was established. An eminent horticulturist Dr. E. Bonavia served as superintendent from 1871 to 1878. The idea of establishing a botanical garden at Lucknow originated in 1929 following the interest aroused by a drug Santonin, obtained from *Artemisia maritima* and Soviet Union was the only supplier of this drug to the world. Department of Botany, Lucknow University undertook the project on cultivation of medicinal plants in Sikander Bagh in 1932. In 1946, it was converted into a botanical garden in 1946 by Professor K.N. Kaul, its 1st director. CSIR took over the garden in 1953 as one of its national laboratories and an excellent library, large herbarium as well as laboratories of botany, horticulture and plant chemistry were erected over 27 acres of land on the bank of river Gomati. Comprehensive collection of bulbous plants and paintings of ornamental plants were established and initiated basic and applied research in areas of plant introduction, floristic botany, plant morphology, cryptogamic botany, algae, fungi, pharmaco botany, aromatics, phytochemistry, plant breeding, radiation technology, palynology, plant propagation, viticulture, virology, entomology and utilization research in various laboratories. Popular attractions of this garden are its Rosarium, palm house, cactus house long with hydroponics, fern house, orchid house, and orchards of mango, citrus, and guava. NBRI conducts basic research on medicinal, horticultural, ornamental, essential oil bearing and other economic plants of agro-industrial value. The garden bears an added experimental research station at Bantha, about 20 km from Lucknow.

12.3.4 Botanical garden of Forest Research Institute, Dehradun.

It has attained a status of one of the principal botanic gardens of the world. It was established in 1934 under the leadership of C.E. Parkinson, F.R.I. botanical garden is now among the famous gardens of the world. It covers an area of about 20 acres in New Forest Estate, Dehradun and almost all the plants in the garden have been artificially introduced from various parts of the world and serves as a centre of research in problems related with plant introduction. There are about 1500 arboreal plants belonging to about 400 genera and about 100 families in this garden. Arboreal plants include trees, bamboos, shrubs and woody climbers. Over half of these 700 species have been introduced from different parts of the world. Main spots of attraction in the garden are greenhouse with a rich collection of orchids, ferns, shade loving plants, water plants in central park, cactus house and a plant introductory nursery. There is an extensive arboretum having more than 1000 ligneous species, both indigenous and exotic ones. The arboretum is primarily meant for the trial of timbers and other economically important trees. Biggest attraction is its renowned herbarium, internationally known as Dehradun Herbarium, which contains more than 3,00,000 sheets including 1200 irreplaceable type or cotype specimens representing new genera and species. This herbarium is the best in South-East Asia and comprises specimens of Indian flora as well as a very fine collection of specimens of plants from all over the world. This is the best maintained and planned botanic garden in India. Buildings, laboratories, workshops, official and residential buildings against hilly backdrops greatly add to its grandeur

12.3.5 Government Botanical Garden, Ootacamund.

Superintendent Mr. Mclvor completed the layout of this garden in 1867 and Lindley's natural system has been adopted in laying out this garden. Dr. S. Krishnamurthy was the 1st Indian superintendent from 1949 to 1951. This garden has made a significant contribution in the economic development of Nilgiri hills by way of introduction of most economic plants like several vegetable crops, fruits plants, etc. This garden has a rich collection of Eucalyptus varieties, Wattle, cinchona which were introduced by Mclvor in 1860. Number of essential oil yielding plants like scented geranium, *Lippia citriodora*, *Mentha piperita*, *Anthems nobilis*, *Rosmarinus officinalis*, *Gaultheria fragratissima*, etc. have been introduced into this garden, Chicory used for blending coffee powder was also introduced, which is now commercially grown in that area. This garden has been well known for introduction of many exotic and commercially important plants and also multiplication and distribution of these plants in that region. Herbarium section is doing great service to the students of botany and research workers. The exotic flora in the garden represents countries like Japan, North America, China, Malaya, Europe, South America, Central Asia, Central America, Burma and Africa. The coniferous section has more than 45 species. Among exotic conifers, names of species of Pinus, Callitris, Cunninghamia, Araucaria, Retinospora, *Cupressus pilea*, *Cephalotaxus*, Thujopsis, may be mentioned. The most remarkable among them is the Australian *Callistris* with its evergreen bluish foliage. There are two Dawn-redwood (*Metasequoia glyptostroboides*) and this beautiful conifer is of immense botanical interest. The genus *Metasequoia* has long been known only from fossil records. This plant is also called as living fossil, because of its occurrence both in living and fossil forms. There are also tulip trees (*Liriodendron elatus*) and collection of exotic oaks (*Quercus* spp.). Exotic maples, with their charming leaves, which changes their colour in autumn, are of great attraction for the visitors.

12.3.6 Botanical Garden, Saharanpur

Now serves as Horticultural Research Station of the UP government. World renowned botanists have worked in this garden and made valuable contributions. The history of this garden goes back to 1779 and Muslim ruler Zabita Khan established. In 1817, East India company introduced medicinal plants and several exotic plants were introduced from America and nurtured there under the leadership of garden superintendent Dr. Govan. Introductions of potato, tobacco, pineapple, chillies, papaya, sapota, logwood, mahogany, *Parkinsonia aculeata*, *Argemone mexicana*, *Asclepias curassavica*, *Martynia diandra*, *Canna glauca* and *Jatropa multifida*, etc. were made on 40 acres land of this garden. Garden has various sections for Linnaean garden, Agricultural and Horticultural nurseries, and in 1820, it became the centre of research on taxonomy. Uttar Pradesh government established as research station here in 1949 with a view to conduct research on tropical and sub-tropical fruits and various laboratories for research work on pomology, chemistry, mycology, biochemistry, entomology and physiology. This institute has a good collection of economic, medicinal and ornamental plants and library containing old and new journals and books.

12.4 Largest Botanical Garden of India

12.4.1 Royal Botanic Garden or Calcutta Botanic Garden or Indian Botanic Garden, Sibpur, Kolkata

Renamed as Indian Botanic Garden in 1950. It was laid in 1787 in the banks of the river Hoogly on 310 acres by Lt. Colonel Robert Kyd of Bengal

Infantry. He founded it "Not for the purpose of collecting rare plants as things of mere curiosity or furnishing articles for the gratification of luxury, but for establishing a stock for disseminating such articles as may prove beneficial to the inhabitants". It is the largest and oldest botanical garden of India. The garden has been a centre of great interest to botanists and horticulturists from India and abroad. Dr. William Roxburgh in 1793, made an outstanding contribution in the establishment of this great garden. The catalogue, Hortus bengalensis was prepared on 3500 plants growing there at that time and was published in 1814. Roxburgh's Flora Indica, Plantae Coromandellianae and painting work of Indian artists formed the basis of Hookers Flora of British India. The credit to use binomial nomenclature for the 1st time in India goes to Roxburgh and he is aptly called the 'Father of Indian Botany'. He was its 2nd director and founded the world famous herbarium of this garden. This garden was destructed by great cyclones in 1864 and 1867. The garden was rehabilitated in 1871 by sir George King and brought back the lost glory of the garden by publishing the Annals of the Royal Botanic Garden in 1887 and organized the Botanical Survey of India in 1890. Dr. K. Biswas was the 1st Indian appointed as superintendent in 1937.

The garden is now noted for potato cultivation and introduction of jute, flax, hemp, vanilla, eucalyptus, paper mulberry, nutmeg, cloves, cinnamon, pepper, mangosteen, tobacco, mahogany, sugarcane, tea and quinine-yielding *Cinchona* from Ceylon. Cultivation of *Aloe*, coffee, India-rubber, cardamom, and henbane are some of the special achievements of this garden.

The 'great Banyan tree', which is one of the largest trees in size in the world, is the main centre of attraction of this garden. It appears like a miniature forest in itself. Over 1,600 of its aerial roots are actually rooted in the ground. The circumference of the canopy of this single tree is more than 1328 feet and height up to 98 feet with 666 aerial roots. It is considered to be over 250 years of age.

There are over 15,000 species of plants in this garden from several countries. Some main attractions of the garden are its palm house, orchid house, pinetum, fernery, cacti-collection, the giant water lily, and the section of medicinal plants.

12.5 Largest Botanical Garden of the World

Royal Botanical Garden, Kew, officially opened in 1841 and originally established in 1760 because of the interest of Princess Augusta of Wales, is the largest botanical garden of world. William Aiton was the 1st curator of this garden. After the death of Princess Augusta in 1772, her Kew garden was united with the gardens of the Palace of Richmond, and the extensive area of the two gardens was named as Royal Botanic Garden, Kew, Richmond, England. At present, the herbarium of this garden contains over 5 million specimens, its arboretum has over 7,000 species, and its glasshouse have over 13,000 species. Its Jordrell laboratory provides best research facilities in the world for cytologists, anatomists, geneticists, and physiologists. A beautiful alpine house, rose garden, bamboo garden, and a lily pond are also attached to this garden. Because of all these facilities and vase developments, Kew gardens are called the 'Botanical Capital of the World'.

Some other remarkable features of Kew are Chelsea Physic Garden and world's best rock gardens. It is also well known for some of its famous publications such as Flora of British India, Index Kewensis and Kew Bulletin.

After 1838, this garden was transferred from British Crown to British Parliament, and Sir William Jackson Hooker took over as its 1st director in 1841.

12.6 Role of Botanical Gardens

1. Botanical gardens provide information of local flora and are used in the preparation of monographs.
2. Several gardens supply seeds and material for botanical investigations.
3. They provide information on food plants, ornamental plants, medicinal plants, etc.
4. Modern gardens supply living plant resources for research in systematics, horticulture, ecology, genetics, etc.
5. They also provide information on the protection of endangered species, and propagation of rare plants.
6. Big botanical gardens provide training facility to younger students by providing them laboratory, classroom, greenhouse, nursery, etc.
7. They supply facility for courses in local flora, horticulture, hybridization, plant propagation, etc. Their educational programme includes workshops and training sessions for teachers, students, naturalists, etc.
8. Gardens practice and provide training for the conservation of nature.
9. They provide instructions for home gardening, and care of plants in winter and summer.
10. Botanical garden provide aesthetically pleasing environment and thus play a major role in providing sound mental health.

11. Botanical gardens play a leading part in achieving conservation of plant life.
12. Botanical gardens can be called as living repository of plants arranged and maintained in some scientific basis

12.7 Other Famous International Gardens

- IABG: The International Association of Botanical Gardens is the official umbrella of more than 800 botanical gardens and arboreta all over the world.
- World famous flower and tulip garden: Keukenhof gardens, Amsterdam.
- Dubai's Miracle garden: World's largest natural flower garden featuring over 50 million flowers and 250 million plants spreading over 7,80,000 sq. ft.
- Circular garden city 'BrondbyHaveby' is a collection of houses nestled inside a grid of perfectly circular greenery is located at Copenhagen, Denmark.
- Atlanta Botanical Garden: Comprises Fuqua Orchid Center, mosaiculture exhibition including giant living sculptures made up of plants.
- South Coast Botanical Garden in California, Los Angel has been declared as the biggest 'dessert garden' in the world. Large plantations of *Euphorbia*, cactus, bonsai have been made and procured from Africa, Mexico and South America. Japanese styles of gardens have been developed alongside with green house cultivation of various species of flower crops

Name of garden	Year of establishment
Yuyuan garden, Shanghai, China	1577
Gardens of Versailles, France	1632
Singapore Botanical Gardens, Singapore	1859
Brooklyn Botanical Garden, USA	1910
Botanical Garden, Berlin	1910
Kirstenbosch National Botanical Garden, South Africa	1913
Montreal Botanical Garden, Canada	1931
Garden of cosmic speculation, Scotland	1931
NongNooch Tropical Botanical Garden, Thailand	1980

12.8 Other famous Indian gardens

Name of garden	Feature(s)	Year of establishment
Shalimar garden, Srinagar, Jammu and Kashmir	Built by Jahangir; fountains; stone pavilion; 1st terrace was *Diwan-e-Aam*& 2nd*Diwan-e-Khas*	1619; renovated & re-laid in 2006
Achabal garden, Anantnag, Jammu and Kashmir (also called as 'The place of the Princes')	Work of Noor Jahan; known for its gushing water	1620
Bijbehara garden, Anantnag, Jammu and Kashmir	Constructed by Dara Shikoh on both sides of Jhelum river; known for oldest *Cheenar* (21 m)	-
Botanical garden, FRI, Dehradun, Uttarakhand	Repositories & arboretum are prime source of conservation, education & research; herbarium consist of 3,00,000 species	1934
Buddha Jayanti park, New Delhi	Created on occasion of 2500th anniversary of Gautama Buddha's enlightenment; Bodhi tree is major attraction brought from Sri Lanka	-
Chasma-e-Shahi, Srinagar, J & K	Developed by Shah Jahan& constructed by Ali Mardan Khan	1632
Deeg Palace garden, Bharatpur, Rajasthan	Founded by Suraj Mal; now it is luxurious summer resort	1772
Indian Botanical Garden, Sibpur, Kolkata, West Bengal (also known as Royal Botanical Garden)	Founded by Col. Robert Kyd; area 109 ha; major attraction is Great Banyan tree (largest tree in the world)	1787
Pinjore garden, Haryana	Known as Yadavindra garden	17th century
Ram bagh	Known as Aarambagh; oldest Mughal garden; water flows in cascade form;	1528
Roshnara park, New Delhi	1st Japanese garden in India	1650
Lalbagh Botanical garden, Bengaluru, Karnataka	Private garden of Hyder Ali & in 1856 attained status of government botanical garden; floral clock; magnificent glass house; Lalbagh rock	1760
Lloyd Botanical garden, Darjeeling, West Bengal	Distinct extension of IBG, Kolkata; conservatory of cacti & succulents	1878
Ganeshkhind Botanical Garden, Pune, Maharashtra	Established by G.M. Woodrow	1873

Name of garden	Feature(s)	Year of establishment
Mandor garden, Jodhpur, Rajasthan	Established by Raja Abhai Singh; slopes on hillocks	1724-1749
Nishatbagh, Srinagar, J & K	Laid on the edge of Dal lake; designed by Asaf khan, brother of Noor Jehan	1633
Verinag garden, Kashmir, J & K	Established by Jahangir & completed by Shan Jahan; water has limpid clarity	-
Government Botanical Garden, Ootacamund, Tamil Nadu	Upper land is scrubby & lower land is swampy; consist 6 major features	1867
Sikandarbagh, Lucknow	Taken over by CSIR in 1953 and renamed as National Botanical Garden; later converted to NBRI	1978
Sim's Park, Coonoor, Tamil Nadu	Main attraction of town; arches, pergola, terraced annual beds; rose garden, etc.	1874
Sayaji park, Vadodara, Gujarat	Established by Maharaja Sayajirao III; comprises toy train, giant wheel, branching palm (*Hybaena thebaica*) avenue; 8000 tres & other plants	1879
Rock garden (Sculpture garden/ Alpine garden), Chandigarh	Developed by Nek Chand; completely built of home & industrial waste; situated near Sukhna lake; humanoid figures made of bottles, glasses, bangles, tiles, etc.	1957 & inaugurated in 1976
University Botanical garden, Coimbatore	Contain formal & informal style; highly pleasant in appearance; garden holds Japanese garden, bog garden, sunken garden, rockery, glasshouse, arboretum, etc.	1908
Byrant park, Kodaikanal, Tamil Nadu	Named after forest officer Mr. H.B. Bryant; area 8.5 ha	1909

Name of garden	Feature(s)	Year of establishment
Brindavan garden, Mysuru, Karnataka	Adjoined to Krishna Raja Sagara dam on river Kaveri; biggest formal garden; garden area 60 ha; one of the best show gardens of India	1927-1932
Rashtrapati Bhavan garden, New Delhi	Combination of Mughal + British style & Sunken garden); constructed by red & cream sandstone; 133.35 ha area;	1929
Red Fort garden, New Delhi	Built by Shan Jahan; 254.67 ha area; designed by Ustad Ahmad Lahauri; 14 gates	1648
Hanging garden, Mumbai (Feroze Shah Mehta garden)	-	1881
Chandrashekhar Azad garden, Prayagraj, Uttar Pradesh (earlier Alfred park)	-	1870
The Agri-Horticultural Society garden, Kolkata	-	1825
Horticultural Research Institute, Saharanpur	-	1750
ZakirHussain rose garden, Chandigarh (Free type; biggest rose garden in Asia)	-	1967
National Cactus and Succulent botanical garden and Research Centre (known as Cactus garden) Panchkula, Haryana	-	1987
Nehru Memorial Botanical Garden, Srinagar, Jammu and Kashmir	-	1969
JayantraoTilak Rose garden, Pune	-	-
Cactus garden, Kevadia, Gujarat	-	2018
Tal katora garden, New Delhi	-	1737
Globe park, Lucknow	-	-
BirbalSahni Institute of Palaeosciences, Lucknow	-	1946
Jawaharlal Nehru Tropical Botanic Garden and Research Institute, Thiruvananthapuram	-	1979

13

The Red Data Book

The Red Data Book is a public document that is created for recording endangered and rare species of plants, animals, fungi as well as some local sub-species that are present in a particular region. It helps us in providing complete information for research, studies and also for monitoring the programs on rare and endangered species and their habitats. Many nations of the world have prepared Red Data Books for their biological resources. Some Red Data Books include both plants and animals whereas, generally the plants are dealt separately from animals. This book is mainly created to identify and protect those species which are on the verge of extinction ("*Extinction is the process of evolution that leads to the disappearance of a population or species*").

Whereas red lists of threatened plants are being highly developed for wild plants and even replaced by green lists (Imboden, 1989) and blue lists (Gigon *et al.*, 2000), ornamental plants still lack similar lists.

13.1 Brief History

The name of this book has its origins in Russia, it was originally known as the Red Data Book of the Russian Federation or the RDBRF. The book was based on research conducted between 1961 and 1964 by biologists in Russia. Hence, it is also called the Russian Red Data Book.

Currently, the International Union for Conservation of Nature (IUCN) was founded in 1948 with an aim to maintain a complete record of every species that ever lived, maintains the Red Data Book. IUCN is the world's most detailed inventory centre of the global conservation status of biological species.

The Red Data Book contains the complete list of threatened species. The main aim behind this documentation is to provide complete information for research and analysis of different species. The Red Data Book contains colour-coded information sheets, which are arranged according to the extinction risk of many species and sub-species.

- Black represents species that are confirmed to be extinct.
- Red represents species that are endangered.

- Amber for those species whose status is considered to be vulnerable.
- White is assigned for species that are rare.
- Green for species that were formerly endangered, but their numbers have started to recover.
- Grey coloured for the species that are classified as vulnerable, endangered, or rare but sufficient information is not available to be properly classified.

In India, in 1980, the National Man and Biosphere Committee asked BSI, Kolkata to prepare an illustrated account of the threatened plants in India and a small State-of-Art Report dealing with 134 species was brought out. Later few seminars were organized by BSI between 1977-82 at Kolkata, Dehradun, Shillong, Delhi, Coimbatore and Port Blair to create wider interest among florists and particularly threatened plants. A Project on Study, Survey and Conservation of Endangered Species of Flora (POSSCEF) supported by U.S. Fish and Wildlife Service has been functioning at BSI since 1980.

In a nutshell, the Red Data Book indexes species as: Threatened, Not threatened and unknown. International Union for Conservation of Nature (IUCN) predicted that 60,000 species of flowering plants could become extinct by 2050. Furthermore, The Red Data Book also has information as to why a species has become extinct along with the population trends and the extent of its range (distribution). Its opposite is the Green Book, which lists rare plants growing in protected areas like botanical gardens. A mimeographed 'Green Book for India' has already been published. It deals with about a hundred rare plant species growing in protected life sanctuary and garden of Botanical Survey of India. The UNEP has complied endangered species of the world under the name Blue Book.

Advantages of the Red Data Book

- It helps in identifying all animals, birds and other species about their conservation status.
- It is used to evaluate the population of a particular species.
- The data available in this book can be used to evaluate the taxa at the global level.
- With the help of this book, we can estimate the risk of taxa becoming globally extinct.
- Provides a framework or guidelines for implementing protective measures for endangered species.

Disadvantages of the Red Data Book

- The information available in the Red Data Book is incomplete. Many species, both extinct and extant are not updated in this book.
- The source of the book's data has been speculated and has been mired in controversy.
- This book maintains the complete record of all animals, plants, other species but it has no information about the microbes.

13.2 Red Data Book of India

The Botanical Survey of India has brought out 'Red Data Book' of threatened plants. Red Data Book of India includes the conservation status of animals and plants which are endemic to the Indian sub-continent. The data for this book is provided through surveys which are conducted by the Zoological Survey of India, the Botanical Survey of India and Forest Research Institute, Dehradun under the guidance of the Ministry of Environment, forest and Climate Change. About 12% Indian fauna has been enlisted in Red Data Book during 2020-21.

• Some endangered plant species found in India

1. Malabar Lily (*Chlorophytum malabaricum*): Found in Tamil Nadu.
2. Musli (*Chlorophytum tuberosum*): Found in Tamil Nadu.
3. Malappuram (*Pterospermum reticulatum*): Found in Kerala and Tamil Nadu.
4. Jeemikanda (*Ceropegia odorata*): Found in Gujarat and Rajasthan.
5. Ebony tree (*Diospyros celibica*): Found in Karnataka.
6. Bird`s foot (*Lotus corniculatus*): Found in Gujarat.
7. Assam catkin yew (*Amentotaxus assamica*): Found in Arunachal Pradesh.

Endangered plants are any living species of plants vulnerable to extinction shortly, either globally or in specific geographic locations. Out of the 1.9 million living species of flora and fauna, about 0.7 million are plants. According to The Guardian newspaper, around 40% of existing plants await extinction. The loss of endangered plant species annually will threaten the supply of food, fuel, medicines, etc.

A living plant species become endangered when

1. Its population shrinks significantly in recent years.
2. The decline is continuous without a check and
3. Its geographic range also sharply shrinks.

13.3 Definitions

IUCN has defined six categories for threatened plants-Extinct, Extinct / Endangered, Endangered, Vulnerable, Rare and Indeterminate. Threatened species is a broad term that is used for species that fall into categories like vulnerable, endangered, rare and extinct.

1. Extinct (Ex): Taxa that are no longer known to exist in the wild after repeated searches of the type localities and other known or likely places.
2. Extinct/ Endangered (Ex/E): Taxa possibly considered to be extinct in the wild.
3. Endangered (E): Taxa in danger of extinction and whose survival is unlikely if the causal factors continue operating. Included are taxa whose numbers have been reduced to a critical level or whose habitats have been so drastically reduced that they are deemed to be in immediate danger of extinction.
4. Vulnerable (V): Taxa believed likely to move into the Endangered category in the near future if the causal factors continue operating. Included are taxa of which most or all the populations are decreasing because of over-exploitation, extensive destruction of habitat or other environmental disturbance; taxa with populations that have been seriously depleted and whose ultimate security is not yet assured; and taxa with populations that are still abundant but are under threat from serious adverse factors throughout their range.
5. Rare (R): Taxa with small world populations that are not at present Endangered or Vulnerable, but are at risk. These taxa are usually localized within restricted geographic areas or habitats or are thinly scattered over a more extensive range.
6. Indeterminate (I): Taxa known to be Extinct, Endangered, Vulnerable, or Rare but where there is not enough information to say which of these four categories is appropriate.

13.4 IUCN Red List Classification of Endangered Plants

The International Union for the Conservation of Nature and Natural Resources (IUCN) devised a plan in 1994 to classify endangered species of plants and animals into the following categories for which adequate data is available.

1. Critically Endangered-There is a 90% decline in population in 10 years.
2. Endangered – There is a 50% decline in population in 10 years.
3. Vulnerable- There is a 30% decline in population in 10 years.

4. Near threatened.
5. Least Concerned.
6. Extinct in wild.
7. Extinction.

The first three categories are collectively called 'Threatened Species' as they are at high risk of immediate extinction without proper safeguards.

The cause of plant species becoming endangered

1. Habitat loss and habitat fragmentation: It is the biggest cause of the endangerment of plants. Human activities like agriculture, deforestation, mining, wars and hydel projects destroy plant habitats. Diminishing plant habitat causes a decline in their population, paving their way to extinction.
2. Invasive species: The plantation of exotic species can adversely affect the local plants' populations. The exotic plant outstrips the local plant in competing for minerals and space, and gradually drives the latter to extinction. For example, the British colonial rulers introduced Congress Grass (*Parthenium hysterophorus*) in India.
3. Commercial use: Commercial overexploitation has endangered several plants, like the exploitation of cherry plants in Africa has made it an endangered plant.
4. Pollution: Some plants are susceptible to pollution, which triggers their extinction.
5. Natural causes: Wildfires, drought, floods and landslides also endanger plants.
6. Diseases: Fungal diseases have wiped several plants at different locations or pushed them towards extinction.
7. Human-wildlife conflict: Due to the ever-increasing human population, farming, industrialization and infrastructure development have expanded gradually, resulting in conflicts with wildlife as these spaces are created at the cost of shrinking the natural habitat of endangered plants.

13.5 Laws Formulated to Protect Endangered Plant Species

I) **Wildlife Protection Act, 1972:** The Indian Parliament enacted the Wildlife Protection Act in 1972 to protect the wild flora and fauna. Through this act, India became one of the signatories of the Convention on International Trade in Endangered Species of Wild Flora and Fauna (CITES) in 1973. The flashpoint of the act is protecting endangered species of plants and animals. It has six schedules, and the sixth schedule is dedicated to protecting endangered plants.

II) **CITES:** CITES, also known as the Washington Convention, was ratified in 1974 by 80 countries. It became effective in 1975 and now has 182 countries as parties to it. CITES aims to impart moral and legal responsibilities to the member countries to protect endangered species in their respective jurisdictions and prevent transborder trades. Species of *Cyclamen*, *Galanthus* and *Sternbergia* are protected under CITES.

III) **Endangered Species Act, 1973 of USA:** The US Congress passed the Endangered Species Act in 1973 that envisages all states to protect endangered species of plants and animals. Also, it prohibits the import, export, and consumption of plants, animals, and fishes listed as threatened or endangered.

Possibly extinct species

A) Gymnosperms

a) **Extinct species**: *Cycas fugax*, *Cycas ophiolitica, C, taiwaniana*, Waxen cycad (*Encephalartos cerinus*), Venda cycad (*Encephalartos hirsutus*), Lydenburg cycad (*Encephalartos inopinus*), and *Podocarpus perrieri.*

b) **Rare plants**: *Cycas seemannii.*

B) Angiosperms

a) **Extinct species:** *Acaena exigua, Acalypha dikuluwensis*, *A. rubribervis* (Stringwood), *A. wilderi*, *Achyranthes atollensis*, *Astragalus robbinsii var. robbinsii, Argocoffeopsis lemblinii, Argyroxiphium virescens, Basananthe cupricola, Begonia eiromischa* (wolly-stalked begonia), *Blutaparon rigidum* (Galapagos amaranth), *Byttneria inovensis, Campomanesia lundiana, Casearia quinduensis, C. tinifolia, Centaurea pseudoleucolepis, Chrysophyllum januariense, Clermontia multiflora, Commidendrum gummiferum, Cupaniopsis crassivalvis, Cyanea arborea* (Tree cyanea), *C. comata, C. cylindrocalyx, C. dolichopoda, C. eleeleensis, C. giffardii (Giffard's cyanea), C. linearifolia, C. mauiensis, C. parvifolia, C. pohaku, C. pycnocarpa, C. quercifolia* (Oakleaf cyanea), *C. sessilifolia, Cyperus rockii, Delilia inelegans, Delissea niihauensis, D. subcordata, D. undulata (Undulata delissea), Dipterocarpus cinereus, Encephalartos woodii, Erica verticillata, Fitchia mangarevensis, Galipea ossana, Gomidesia cambessedeana, Heliotropium pannifolium* (St. Helena heliotrope), *Hibiscadelphus bombycinus* (Kawaihae hibiscadelphus), *H. crucibracteatus, H.*

wilderianus, Holarrhena pubescens, Lepidium amissum (Waitakere scurvy grass), *L. obtusatum, Madhuca insignis, Melicope cruciata (Cross-bearing pelea), M. haleakalae, M. macropus, M. nealae, M. obovata* (Obovate melicope), *Musa fitzalanii, Myosotis laingii* (Waiautoa forget-me-not), *Oldenlandia adscensionis, Ormosia howii, Pitcairnia undulata, Pluchea glutinia, Podophorus bromoides, Pouteria stenophylla, Pradosia glaziovii, P. mutisii, Psiadia schweinfurthii, Santalum fernandezianum, Sporobolus durus, Stenocarpus dumbeensis, Tacitus bellus, Trochetiopsis melanoxylon* (St. Helena ebony tree), *Vernonia sechellensis, Viola cryana, Wikstroemia spiraeoides, W. Hanalei, W. skottsbergiana*, and *Xanthostemon sebertii.*

b) **Possibly extinct (Extinct/ endangered) species:** *Aechmea cymosopaniculata, Aeschynomene ruspoliana, A. hypoleuca, A. ledifolia, A. pichinchensis, Agelanthus rondensis, Alnus henryi, Amsinckia marginata, Andropogon benthamianus, Angraecum mahavavense, A. muscicolum, A. perhumile, A. potamophilum, A. rigidifolium, Argyroxiphiun sandwicense, A. rubellum, A. serpens, Anisopappus burundiensis, Artemisia kauaiensis, Aster quitensis, Bulbophyllum erythroglossum, B. hirsutiusculum, B. minax, B. sanguineum, B. tampoketsense, B. zaratananae, Canavalia veillonii, Clitoria andrei, Corypha taliera, Cyanea kolekoleensis, Cynorkis bimaculata, C. catatii, C. rolfei, C. sylvatica, Cyperus chionocephalus, Cyrtandra crenata, Dalea sabinalis* (Sabinal prairie-clover), *Delonix tomentosa, Devillea flagelliformis, Disperis bosseri, D. egregia, Eleocharis lepta, Eugenia albida, E. guayaquilensis, E. scheffleri, Euphorbia alcifornis, E. boinensis, E. neospinescens, E. pirahazo, Exacum bicolor, Franklinia alatamaha, Guzmania lepidota, G. poortmanii, G. striata, Hymenodictyron seyrigii, Lijndenia brenanii, Magnolia emarginata, Manetti canescens, M. holwayi, Maytenus manabiensis, Miconia scabra, Mikania seemannii, Mimosa kitrokala, Peperomia dauleana, P. petraea, Philodendron balaoanum, Phyllostegia kahiliensis, P. knudsenii, Polystachya acuminata, P. canaliculata, P. porphyrochila, P. rugosilabia, Prestonia quitensis, Pritchardia affinis, Rhynchosia ladermannii, Senecia navugabensis, Solanum semicoalitum, Tournefortia obtusiflora, Tridactyle sarcodantha, Uvaria decidua, Voburnum divaricatum, Xylosma serrata.*

c) **Extinct in the wild species:** *Betula szaferi, Bromus bromoideus, Brugmansia arborea, B. aurea, B. insignis, B. sanguinea, B. suaveolens, B. versicolor, B. vulcanicola, Corypha taliera, Cryosophila williamsii* (Lago Yojoa palm), *Cyanea pinnatifida, Euphorbia mayurnathanii,*

Franklinia alatamaha (Franklin tree), *Lachanodes arborea* (She cabbage tree), *Nymphaea thermarum, Rhododendron kanehirai, Senecio leucopeplus, Terminalia acuminata, Trochetiopsis erythroxylon* (St Helena redwood).

d) **Possibly extinct in the wild species:** *Cyanea kuhihewa, Kanaloa kahoolawensis, Magnolia wolfii, Mirabella estevesii, Silene perlmanii, Stenogyne bifida, S. campanulata, S. kanehoana.*

e) **Extinct sub-species:** *Costus vinosus*, *Cyanea copelandii* ssp. *copelandii*, *C. superba* ssp. *regina*, *Delissea subcordata* ssp. *obtusifolia*, *D. subcordata* ssp. *subcordata*, *Tetramolopium consanguineum* ssp. *consanguineum*.

f) **Extinct varieties:** *Cenchrus agrimonioides var. laysanensis*, *Euphorbia celastroides* var. *tomentella*, *E. remyi* var. *hanaleiensis*, *Ryania speciosa* var. *mutisii*.

g) **Endangered species:** *Agave wercklei, Alphonsea zeylanica, Aloe albiflora, Aloe ballii, Aloe bellatula, Araucaria rulei, Araucaria cookii, Areca concinna, Astrophytum asterias, Atriplex canescens, Balfourodendron riedelianum, Beccariophoenix madagascariensis, Brahea edulis, Brighamia insignis, Butia campicola, Calanthe pachystalix, Chamaedorea brachypoda, Coccothrinax crinita, Columnea allenii, Cupressus goveniana, Dillenia retusa, Dracenna synbarry, Dypsis decipiens, Encephalartos arenarius, E. cerinus, E. chimanimaniensis, E. concinnus, E. cupidus, E. dolomiticus, E. aevifolius, E. pterogonus, Geranium maderense, Goniothalamus wynaadensis, Haemanthus pumilio, Impatiens langeana, I, podocarpa, I. polysciadia, I. reidii, I. stoliczkai, Iris duthiei, Juniperus barbadensis, J. bermudiana, Latania loddigesii, Lavatera phoenicea, Limonium dufourei, Livistona carinensis, L. jenkinsiana, Lotus berthelotii, Lotus maculates, Malus hupehensis, Malvaviscus arboreus, Mammillaria carmenae, Melocactus matanzanus, Metasequoia glyptostroboides, Miliusa nilagirica, Neoveitchia storckii, Orania trispatha, Orophea thomsoni, Paphiopedilum armeniacum, P. micranthum, Pinus maximartinezii, P. muricata, P. radiata, P. torreyana, Pleione lagenaria, Pritchardia remota, Puya laxa, P. macrura, Rhododendron concinnoides, R. elliottii, R. falconeri, R. santapaui, R. subansiriense, R. wattii, Sabal bermudana, Sedum obtusatum* ssp. *paradisum, Tillandsia balsasensis, T. califanii, T. hildae, T. hondurensis, T. ignesiae, T. ixioides, T. kammii, T. lindenii, T. plumosa, Veitchia montgomeryana, Widdringtonia cedarbergensis, W. schwarzii, Zamia vasquezii.*

h) **Vulnerable plants:** *Acacia flocktoniae, A. koaia, Acanthophoenix rubra, Aeonium sedifolium, Ageratum houstonianum , Araucaria heterophylla, Argyranthemum broussonetii, Ariocarpus fissuratus, Armeria welwitschii, Astrophytum capricorne, Begonia cubensis, Caesalpinia echinata, Carpenteria californica, Ceanothus cyaneus, C. dentatus, Cephalocereus senilis, Chamaedorea graminifolia, C. microspadix, C. radicalis, Cupressus bakeri, C. cashmeriana, Dactylorhiza hatagirea, Dianthus serotinus, Dioscorea elephantipes, Diplomeris hirsuta, Dypsis decaryi, D. hildebrandtii, D. jumelleana, D. rivularis, Erythronium tuolumnensa, Eucalyptus argophloia, E. burdettiana, E. nicholii, E. pulverulenta, Gastrochilus japonicus, Haemanthus amarylloides, Heliconia angusta, Hyphaene dichotoma* (The Indian Doum palm or Branched palm), *Impatiens jaeschkei, Jasminum azoricum, Jubaeopsis caffra, Juniperus recurva, Laelia furfuracea, Leucadendron daphnoides, L. galpinii, Leucospermum formosum, L. fulgens, L. grandiflorum, Limonium perezii, Livistona drudei, Lyonothamnus floribundus, Lythrum flexuosum, Mammillaria bocasana, Masdevallia instar, Michelia velutina, Nephrosperma vanhoutteanum, Nymphaea tetragona, Oncidium phalaenopsis, Opuntia whipplei, Phalaenopsis schilleriana, Paphiopedilum fairieanum, Phoenix rupicola, P. theophrasti, Pinus muricata, Prosopis tamarugo, Roystonea elata, Salix magnifica, Serruria florida, Sparaxis elegans, S. tricolor, Stanhopea hernandezii, S. tigrina, Tanacetum ptarmiciflorum, Vanda coerulia, Verschaffeltia splendida, Zamia splendens*, etc.

i) **Rare plants**: *Abies pinsapo, Abutilon ramosum, Acacia howittii, A. iteaphylla, A. jonesii, A. quornensis, A. robynsiana, Aconitum deinorrhizum, A. heterophyllum, A. leave, A. nagarum, Adansonia za, Agathis atropurpurea, A. microstachya, Alloxylon pinnatum, Aloe forbesii, Alyssum wulfenianum, Anthemis sancti-johannis, Aphyllorchis montana, Aporocactus flagelliformis, Aquilegia eximia, A. longissima, Araucaria angustifolia, A. araucana, Archineottia microglottis, Aruncus dioicus, Bauhinia bowkeri, Begonia dregei, Bombax insigne, Bowkeria citrine, Brunfelsia undulate, Bulbophyllum ornatissimum, Calanthe alpina, C. mannii, C. plantaginea, Calothamnus rupestris, Calycanthus occidentalis, Calyptronoma occidentalis, Campanula davisii, C. elatinoides, C. incurva, C. portenschlagiana, Caralluma edulis, Cassia splendida, Catamixis baccharoides, Chamaedorea klotzschiana, Cheirostylis griffithii, C. pusilla, Clarkia purpurea, Clematis bourdillonii, C. theobromina, Coelogyne cristata, Coreopsis maritima, Corypha umbraculifera, Corybes purpureus, Crinum* *campanulatum, Cyathocalyx martabanicus, Cymbidium mackinnoni, C. tigirinum, Cypripedium elegans, Dendrobium wassellii, D. primulinum, D. peguanum, Desmos viridiflorus, Dianthus gallicus, D. knappii, D. spiculifolius, Dietes bicolor, Drymophloeus pachycladus, Echium wildpretii, Erica propendens, Eucalyptus caesia* ssp. *caesia, E. caesia* ssp. *magna, E. dunnii, E. ansdowneana, E. eptoloma, E. luehmanniana, E. macarthurii, E. risdonii, E. rudderi, E. yarraensis, Eugenia zeyheri, Geranium canariense, Ginkgo biloba, Gladiolus oppositiflorus, G. varius* ssp. *varius, Goniothalamus simonsii, G. thwaitesii, Horkelia frondosa, Howea belmoreana, Hybanthus travancorius, Impatiens flanaganiae, Isoplexis canariensis, Jacaranda mimosifolia, Kolkwitzia amabilis, Lavatera acerifolia, Leucadendron argenteum, L. nobile, Leuchtenbergia principis, Leucospermum saxosum, Lithops lesliei, Livistona alfredii, Magnolia griffithii, M. gustavi, Mammillaria matudae, Michelia mannii, Musschia aurea, Nemesia strumosa, Nerine pudica, Pancratium canariense, Paphiopedilum hirsutissimu, P. insigne, P. spicerianum, P. philippinense, P. venustum, P. villosum, Pereskia bahiensis, Phoenix rupicola, Pimpinella anisetum, Pinus canariensis, P. chihuahuana, P. lawsonii, Picrorhiza kurrooa, Pitcairnia andreana, Polyalthia rufescens, P. punicea, Orophea uniflora, Raphia australis, Renanthera imschootiana, Rheum rhaponticum, Rhododendron johnstoneanum, R. macabeanum, R. nuttallii, R. triflorum, Roystonea borinquena, Sabal uresana, Sedum hispanicum, Sparaxis grandiflora* ssp. *grandiflora, Strelitzia juncea, Tanacetum ferulaceum, Tecoma guarume, Tillandsia heteromorpha, Trachycarpus takil, Trivalvaria dubia, Washingtonia filifera, Zantedeschia pentlandii.*

j) **Indeterminate plants:** *Aechmea orlandiana, Aerides vandara, Ageratum houstonianum, Amherstia nobilis, Aristolochia brevilabris, Astrophytum ornatum, Butia eriospatha, Caryota no, Cattleya trianae, Ceiba insignis, Chamaedorea geonomiformis, Cotoneaster simonsii, Crocosmia masonorum, Delonix regia, Embreea rodigasiana, Epithelantha micromeris, Hatiora gaertneri, Hatiora rosae, Lobelia valida, Oncidium papilio, Paphiopedilum philippinense, P. randsii, P. sukhakulii, P. aff. scandens, Pseudophoenix sargentii, Renanthera imschootiana, Rhipsalis pilocarpa, Siphonochilus aethiopicus, Veitchia merrillii.*

13.6 Conclusion

After finishing the first comprehensive work on threatened crop plants (Hammer and Khoshbakht, 2005) the question arose concerning threatened ornamental plants. Ornamental plants are not included in Mansfeld's Encyclopedia of Agricultural and Horticultural Crops (Hanelt and IPK, 2001), which has been used as the world-wide basis for crop plants (excluding ornamentals). Mansfeld's Encyclopedia was checked against the Red List of Threatened Plants (IUCN, 2001). In principal, the same procedure was planned for the ornamental plants but for them no world-wide Encyclopedia is available.

Plants are becoming endangered at an alarming rate, mostly due to human activities like destruction and fragmentation of their habitat, excessive commercial exploitation, the introduction of invasive species that disturb their typical ecosystem and pollution. Some of the natural factors that also contribute are diseases, natural calamities, rare populations because of low fertility and genetic vulnerability. It is now crucial to collect information about such endangered plant species, classify them based on their vulnerability indices, enact laws and policies to protect their natural habitat and devise plans to save them through captive breeding and habitat restoration.

Each plant species plays a definitive role in their habitat, which forms a part of the food chain and balances the functioning of the ecosystem. Any loss of endangered species will disrupt our ecosystem and also deprive us of essential requirements like medicines, food and biofuel.

As of September, the International Union for Conservation of Nature (IUCN) lists 116 extinct species, 132 possibly extinct species, 35 extinct in the wild species, 13 possibly extinct in the wild species, five extinct subspecies, one extinct in the wild subspecies, and four extinct varieties of plant.

Many efforts have been done to find effective methods for the protection of rare ornamental

plants. In Great Britain "The Pink Sheet" (Anon, 2000) is published for rare and endangered garden plants (see also Hammer and Khoshbakht (2005b)). As already stated, the numbers of garden plants comprises mostly ornamentals. Ornamental plants are often taken in the gardens and are protected there.

References

Anonymous, 2000. The National Plant Collections Directory; National Council for the Conservation of Plants and Gardens. The Stable Courtyard, Wisley Garden, Woking, Surrey.

Hammer, K. and Khoshbakht, K. 2005. *Genetic Resources and Crop Evolution,* **52**: 249-265.

Hanelt, P. and IPK, 2001. (Eds.) *Mansfeld's Encyclopedia of Agricultural and Horticultural Crops, 6 vols.*; Institute of Plant Genetics and Crop Plant Research (IPK), Springer, Berlin.

Imboden, C. 1989. From the Directors desk: green lists instead of Red Books? *World Birdwatch*, **9**(2): 2.

Gigon, A., Lagenauer, R., Meier, C. and Nievergelt, B. 2000. Blue lists of threatened species with stabilized or increasing abundance: a new instrument for conservation; *Conservation Biology*, **14**: 402-413.

IUCN, 2001. IUCN Red List Categories: Version 3.1. Prepared by the IUCN Species Survival Commission; IUCN, Gland, Switzerland.

14

Herbarium

14.1 Herbarium and Its Limitations

A collection of dried and pressed plants arranged according to a classification system and available for study or reference is known as herbarium (plural, herbaria), a name first applied by Linnaeus.

Plant specimens are usually mounted on a sheet of high quality paper. Properly dried, pressed and identified plant specimens are placed in thin paper folders (species covers) which are kept together in thicker paper folders (genus covers), and finally they are incorporated into the herbarium cupboards in their proper position. Some fleshy members (e.g. Cactaceae) are preserved in liquid preservatives instead of being pressed and dried. Bulky plant parts (e.g. dry fruits, cones of several gymnosperms, etc.) are dried without pressing and are stored in special bags.

A herbarium may contain a few hundred locally collected plant specimens kept in a small place, or it may contain millions of them collected from different parts of the world and housed in a very big building. Herbaria in different countries remain associated with colleges, universities, scientific societies, research institutes, botanical gardens or flora of a district, state, country, continent or several continents. There are also private herbaria.

The world's largest herbarium is at Royal Botanic Garden, Kew, Richmond Surrey, England containing over 5 million specimens. But findings of Holmgren *et al.* (1981) suggest that the Herbarium of the Museum of Natural History, Paris contains over 6.5 million plant specimens. The biggest Indian herbarium is of Botanical Survey of India at Kolkata holding over 1.3 million specimens.

Over 1,600 world's most important herbaria are listed in 'Index Herbariorum' of Holmgren *et al.* (1981). Missouri Botanical Garden publishes a valuable monthly newsletter entitled 'Herbarium News' covering the news of different herbaria of the world.

According to Shelter (1969), the herbarium resources of the world include as many as 250 million specimens. However, the total number of actually reported from different institutions of the world was only 148 million specimens.

Online Herbarium (museum) www.ivh.bsl.gov.in has been established at Kolkata with its 11 branches throughout India. Online Herbarium Museum contains list of plants collected (52%) from 82 foreign countries. At present, digital herbarium comprises state wise list of flora and fauna and also information on effects of climate change on bio-diversity.

14.2 History of Herbarium Development

Luca Ghini (1490-1556) of Italy has been the initiator of the art of the herbarium. He collected plants, dried, and affixed them on a paper in the form of herbarium specimens. Gherards Gibo, a student of Ghini, Started collecting and preserving plant specimens in 1532. John Falconer, an Englishmen, learned the art of preparing a herbarium, perhaps from Luca Ghini, and prepared one as early as 1553. Holmgren *et al.* (1981) mentioned that the great herbarium of the Museum of Natural History, Paris was founded in 1653.

Up to the time of Linnaeus (1707-1778), the usual practice of preparing a herbarium was mounting the specimens and binding them into volumes. The plants were usually sewn by thread. Linnaeus started the current practice of mounting specimens on single sheets of paper and storing them horizontally. By such a simple beginning of Linnaeus, herbaria have developed into facilities of housing millions of specimens, usually in steel cases. The Herbarium of the Indian Botanical Gardens, Kolkata (now known as Central National Herbarium) was established in 1832.

In USA, the oldest herbarium was started at Salem College in 1772. The herbarium of Academy of Sciences, Philadelphia was started in 1812. University of Michigan in 1838, Missouri Botanical Garden in 1857 and U.S. National Herbarium in 1868. The largest herbaria of U.S.A. are those of the Harvard University, the New York Botanical Garden and the Missouri Botanical Garden.

In the earlier days the herbaria existed only in the form of a few pressed and dry specimens of local or regional significance. But now, many of them are the centres of advanced research in the field of taxonomy. Herbaria of Paris, Kew, Leningrad, Cambridge, Geneva and Kolkata are big centres of taxonomic research.

14.3 A Modern Herbarium

According to Fosberg (1946), a modern herbarium is a "great filling system for information about plants, both primary in the form of actual specimens of the plants and secondary in the form of published information, pictures and recorded notes". However, according to Radford (1986), a modern herbarium "is a research training and service institution that serves as a reference centre, documentation facility, and data storehouse". It includes "diverse collections of flowering plants, gymnosperms, ferns, mosses, liverworts, lichens, fungi, algae, and fossils". It also contains "microscope slides, photographs, photomicrographs, wood specimens, camera lucida drawings, field notebooks, diaries, letters, unpublished reports, manuscripts, reprints and botanical illustrations" according to Radford (1986).

Modern herbaria are utilized as reference centres for identification of plants by the botanists doing research work in taxonomy, ecology, agriculture, pharmacy, etc.

Modern herbaria are the documentation centres because they contain the specimens of new taxa, plants belonging to new discoveries, plants of economic importance, voucher specimens of cytological and cytogenetic studies, photographs of important plants, and all documents related with systematic research.

Geneticists, chemists, pharmacists, etc. use the modern herbaria as a data storehouse. They store data on ecology, habitat, distribution of plants, etc.

Modern herbaria provide training for young undergraduates, solve queries of scientists, environmentalists, doctors, public school students, etc. and thus work as service institutions.

14.4 Twenty Major Herbaria of the World

Under mentioned is the list of 20 major herbaria of the world, along with the number of their specimen holdings as worked out by Holmgren *et al.* (1981).

1.	Museum of Natural History, Paris	6.5 million
2.	Royal Botanic Garden, Kew	>5 million
3.	Komarov Botanical Institute, Leningrad	> 5 million
4.	Conservatory and Botanical Garden, Geneva	5 million
5.	Combined Herbaria, Harvard University, Cambridge	4.5 million
6.	New York Botanical Garden, Bronx	4.3 million
7.	U.S. National Herbarium, Washington	4.1 million
8.	British Museum of Natural History, London	4 million
9.	University of Lyon, Lyon	3.8 million
10.	Natural History Museum, Vienna	3.5 million
11.	Missouri Botanical Garden, Saint Louis	2.9 million
12.	Field Museum of Natural History, Chicago	2.4 million
13.	University of Uppasala, Uppasala	2.2 million
14.	National Botanical Garden of Belgium, Brussels	2.1 million

15.	Botanical Garden and Botanical Museum, Berlin	2 million
16.	Academy of Natural Science, Philadelphia	2 million
17.	Royal Botanic Garden, Edinburgh	1.7 million
18.	University of California, Berkeley	1,5 million
19.	University of Michigan, Ann Arbor	1.4 million
20.	Botanical Survey of India, Kolkata	1.3 million

However, the latest estimates show that in India, there exist two more herbaria having more than a million specimens. According to Tiagi and Kshetrapal (1988) these are (1) The Central National Herbarium, Indian Botanical Gardens, Kolkata (2,50,000 specimens), and (2) Herbarium of the Forest Research Institute, Dehradun (1,50,000 specimens). But in majority of the other Indian books, it is mentioned that herbarium of F.R.I. Dehradun has only about 3,00,000 specimens and not 15,00,000.

14.5 Major Indian Herbaria

In India, the biggest herbarium is at Indian Botanic Garden, Kolkata. Under mentioned is the list of some major Indian herbaria, along with the approximate number of their specimen holdings as mentioned by Naik (1984) and Tiagi and Kshetrapal (1988).

1.	The Central National Herbarium, Kolkata	25,00,000 specimens
2.	Herbarium of Forest Research Institute, Dehradun	3,00,000 specimens
3.	Botanical Survey of India i) Eastern circle herbarium, Shillong ii) Southern circle herbarium, Coimbatore iii) Western circle herbarium, Pune iv) Northern circle herbarium, Dehradun v) Central circle herbarium, Allahabad	 10,00,000 1,75,000 50,000 42,000 45,000
4.	Blatter Herbarium of St. Xavier's College, Mumbai	1,00,000
5.	National Botanical Garden Herbarium, Lucknow	80,000
6.	Herbarium of Industrial Section, Indian Museum, Kolkata	50,000
7.	Herbarium of Rajasthan University, Jaipur	30,000
8.	Herbarium of School of Plant Morphology, Meerut college, Meerut	25,000
9.	Herbarium of Delhi University, Delhi	15,000
10.	Herbarium of Jiwaji University, Gwalior	15,000

14.6 Functions of Herbaria

Some of the general functions of herbaria are mentioned below:

1. Plant specimens are permanently stored in herbaria and therefore they are the major sources of information about plants and vegetation.

2. Preserved specimens of herbaria are used in almost all types of taxonomic research.
3. A picture of all species of a genus, or all the genera of a family may be gathered only in the herbarium.
4. The classification of the world's flora is based mainly on the herbarium material.
5. List of the endangered species of any region may be prepared only by herbarium specimens.
6. Limited individual collections are identified only with the help of herbaria.
7. Monographs of genera or families are prepared only by the herbarium specimens.
8. Our knowledge of the distribution of plants, evolution, and several taxonomic problems, etc. is based mainly on the herbarium specimens.
9. Big herbaria provide training to young students in herbarium practices.
10. Exact area, region or location of the occurrence of important plants may be gathered from the herbarium specimens.
11. Herbaria preserve type specimens, and thus serve as a repository of chromosome, chemotaxonomy, and experimental voucher specimens.
12. Herbarium material is used in studying the palynology, anatomy, and chemical aspects of desired plants.
13. Herbaria provide loan of specimens for study at other institutions.
14. Herbaria preserve the natural plant wealth, and provide scientific information to the public regarding the plants.

14.7 Precautions for Using Herbarium

1. Handle the plant specimens with extreme care because they are of scientific value and generally irreplaceable.
2. Dry specimens are fragile and easily damaged, and so keep the specimen sheets flat.
3. Store the specimens always in herbarium cases.
4. Keep the materials in folders when not in use.
5. Never keep heavy books or other heavy object on the specimens.
6. Support the specimens with a ventilator when carrying them even for short distances.
7. If you, notice that specimens are being damaged by insects, etc. inform

the curator immediately.

8. Do not write anything on sheets of herbarium, unless permitted to do so.
9. Do not reshelve specimens in the herbarium cases.
10. If you have taken specimens on loan from another institution, return them before the loan period is over.
11. The specimens, borrowed from other institutions, should be properly housed in fire-proof, dust-proof and pest-free cases.

References

Fosberg, F.R. 1946. *Scientific Monthly*, **63**: 429-434.

Holmgren, P.K., Keuken, W. and Schofield, E.K. 1981. *Index Herbariorum*, Part I. The herbaria of the world, 7th ed. Bohn, Scheltema and Holkema, Utrecht.

Naik, V.N. 1984. Taxonomy of Angiosperms. Tata McGraw-Hill, New Delhi.

Shelter, S.G. 1969. The Herbarium: past, present and future, *Proceeding of Biological Society of Washington*, **86**: 687-758.

Tiagi, Y.D. and Kshetrapal, S. 1988. An Introduction to Taxonomy of Angiosperms. Hill publishing.

15

Floral Formula and Floral Diagram

A floral formula is a numerical representation of various parts of the flower. Such notations use numbers, letters and various symbols to convey significant information in a compact form. They may represent the floral form of a particular species, or may be generalized to characterize higher taxa, usually giving ranges of numbers of organs. It also depicts the symmetry, interrelationship of various floral parts, unisexuality as well as bisexuality in the flower.

Floral formulae were developed at the beginning of the 19th century (Prenner *et al.*, 2010). The first authors using them were Cassel (1820) who first devised lists of integers to denote numbers of parts in named whorls; and Martius (1828). Grisebach (1854) used 4-integer series to represent the 4 whorls of floral parts in his textbook to describe characteristics of floral families, stating numbers of different organs separated by commas and highlighting fusion. Sachs (1873) used them together with floral diagrams, he noted their advantage of being composed of "ordinary typeface".

15.1 Symbols Employed In Floral Formulae

There has been no attempt to standardize the symbols used in floral formulae on a universal basis. Different authors have used different symbols for the same part, e.g. calyx is variously designated as K, C, and CA and corolla as C, CO, and COZ. However, the generally accepted symbols, used in the floral formula presentation by the taxonomists of Indian sub-continent, are presented in following Table 15.1, and the same usage has been followed in the present book.

Table 15.1: Floral symbols

Br	Bracteate	G	Gynoecium
Ebr	Ebracteate	( )	Cohesion of floral parts in a whorl
Brl	Bracteolate	C A	Epipetalous stamens
Epik	Epicalyx	P A	Epiphyllous stamens
0 (zero)	Absence of particular whorl	Std	Staminodes
∞	Indefinite number	G_4	Tetracarpellary, free carpels
⊕	Actinomorphic	$G_{\underline{(4)}}$	Tetracarpellary, syncarpous, ovary superior (hypogynous)

·|·	Zygomorphic	G(4)-	Tetracarpellary, syncarpous, ovary semi-inferior (perigynous)
♂	Male flower	G(4)	Tetracarpellary, syncarpous, ovary either inferior (epigynous)
♀	Female flower	G(4)	Tetracarpellary, syncarpous, ovary either superior or inferior
⊠	Bisexual flower or hermaphrodite condition	Pistd.	Pistillode
K	Calyx, sepals	A G	Androecium and gynoecium united
C	Corolla, petals	2+4	2 in one set and 4 in another
P	Perianth	2-4	2 or 4
A	Androecium	X	Variable

The different whorls of a flower, their number, cohesion and adhesion may be represented by a formula known as the floral formula (Rangaswamy, 1966). The floral formula, K stands for calyx. C for corolla, P for perianth, A for androecium and G for gynoecium. The figures following the letters K, C, P, A and G indicate the number of parts of those whorls. Cohesion of a whorl is shown by enclosing the figure within brackets, and adhesion is shown by a line drawn on top of the two whorls concerned. In the case of the gynoecium, the position of the ovary is shown by a line drawn above or below G on the figure. If the ovary is superior, the line should be below it, and if inferior, the line should be on top. Thus, all the parts of a flower are represented in a general way by a floral formula. The floral characteristics of a family may also be represented by one or more formulae, as follows.

Ranunculaceae	$\oplus$ ⊠ $K_5 C_5 A_\infty \underline{G}_\infty$
Nymphaeaceae	$\oplus$ ⊠ $K_4 C_\infty A_\infty \underline{G}_{(\infty)}$ or $_\infty$
Malvaceae	$\oplus$ ⊠ $K_{(5)} C_5 A_\infty \underline{G}_{(5-\infty)}$
Solanaceae	$\oplus$ ⊠ $K_{(5)} C_5 A_5 \underline{G}_{(2)}$
Labiatae	% ⊠ $K_{(5)} C_{(5)} A_4 \underline{G}_{(2)}$
Liliaceae	% $P_{(3+3)} A_{(3+3)} \underline{G}_{(3)}$

Besides, some symbols are used to represent certain features of flowers. Thus, ♂ represents for male (staminate), ♀ female (pistillate), ⊠ for hermaphrodite (bisexual), ♂♀ dioecious, ♂-♀ Monoecious, ♂♀ ⊠ polygamous, $\oplus$ actinomorphic, % zygopmorphic, ∞ indefinite umber of parts, etc.

15. 2 Floral Diagram

The number of parts of a flower, their general structure, arrangement and the relation they bear to one another (aestivation), adhesion, cohesion, and position with respect to the mother axis may be represented by a diagram known as the floral diagram. The floral diagram is the ground plant of flower. A floral diagram visually depicts the essential features of a flower in cross section. The parts of the flower are drawn by semi-diagrammatic symbols or ideographs. The number of the whorls and floral parts are both illustrated in a floral diagram. This also illustrates the fusion of the floral parts as well as the floral symmetry.

A floral diagram is thus the diagrammatic representation of the ground plan of flower depicting the arrangement of various floral parts as projected in the transverse plane. In brief, a floral diagram is an ideal ground plan of a flower. In the diagram, the calyx lies outermost, the corolla internal to the calyx, the androecium in the middle, and the gynoecium in the centre. Adhesion and cohesion of members of different whorls may also be shown clearly by connecting the respective parts with lines (Schaffner, 1916). Fig. 15.2 (Rubiaceae) shows that the petals and stamens are united (adhesion). The black dot on the top represents the position of the mother axis (not the pedicel), which bears the flower. The axis lies behind the flower and, therefore, the side of the flower nearest to the axis is called the posterior side, and the other side away from the axis the anterior side. The floral characteristics of species may be well represented by a floral diagram, while more than one diagram may be necessary to represent a genus or family.

Information that will get from floral diagram:

A floral diagram furnishes the following information.

1. Sex and symmetry of a flower.
2. Bracteate or ebracteate, and bracteolate or ebractrolate nature of a flower.
3. Number of floral cycles of a flower, i.e. its tetracyclic or pentacyclic nature.
4. Number of a floral parts in each cycle of a flower.
5. If the number of sepals or petals is odd, the floral diagram informs about the anterior or posterior position of the odd sepal or petal.
6. Type of aestivation of sepals and petals.
7. Number of whorls of stamens, their position in relation to the petals, attachment with the petals or free nature, are all shown in a floral diagram.
8. Carpel number, their free or fused nature, number of locules, number of ovules in each locule, and the type of placentation are all shown by a floral diagram.
9. Position of all floral parts in relation with mother axis.

15.3 Drawing of Floral Diagram

The floral diagram is always drawn circular in outline. The different floral whorls are represented in concentric circles, the sepals on the outermost circle, then the petals, the stamens and carpels towards the inner side.

The first step to draw a floral diagram is to examine mature floral buds which are due to open shortly but have not yet opened. Pluck the floral bud from the mother axis only after you have noted down the anterior and posterior sides. Floral parts are drawn in floral diagram as they would be seen in their transverse sections below the mother axis. Make the floral diagram in the following sequential stages:

1. A very small circle is drawn above the floral diagram. This circle represents the mother axis. In actinomorphic flowers the mother axis circle is denoted as ⊕, but in zygomorphic flowers it is drawn as %. If the flowers are terminal, the mother axis is not drawn.
2. In bracteate flowers, a section of bract is drawn below the floral diagram. In flowers without any bract, such a section is not drawn.
3. In bracteolate flowers, bracteoles are drawn in section on the left and right sides of the diagram.
4. Note the number of sepals, their arrangement in relation to the mother axis and their aestivation. Draw transverse sections of sepals between the mother axis and the bract, keeping all these points in view. In case of odd number of sepals, the odd sepal would be drawn either posterior or anterior to the flower, i.e. opposite the mother axis or opposite the bract, respectively.
5. The same procedure is repeated for petals as for sepals mentioned above. However, petals should be drawn alternate with the sepals.
6. If the flower is zygomorphic, petals are drawn of unequal sizes. Same may be the case with the sepals also in zygomorphic flowers.
7. If any sepal or petal is spurred, it is shown by drawing a loop at the back of that particular part in a floral diagram.
8. If parts of sepals or of petals are fused, draw lines to connect their edges together in the floral diagram.
9. In the epipetalous condition, (i.e. when stamens are joined with petals), link the stamens and petals with small radial lines.
10. In case of bilabiate calyx or corolla, the two lips are joined by bulging lines.

11. Count the number of stamens, the number of whorls in which they are arranged, their cohesion and adnation to other floral parts, their position in relation to petals, their introrse or extrorse position, and draw them inside the petals in the floral diagram. Stamens are represented through transverse sections of the anthers. In the obdiplostemonous condition, the stamens of the outer whorl are drawn opposite to the petals. Introrse stamens face towards the centre whereas the extrose towards the petals. Staminodes are represented either by an asterisk (*) or by a cross (X).
12. The gynoecium is presented by a transverse section of the ovary. Also draw the number of locules and the number ovules in each locule, in the ovary. Type of placentation is also drawn.

References

Cassel, F. P. 1820. Morphonomia botanica: sive observationes circa proportionem et evolutionem partium plantarum. Colonia Agrippina [Cologne]: M. DuMont-Schauberg.

Grisebach, A. 1854. Grundriss der systematischen Botanik. Göttingen: Verlag der Dieterichschen Buchhandlung.

Martius, C. F. 1828. "Über die Architectonik der Blüthen". Isis (Oken) **21**: 522-529.

Prenner, G., Bateman, R.M. and Rudall, P.J. 2010. Taxon, **59**(1): 241-250.

Rangaswamy, N.G. 1966. Science and Culture, **31**(1): 33-34.

Sachs, J. 1873. Lehrbuch der Botanik nach dem gegenwaertigen Stand der Wissenschaft. Leipzig: Engelmann.

Schaffner, J.H. 1916. Ohio Journal of Science, **16**: 300-360.

16

Conservation and Tissue Culture Repository

Collection and conservation of genetic resources in horticultural crops is very much essential in regions where local seed sources are still used in genetic enhancement and cultivar production. Domesticated crops grow well adapted to both tropical and temperate regions. In particular, if the crop is domesticated over centuries to be grown in different latitudes, the genetic diversity estimates are essential for a breeder to develop location specific cultivars. Genetic vulnerability, erosion and extinction of some wild progenitors and species has been altered, warranting the need for *ex situ* conservation measures. The material is conserved in various forms, ranging from botanical gardens, clonal and seed populations, vegetative methods, *in vitro* active repositories, pollen cryobanks and DNA banks. The objective, in genetic resource context, is to conserve for prolonged duration with minimal loss of viability and genetic diversity. Most *ex situ* conservation methods are technology dependent. Seed and pollen banks are dependent in refrigeration technology and cryogenic technology.

The conservation of plant genetic resources in ornamental species is generally driven by demand, focusing on preserving a wide range of genetic diversity. This diversity can include individual genes and coadapted gene complexes at the micro level, as well as entire genotypes, elite clones, and plant populations at the macro level. The distribution of genetic diversity (GD) in plant species is predominantly dependent on the reproductive mechanisms they adopt and the mode of inheritance of the characters targeted for collection. The main reasons for collection and conservation genetic resources of an ornamental species gene pool in a given ecosystem are:

1. Status of the species-extensive loss due to genetic erosion leading to extinction.
2. Need based, for economic causes at a national or global level.
3. Insufficient representation or total absence of GD in *ex situ* collections.

Conservation Strategies

Genetic conservation strategies for ornamental species ensure preservation and use of value-added traits contained within the gene pool for future commercial exploitation. The primary objective of this conservation program focuses on collection, maintenance and conservation of species-specific core collection with minimum samples of gene pool components by containing maximum genetic diversity within the collections. Especially this ornamental consists of diverse habitat i.e., herbaceous annuals to tree species and acclimatized to different climatic conditions i.e., tropical, sub-tropical and temperate regions. Because of this diversified growth habit and found in different regions their exhibit enormous quantum of variations in the plant species.

Gene sanctuaries as well as *ex situ* and *in situ* methods are practiced to conserve variety of tree species. In botanical gardens, tree species are conserved and maintained from horti-botanical aspects. Germplasm conservation of rare/ endangered plant seeds and tissues is done through using meristem culture, embryo culture or recalcitrant seeds stored in liquid nitrogen.

In India, maximum fruit germplasm is maintained through *in situ* conservation. Germplasm conservation is attempted in three forms viz. storage of recalcitrant seeds, vegetative propagules and clonal propagules. Most common method of germplasm conservation is seed collection and it is done via-

1) *Ex situ* conservation: Conservation of material outside its natural habitat e.g. cryobank, seed gene banks, field gene banks, natural monuments, cultural landscapes, botanical gardens, arboretum, *in vitro* repositories, DNA bank, herbal seed/ garden, long term storage of seeds, perennial fruit crops.

   a) Seed conservation: Seeds are categorized as orthodox, where seed longevity is enhanced with increase in moisture content or storage temperature. The preferred conditions for long term storage with 5 $\pm$ 1^{0}C moisture content is at -18^{0}C or below, where a progressive decline in viability is observed, a much slower rate than compared to ambient conditions. There is a close relationship between viability and accumulation of chromosomal damage. There exists a relationship between longevity, seed moisture content and storage temperature. Some disadvantages in a seed conservation program could be a progressive decline in viability over successive generations in relation to the rate of decline or erosion in diversity due to regeneration and pollination methods followed. Seed banks are encountered with practical problems in managing controlled pollination programs especially for out-breeding species. Seed conservation strategies using cryogenic technology has been proved ideal for most angiosperms, especially in species where seed is the unit propagation. In clonally propagated crops, trees, etc. the same is not desired due to loss of clonal identity. The seeds of most plants withstand desiccation and low temperatures.

b) Conservation of pollen: Conservation of nuclear genetic diversity using pollen is desirable in horticultural species for a variety of reasons. Cryopreserved pollen can be a major access point for pre-breeding germplasm lines, hybrid seed production, biotechnological and other basic studies. In case of tree species, germplasm can be easily received and exchanged through pollen, eliminating a long juvenile phase. The objective of a useful pollen cryostorage protocol is to collect mature pollen from plant and treat it so as to retain its normal function, ultimately assessed by its ability to germinate *in vivo* and effect normal fertilization. The international transfer of germplasm in the form of dry pollen is not generally restricted. Moreover, this will eliminate the need of growing plant populations to produce pollen. Pollen is subjected to less stringent quarantine restrictions. So, it can be easily shipped and used. Through exchange of pollen, desired crosses can be made directly on the seed parent, allowing introgression of characters at a much faster rate. This would find favour especially in breeding of tree species with long juvenile phase.

2) *In situ* / On farm conservation: Conservation of diversity in its natural habitat by NBPGR and NAGS e.g. Wild species, land races, bio-sphere reserve, gene sanctuaries, sacred groves, national parks, sanctuaries with legal protection and traditional farmers' varieties/ genotypes. National Council for the Preservation of Plants and Gardens of UK conserves species of *Narcissus* (maintain 2,000 taxa).

3) *In vitro* conservation: The role of *in vitro* conservation techniques in the overall conservation strategies should be indicative of the fact that it should complement other conservation strategies within the total program of a given species or population. The methods chosen should be carefully considered taking into account the feasibility, practicality, economy and security.

It is a high-tech technique for germplasm conservation. It has very high multiplication rate, and free from fungi, bacteria, viruses and insect pests, thereby producing pathogen free stock. However, it is labour intensive (in regular sub-culturing), carry risk of infection and there is a danger of breakdown of equipments.

- *In vitro* storage techniques have been established for those crops for which *ex situ* storage as seed is not viable option e.g. a) crops that are vegetatively propagated e.g. potato, yam, cassava; b) fruit species with recalcitrant seeds (coconut, mango).
- Comprises storage at low temperature i.e. cryopreservation (-154 to -196°C) which arrest cell division, all metabolic activity without modification or alteration for unlimited periods of time. At this temperature, due to lack of liquid water, chemical reactions hardly occur inside the cells. It is most successful for cell cultures e.g. Germplasm of apple, *Prunus* spp., *Pyrus* spp., blue berry, goose berry, currents, *Elaeis guineensis* and *Solanum* spp.
- Pollen cryopreservation is used for hybridization purpose of plants which do not flower at the same time e.g. gladiolus, carnation, gerbera, coconut and other fruit crops was attempted at IIHR.
- DNA storage: DNA libraries store total genomic information of germplasm.

Cryopreservation

Word 'cryo-preservation' was derived from Greek words '*cryo'* meaning 'cold' and 'preserve' to 'maintain'. It is the non-lethal storage of biological tissues at low or ultra-low temperatures without losing the viability and genetic integrity of the material preserved. It is a good method to store live material. It ensures safety of rare or endangered plant seeds and tissues. At IIHR, Bengaluru, cryopreserved pollen grains of gladiolus, carnation and gerbera retain the viability, as confirmed by fertilization and seed setting tests, for several years.

i) Seeds and pollen grains of many crop species tolerate desiccation and ultra-low temperatures and can be preserved for extended periods without any negative effect on cell viability.

ii) Such extremes of environmental conditions are lethal for many other propagules.

iii) Shoot tips and meristems from various species including those producing recalcitrant seeds are ideal material for germplasm conservation by adopting cryopreservation techniques.

iv) Cryopreservation of callus and suspension cultured cells is undesirable due to high incidence of somaclonal variations.

v) Cryopreservation protocols for cold-hardy and non-cold hardy species are distinctly different.

vi) Cold hardy temperate species can tolerate -25^{0}C or less freezing stress and these plants have natural ability to tolerate freezing stress and can be preserved for extended periods without any negative effect on cell viability due to slow cooling @ 2^{0}C / hour to temperatures between -30 to 50^{0}C before exposing them to liquid nitrogen.

vii) Exposure to liquid nitrogen is used for non-cold hardy plants.

viii) Practically, all the non-cold hardy plants require some form of cryoprotective treatments for their survival at ultra-low temperatures.

ix) Both slow cooling and rapid freezing by direct exposure to liquid nitrogen can be used for non-cold hardy plants.

Cryoprotectant chemicals like Di-methyl Sulphoxide (DMSO), ethylene glycol & propylene glycol which exert the protective effects through colligative properties. And glycerol, proline, Mannitol/ sorbitol, sucrose & polyethylene glycol (PEG) protect the tissue from ice damage by osmotic dehydration. These cryoprotectants alter the form of ice crystals in the vacuoles and lower the initial freezing temperature leading to slow freezing.

For cryopreservation, shoot tips and axillary buds, measuring about 0.3-1.0 mm in length, isolated from field frown or *in vitro* plants can be used and soaked in 0.5-5% DMSO to enhance survival. Cryopreservation methods have been reported in chrysanthemum, carnation, rose, camellia, etc.

Cryobanking of flower seeds

- 155 species and 455 selections have been cryobanked at the National Seed Storage Laboratory, Fort Collins, USA, includes around 20 ornamental species seeds.
- The USDA-ARS-NSSL Fort Collins, Colorado, USA maintains over 30,000 seed accessions through cryopreservation facility.

Tissue Culture Repository

A tissue culture repository is a specialized facility or collection that stores, maintains, and distributes various types of tissue cultures. These cultures include various plant parts, animal, or microbial cells, tissues, or organs that are grown and sustained under controlled laboratory conditions. Tissue banks and genetic revolutionary race histories of bougainvillea and verbena are maintained at NBRI, Lucknow.

Role of Tissue Culture Repository

- Preservation of genetic resources i.e., conservation of biodiversity and plant germplasm which helps to conserve the genetic materials of rare, endangered and threatened species and preserved germplasm can be used for research and reintroduction of such species into the wild, respectively (Reddy *et al*., 2004).

Tissue Culture Repository Centers

1. The American Type Culture Collection (ATCC) located at Rockville, Maryland, United States, around 76 plant tissue cultures are preserved.

Micropropagation Technique Followed In Ornamental Crops

- Species cultured easily through mass production. Plants are free from known pathogens, sustain year round production, high quality, improved habits and many desirable horticultural characteristics e.g. Alstroemeria, Anthurium, Caladium, Chrysanthemum, *Dieffenbachia*, Drosera, Gerbera, Gypsophila, Heliconia, Freesia, Banana, Nepeta, Nephrolepis, Orchids, Philodendron, Rhododendron, *Rosa* and Saintpaulia.
- Plants are amenable for in vitro culture, pathogen tested, produce superior clones, breeding stocks and germplasm collection e.g. *Begonia*, *Dianthus*, *Gladiolus*, *Haemanthus*, *Hemerocallis*, *Lilium*, Hyacinth, *Iris*, *Pelargonium* (virus tested stocks) and *Petunia* (hybrids & male sterile lines).
- Plants cultured with some difficulty e.g. Acer, Junipers, Paeonia, Potentilla, Sequoia, and Texus. Some of the ornamental palms e.g. Howeia and many Australian native species. Grevillea.

Methods followed	Crops
Meristem culture	Carnation, Gladiolus, Lilium, Orchids,
Embryo rescue/ Ovary culture	Amaryllis, Balsam, Calla lily, Carnation, Cosmos, Hibiscus, LA, LO hybrids of lilium, Orchids, Pelargonium, Rose, Statice, Tulip
Ovule culture	Alstroemeria, Chrysanthemum, Lilium, Pelargonium, Tulip,
Anther/ Double haploid culture	African violet, Anthurium, Calla lily, Carnation, Chrysanthemum, Easter lily, Geranium, Gerbera, Orchids, Pelargonium, Petunia, Phlox, Rose, Tuberose, Tulip
Somatic embryogenesis	Alstroemeria, Anthurium, Carnation, Chrysanthemum, Cyclamen, Day lily, Geranium, Gladiolus, Gerbera, Lilium, Marigold, Orchids, Pansy, Periwinkle, Poinsettia, Rose, Tulip
Somatic hybridization	Geranium, Hibiscus, Lilium, Petunia,

Methods followed	Crops
Protoplast culture	Anthurium, Carnation, Chrysanthemum, Day lily, Gypsophila, Lilium, Lotus, Orchids, Pelargonium, Petunia, Snapdragon, Stock, Rose
Callus culture	Dieffenbachia, Rose, Gladiolus, Gerbera, *Ficus benjamina*, *Begonia rex*, *Pepromia obtusifolia*, *Philodendron,* Spathiphyllum, Syngonium

Reference

Reddy, B. S., Janakiram, T., Kulkarni, B. S., Narayanaswamy, P., 2004, Emerging trends in ornamental horticulture. Indian Society of ornamental horticulture, pp- 24-27.

17

Use of Molecular Techniques in Modern Plant Systematics

Taxonomy, also known as systematics, is the study of classifying organisms into logically related groups. Historically, taxonomy was performed by examining physical characteristics of organisms and classifying species based on the most commonly held traits. Unfortunately, this method of classifying plants and animals assumes that because they share physical characteristics, they must have shared ancestors. Markers are used over the years for classification of plants. Markers are any trait(s) of an organism that can be identified with confidence and relative ease, and can be followed in a mapping population. They can be defined as heritable entities associated with the economically important trait under the control of polygenes. Molecular markers are any kind of molecule indicating existence of a chemical or a physical process. Molecular markers include biochemical constituents (e.g. secondary metabolites in plants) and macromolecules (e.g. proteins and deoxyribo-nucleic acid). These macromolecules show easily detectable differences among different strains of a species or among different species.

The prerequisite for the improvement of any plant species are germplasm collection and its characterization to find out the diversity among the accessions. Characterization can be done by several ways which include morphological, enzymatic and genetical methods. The best way to assess the diversity is based on DNA markers which allow differentiation among genotypes without any environmental influences. These markers have a great potential to identify genetic diversity within and among accessions which help in successful implementation of pre-breeding approaches. Molecular markers are considered the best tools for determining genetic relationships/ diversity as they are unlimited in number, show high polymorphism, provides more reliable information, detect more variation than morphological and biochemical markers; and are stage independent, not affected by environment and simply inherited. Molecular markers are 'tags' that can be used to identify specific genes and locate them in relation to other genes. They help in ensuing a desired beneficial trait to be maintained or an undesirable trait to be eliminated.

Previous century belonged to 'Green' revolution and present century belongs to 'Gene' revolution.

Morphological markers	Molecular markers
Phenotypes of most morphological markers can only be distinguished at plant level	Using molecular markers, genotypes can be distinguished at plant, tissue or even at cellular level
Distinguishable alleles for morphological marker loci occur less frequently	A relatively large number of naturally occurring alleles can be found at molecular marker loci
Morphological markers are usually associated with undesirable phenotypic effects	Usually, no undesirable effects are associated with alternate alleles of molecular marker loci
Alleles usually interact in a dominant recessive manner, causing difficulties in selecting recessive genotypes	Alleles usually exhibit codominance; therefore, all possible genotypes can be distinguished in the segregating population
Strong epistatic effects limit the number of markers that can be scored in the same segregating population	Fewer epistatic or pleiotropic effects allow limitless number of markers that can be scored in the same segregating population
These are stage specific & affected by environment	These are not stage specific & not affected by environment

Molecular systematics is the discipline of classifying organisms based on differences in protein and DNA in order to make fine taxonomic classifications that are not solely based on morphology. Molecular systematics provided a method for objectively testing the robustness of the major clades of ciliates established based on ultrastructural features. The *SSU rRNA* gene, which specifies the RNA skeleton of the small subunit of the ribosome, has been the gene of choice for ciliates. This molecule contains both highly conservative and more variable regions, allowing it to be used to test both deep (ancient) and recent phylogenetic relationships.

There are now hundreds of *SSU rRNA* sequences for ciliates, with multiple representatives for each of the phylum's major subdivisions. Analyses of these data have confirmed the basic integrity of Small and Lynn's eight major classes, with some changes in assigning subclasses.

Molecular breeding is a term that is used to describe the application of variety of tools for manipulating the DNA of plants, which may or may not involve rDNA. Marker Assisted Selection (MAS) denotes the use of molecular techniques in identification of parents, cultivar identification, genetic diversity studies and cluster analysis. Using molecular markers, genotypes can be distinguished at plant, tissue or even at cellular level.

DNA Based Molecular Marker Techniques	Abbreviations
Amplified Fragment Length Polymorphism	AFLP
Amplicon Length Polymorphism	ALP
Cleaved Amplified Polymorphic Sequences	CAPS
DNA Amplification Fingerprinting	DAF
Inter-SSR Amplification	ISA
Random Amplified Polymorphic DNA	RAPD
Restriction Fragment Length Polymorphism	RFLP
Simple Sequence Repeat Markers	SSR
Sequence Tagged Microsatellite Markers	STMS
Single Nucleotide Polymorphism	SNP

The molecules used in molecular taxonomy are as follows:

- Nucleotides are monomeric units made up of single nucleotides that make up DNA. Each nucleotide is composed of three components: sugar, a nitrogenous base, and a phosphate group. Deoxyribose is the sugar found in DNA, and it has a phosphate group on one side and a nitrogenous base on the other.
- Nucleotides with a nitrogenous base, a pentose sugar, and one or more phosphate groups make up RNA. Deoxyribose has a structure similar to ribose, except it has a H instead of an OH at the 2′ position.
- Genes aid in the determination of genetic information.

17.1 Use of Molecular Markers in Ornamentals and Flowers

Molecular markers are widely used for cultivar identification and mapping, phylogenetic studies, genetic diversity studies, DNA finger printing, genetic diagnostics, expression profile analysis, study of genome, gene mapping / gene tagging, cluster analysis, seed testing, identifying location of QTL's and Marker Assisted Selection (MAS).

Marker-assisted selection

It is the indirect selection for desirable traits based on their molecular genotype. It can be used alone or in combination with the classical methods of selection. Marker-assisted selection refers to such a breeding procedure in which DNA marker detection and selection are integrated into a traditional breeding program. The fundamental advantages of MAS compared to conventional phenotypic selection are, 1) They simpler compared to phenotypic screening, selection may be carried out at seedling stage and 2) Single plants may be selected with high reliability.

Amaryllis

Considering the importance of protecting plant breeders' rights for commercial exploitation of hybrids, a PCR based technique (DNA-RAPD) was used to correctly identity known and unknown hybrids and to determine cultivar relatedness. RAPD profiles were very successful in tracing and confirming the parentage of all hybrids and determining molecular relationship among varieties at NBRI (Chakrabarty *et al.*, 2007).

Anthurium

After conducting analysis of floral fragrance compounds on five *Anthurium* species (*A. armeniense, A. fragrantissimum, A. lindenianum, A. orchranthum* and *A. roseospadix*) and three hybrids of *A. armeniense*, Kuanprasert *et al.* (1998) suggested that concentrations (%) and chemical profiles of 28 compounds can be considered as useful taxonomic markers to clarify phylogenetic relationships in *Anthurium* spp.

A number of genomic markers based on DNA polymorphism such as RAPD and ISSR have been developed and used for genetic diversity analysis. However, RAPD and ISSR markers are dominant, limiting their use for population genetic studies. The RAPD markers determined the genetic relationships of *Anthurium* species. Lady Ann and Lady Beth, pot plant cultivars exhibit high degree of genetic similarity or closest relationship with genetic distance (0.06) and similarity coefficient values (0.98). Nowbuth *et al.* (2005) assessed the genetic diversity of 24 anthurium cultivars using 52 RAPD markers. Nei and Li's genetic distance (GD_{NL}) coefficient values among the cultivars ranged from 0.018 to 0.163 with an average genetic similarity of 91.34 per cent. Wang and Chuang (2013) developed and characterized six microsatellite markers for use in anthurium fingerprinting and genetic diversity studies for a set of 28 diverse genotypes. From the selected six polymorphic SSR loci, 52 alleles were amplified and 27 distinct genotypes were found, except for cv. Tropical and its mutant, with a mean number of 8 alleles per locus.

Antirrhinum

The RAPD analysis was carried out to investigate genetic diversity and pigmentation structure of endangered species *A. subbaeticum* and *A. microphyllum*, and to molecular cytogenetic characterization of *A. majus* genome and to establish relationship between chromosomes and linkage groups (Jinenez *et al.*, 2002 and Torres *et al.*, 2003).

Balsam

The RAPD markers have proven their usefulness in evaluating genetic variability among Lithuanian populations of *I. glandulifera* (Zybartaite *et al.*, 2011).

Calla lily

Different molecular methods are efficiently used to study gene pool of species. Wei *et al.* (2012) conducted EST-SSR marker study to evaluate the level of genetic variation present in 24 accessions of *Z. aethiopica* from Asia, Europe and Africa. Total of 43 loci were polymorphic with number of alleles per locus ranging from 2-10.

Bougainvillea

The RAPD, EST-SSR and AFLP markers were used for identification of cultivars, determining diversity and genetic relationship. Study of Srivastava *et al.* (2009) deals with authenticating knowledge about 21 bougainvillea cultivars comprising of 9 hybrids and their parents through RAPD analysis. The 19 degenerate primer sets generated 234 bands from which 158 (67.5%) were polymorphic. The UPGMA based dendrogram divided 21 cultivars into two major groups with Jaccard's similarity coefficient ranging from 0.51 to 0.942. Group A had three cultivars namely Trinidad, Formosa and Dr. H.B. Singh in which Dr. H.B. Singh was confirmed as a hybrid of other two cultivars. Group B was sub-divided into 8 clusters. The parentages of 7 out of 8 hybrids have been confirmed based on clusters. The study concluded that the RAPD technique is suitable for confirmation of parent-hybrid relationship.

Kumar *et al.* (2014) studied the molecular characterization and cultivar identification in *Bougainvillea* spp. using SSR markers to determine the genetic relatedness of fifty cultivars to the four major species namely, *B. glabra, B. spectabilis, B. peruvuiana* and *B.* x *buttiana.* In all, 28 alleles were detected at an average number of alleles of 5.6 alleles/locus. Primer *BOUG*-1 showed highest polymorphism index content of 0.891 which reflects its ability to differentiate these cultivars better at molecular level. Around 18 rare alleles were identified among them Blondie had more rare alleles (3).

Calendula

RAPD analysis performed for six diversified species and preparing their phylogenetic tree indicated their usefulness in showing relationship between genotypes, species, gender and family (Baciu *et al.*, 2012). DNA fingerprint markers were used for identification purposes, genetic analysis, detecting genetic association between DFP bands and agronomically important traits. AP-PCR technique was also applied for precise characterization cultivar lineages and relationships.

Carnation

Isozymes, as phytochemical compounds, are considered as useful taxonomic markers to clarify phylogenetic relationship in carnation species. Thirteen commercial cultivars were assessed using isozymes as markers (Patial *et al.*, 2007). Polymorphism was found to be 13.8 per cent. Cultivars were grouped into two clusters and three sub-clusters on the basis of genetic distance. Information on genetic diversity and relationship within and among species and their wild relatives is essential for the efficient utilization of plant genetic resources. Microsatellite markers were studied by Arens *et al.* (2009).

Chrysanthemum

Identification and characterization of new cultivars is extremely important to meet the DUS testing, address IPR issues and their utilization and conservation. Traditionally, morphological markers are used for germplasm characterization as being simple and irreplaceable. Though vast diversity exists in chrysanthemum, there are only few reports on the use of molecular markers for diversity analysis in chrysanthemum. Wolff *et al.* (1995) developed DNA fingerprinting key for identification of chrysanthemum cultivars. The RAPD, AFLP and SRAP markers can be utilized to examine phenotypic variation, genetic distances in different cultivars and/or species of chrysanthemum. Analysis of genetic diversity through RAPD is technically simpler and quicker.

Wide range of genetic diversity/ genetic distance among mini cultivars gives opportunity to introgress various desirable traits by further breeding. Systematic taxonomical studies at molecular level of mini chrysanthemum group are possible along with improvement in flower quality by broadening genetic pool through introgression of foreign genes. Studies on genetical similarity gives more accuracy for classification of mini cultivars instead of morphological parameters (Chatterjee *et al.*, 2006). RAPD profile serves as way for identification of species specific markers and also produce reliable data to construct polygenetic relationship of genus *Chrysanthemum* (Chatterjee *et al.*, 2005).

The development and utilization of biochemical and molecular markers are widely used for different purposes. The use of molecular markers for genetic diversity analysis and as a selection tool is a high priority for new efforts in the development of ornamental cultivars and exploitation of species diversity.

DNA profiling and diversity analysis of chrysanthemum was done by RAPD markers (Kumar *et al.,* 2017).

Gladiolus

Use of RAPD (Takatsu *et al.*, 2001) and AFLP (Ranjan *et al.*, 2010) are discussed as strong tools for dissection of genetic relatedness in gladiolus species, cultivars and their interspecific crosses. The assessed genetic relationships may enhance the efficiency of breeding program by selecting desirable parents with reduced breeding cycle.

Hibiscus

Molecular characterization using SSR and ISSR markers exhibited high diversity among hybrid plants and could clearly distinguish hybrids from parents. Both these markers were efficient for hybrid identification and species discrimination. Cluster analysis revealed that hybrids formed distinct groups and exhibited high genetic diversity. Molecular markers are exploited for characterization of species, varieties and advanced hybrids. The genomic repeat specific markers like ISSR and SSR have been found to be suitable for genetic diversity analysis, genome association studies and hybrid identification in hibiscus crop due to their high polymorphism and resolving power. Genetic diversity of *H. tiliaceus* using AFLP markers was assessed by Tang *et al.* (2003). Out of total 145 samples, 20 samples showed no polymorphism while 125 samples showed 501 of 566 polymorphic bands (88.5%). Estimates of genetic diversity agreed with life history traits and geographic distribution. Use of four RAPD primers (AO-1, AO-7, AO-8 and AO-9) generated informative bands to confirm hybridity of hardy *Hibiscus* species at the molecular level (Kuligowska *et al.*, 2016). Molecular diversity and genetic relationship in two *Hibiscus* species and 16 genotypes of *H. rosa-sinensis* was revealed by DNA (RAPD) markers (Barik *et al.*, 2006). In *Hibiscus* species, Kadve *et al.* (2012) conducted genetic analysis studies using 9 RAPD primers. Out of total 9 primers used, the primers RPI-07, RPI-08, RPI-10, RPI-19 and RPI-21 were more suitable for genetic variation analysis as it has shown maximum polymorphic bands.

Jasmine

In *Jasminum* species, very little work has been reported regarding its genetic aspects. Joy and Raja (2010) reported the use of RAPD technique to detect somaclonal variation in *J. auriculatum*. Genetic divergence in 32 varieties of jasmine belonging to four species was studied using RAPD markers (Mukundan *et al.*, 2008). The study showed low to moderate genetic diversity among the

Jasminum species. RAPD marker study demonstrated that 21 accessions of *Jasminum* spp. had largest genetic diversity. Similarity matrix index values ranged from 0.263-0.776 among the genotypes (Mahmood *et al.*, 2013).

Lilium

Molecular markers are effectively used in lily to trace parentages, identify diversity, for marker assisted selection, classification of species (Haruki *et al.*, 1997), phylogenetic relationships (Lee *et al.*, 1996), genetic variation within and among populations of *L. martagon* (Person *et al.*, 1998).

Lotus

In *N. nucifera*, EST-SSR and AFLP marker analysis was carried out to investigate genetic diversity and genetic relationships between cultivated and wild lotus cultivars (Hu *et al.*, 2011). These studies demonstrated that sacred lotus had low genetic diversity which was attributed to asexual reproduction.

Orchids

Genetic relationship among 43 *Dendrobium* hybrids was determined by Xiang *et al.* (2003). Ninety bands were polymorphic. Genomic markers find application in the area of plant patenting and breeder's rights and also for fast-track gene screening of traits such as novel flower colour and dwarfing pertinent to potted *Dendrobium*.

Feng *et al.* (2013) developed genetic linkage maps by the double pseudo-test cross mapping strategy based on RAPD and ISSR markers in 90 F_1 progeny individuals derived from interspecific cross between *D. nobile* as female and *D. moniliforme* as male; both 2n=38, having large divergence in terms of phenotype, such as shapes and colour of leaf, flower and labellum. The resulting frame map of *D. nobile* was 1474 cM in total length with 116 loci distributed in 15 linkage groups and *D. moniliforme* linkage map had 117 loci placed in 16 linkage groups spanning 1326.5 cM. Both maps showed 76.91 per cent and 73.59 per cent genome coverage for both species, respectively. These primary maps provide an important basis for genetic studies and further medicinal and horticultural traits mapping and marker assisted selection (MAS) in *Dendrobium* breeding programmes.

Pansy

Molecular markers are useful in elucidating diversity relationships between species. Using DNA sequencing of the ITS1-5.8S rDNA-ITS2, Hildebrandt *et al.* (2006) found that two blue zinc violets are closely related and have strong affinity to *V. lutea*, but not to *V. tricolor*. Using scanning electron microscopy, Kuta *et al.* (2012) characterized the reproductive organs of two blue zinc violets and their relatedness to solve the controversial problem of their descent from either *V. lutea* or *V. tricolor*.

Yockteng *et al.* (2003) studied phylogeny among pansies using internal transcribed spacer (ITS) and ISSR markers. It was confirmed that sect. *Melanium* species are very closely related, showing reduced genetic variation compared to other sections of *Viola* and probably monophyletic. ISSR markers have been used by (Cortés-Palomec *et al.*, 2006) as a means for identification and analyzing genetic diversity in *V. striata* (chasmogamous) and *V. grahamii* (cleistogamous). Genetic diversity is maintained in pansy even though presence of high level of cleistogamy and vegetative reproduction.

Periwinkle

Genetic diversity assessment in nine cultivars of *C. roseus* from Central Gujarat was attempted by Lal *et al.* (2011). Very high polymorphism RAPD (78.71 %), ISSR (78.94%) and SSR (79.62%) was observed among cultivars and this technique proved to be sensitive and effective for genomic analysis. SSR and EST markers have been used by to analyze genetic diversity, variability, linkage mapping and evolutionary relationships (Joshi *et al.*, 2011).

Petunia

Molecular markers *viz.*, RAPD, AFLP, RFLP, DAF, and FISH are useful in elucidating diversity relationship between cultivars and species, preparing linkage maps, genetic dissection of genome and identification of clones containing *Rf* genes in petunia. Using rDNA, Kabbaj *et al.* (1995) suggested that *P. integrifolia* is phylogenetically distant to *P. violacea*. DNA-amplified fingerprinting was compared to distinguish between relationship of five petunia species and ten cultivars of *P.* x *hybrida* (Cerny *et al.*, 1996).

Rose

The RAPD, ISSR, EST, SSR, STMS, AFLP markers can be used for characterization of genetic diversity/variation/ relationships and for dissection of rose genome particularly for genetic linkage maps of diploid and tetraploid population. DNA-RAPD technology can be used to identify genetic diversity in rose cultivars and species (Sasikumar *et al.*, 2007). Vainstein *et al.* (1993) reported that the genetic similarities are small within the cultivated rose groups by using 28 DNA fragments from microsatellite fingerprints. They also indicated a higher genetic similarity between the hybrid tea and the floribunda group than within each group. Cultivars were grouped into three major clusters and species formed a separate cluster.

The double flowered trait is associated with increased number of petals and larger overall flower size and is strongly associated with the attractiveness of the flower for ornamental purposes. Crespel *et al.* (2002) reported genetic mapping approach and their potential use for double flower trait in rose. The map of diploid *Rosa* species showed that *R. hybrida* AGAMOUS gene (*RhAG*) which played a key role in the formation of double flower forms, did not co-localize with the d6 locus or with the QTL for petal number (Spiller *et al.*, 2011; Bendahmane *et al.*, 2013). These findings have triggered an interest in identifying those still unknown genes that relate to the formation of double flowers.

The distinguishable genetic background and high degree of variation revealed by molecular markers, specifically ISSR and SCoT, offer valuable insights that can contribute to the strategic improvement of roses (Chettri *et al.,* 2024).

Salvia

Guofu *et al.* (2007) studied ornamental inherit relations in *S. splendens* via cluster group analysis using RAPD combined with palynology classification. Through clustering diagram four kind of flower colours were divided in two lineages i.e. red and white grouped in one lineage and gathered with purple. White colour is alone one lineage.

Tuberose

Sarkar *et al.* (2010) developed two RAPD primers OPC-13 and OPD12 for identifying genetic diversity of tuberose cultivars. RAPD primers OPC-13 and OPD12 were identified as efficient primers for diversity analysis of tuberose. A total of 157 RAPD bands were generated by the 17 random decamer primers. The RAPD analysis also confirmed their relatedness as they grouped in the same cluster. The suitability of this technique for genotyping and diversity analysis was also established. Genotypes, Vaibhav and Pune Single were found to have least pair-wise similarity although they had a greater morphological similarity with each other.

Six ISSR primers *viz*., 808, 836, 840, 842, 855 and 857 were used for genetic diversity studies. A total of 62 bands were produced across seven genotypes, of which 53 bands were polymorphic, accounting for 85.48 per cent polymorphism. The ISSR dendrogram grouped the genotypes into two major clusters. The genotype Calcutta Single and Prajwal recorded the highest similarity index value of 0.706 which were considered closely related genotypes (Kameswari *et al*., 2014).

Tulip

Heterogeneity or DNA within individuals of *T. gesneriana, T. kaufmanniana* and their inter-specific hybrid was analyzed by using RFLP markers (Booy *et al*., 2000). Use of AFLP markers was effective in making pre-selection feasible since tulips have long juvenile period. AFLP markers can be used to detect 12 linkage groups in entire tulip genome associated with TBV resistance (Van Heusden *et al*., 2002).

Verbena

SCAR marker could distinguish *V. officinalis* from all other *Verbena* species except its closest relative *V. hastata* (Ruzicka *et al*., 2009).

Zinnia

RAPD analysis was carried out among and between zinnia varieties (Alderuccio *et al*., 2007). This marker study demonstrated that *Z. elegans* cv. Dreamland had the largest genetic identity when compared to *Z. haageana*. The three analyzed species *Z. angustifolia, Z. haageana* and *Z. ellotiana* could be clearly distinguished by their DNA content.

Name of the plant	Molecular markers used	Result	Reference
Camellia nitidissima.	RAPD, AFLP	Phylogenetic relationships	Tang *et al*. (2006)
Camellia sinensis	RAPD, AFLP	Genetic variation among germplasm	Wachira *et al*. (2001)
Cephalotaxus spp.	AFLP	Genetic diversity	Wang *et al*. (2016)
Chrysanthemum cvs.	RAPD	Genetic analysis of hybrids	Huang *et al*. (2000)
Chrysanthemum genotypes	AFLP	Genetic structure	Roein *et al*. (2014)
Cornus florida	DAF	Determination of parentage	Ament *et al*. (2000)
Cupressus sempervirens var. *horizontalis*	RAPD	Genetic analysis	Cvjetan *et al*. (2000)
Cymbidium cvs.	RAPD	Identification of cultivars; genetic diversity study	Obara-okeyo and Kato (1998)
Dianthus caryophyllus	RAPD, Ministellite	Genetic relatedness between genctypes	Vainstein *et al*. (1991)

Name of the plant	Molecular markers used	Result	Reference
Eucalyptus erythroma	RAPD	Molecular analysis of an interspecific hybrid	Delaporte *et al.* (2001)
Eustoma grandiflorum	RAPD	Somaclonal variant identification	Cheong *et al.* (1999)
Gladiolus spp.	RAPD	Genetic relationship	Takatsu *et al.* (2001)
Gymnnadenia conopsea	Micro-satellite	Genetic variation	Gustafsson (2000)
Hibiscus syriacus, H sinosyriacus, H paramutabilis	AFLP	Genetic relationships	Van Huylenbroeck *et al.* (2000)
Ixora coccinea, I javanica	RAPD	Genetic diversity study	Rajaseger *et al.* (1997)
Lilium lomgiflorum var. *formosanum*	RAPD	Genetic diversity	Wen and Hsiao (2001)
Lilium xelegans,	RAPD	Phylogenetic	Lee *et al.* (1996)
Rosa odorata	SNVs	Phylogenetic advances	Li *et al.* (2020)
Rosa spp.	RAPD markers	Genetic diversity	Visalakshi *et al.* (2023)
Flower Plants	Markers	MAS, QTL	Hasan *et al.* (2021)
Rosa damascena	URP & SCoT markers	Genetic diversity analysis	Mostafavi *et al.* (2021)

Conclusion

Molecular systematics is the discipline of classifying organisms based on differences in protein and DNA in order to make fine taxonomic classifications that are not solely based on morphology. Taxonomy is the naming of science describing, and classifying organisms, which encompasses all plants, animals, and microorganisms on the planet describe, and taxonomists identify and classify species using morphological, behavioral, genetic, and biochemical observations, including those that are new to science. The study of the diversification of living forms, both past and present, as well as the relationships between living things over time, is known as systematics.

References

Alderuccio, F., Chan, J. and Toh, Ban-Hock. 2007. Advances in Gene, Molecular and Cell Therapy-Floriculture and Ornamental Biotechnology (Eds.) Anantasarn, J., Schroder, M.B., Eimert, K. and Kanchanapoom, K.), pp. 82-90.

Ament, M.H., M.T. Windham and R.N. Trigiano. 2000. Journal of Arboriculture, **26**: 206-212.

Arens, P., Esselink, D., Noordijk, Y., Kodde, l., Hof, L., Wietsma, W. and Vosman, B. 2009. *Acta Horticulturae,* **836**: 456-460.

Barik, S., Senapati, S.K., Aparajita, S., Mohapatra, A. and Rout, G.R. 2006. *Journal of Biosciences*, **61**(1-2): 123-128.

Baciu, A.D., Pamfil, D., Mihalte, L., Sestras, F. and Sestras, R.E. 2012. Bulletin of University of Agricultural Sciences and Veterinary Medicine Cluj-Napoca Horticulture, **69**(1): 40-47.

Bendahmane, M., Dubois, A., Raymond, O. and Le Bris, M. 2013. Journal of Experimental Botany, **64**: 847-857.

Booy G., Schoot, J. and Vosman, B. 2000. *Plant Systematics and Evolution*, **225**: 29-41.

Cerny, T.A., Caetano-Anollés, G., Trigiano, R.N. and Starman, T.W. 1996. Theoretical and Applied Genetics, **92**(8): 1009-1016.

Chakrabarty, D., Gupta, V.N. and Dutta, S.K. 2007. Plant Biotechnology Reports, **1**(4): 211-217.

Chatterjee, J., Mandal, A.K., Ranade, S. A. and Datta, S.K. 2006. *Pakistan Journal of Biological Sciences*, **8**(4): 546-549.

Chatterjee, J., Mandal, A.K., Ranade, S.A., Teixeira, J.A. and Datta, S.K. 2005. *Scientia Horticulturae*, **110**: 373-378.

Chettri, K., Majumder, J., Mahanta, M., Mitra, M. and Gantait, S., 2024. *Scientia Horticulturae*, **332**: 113243.

Cheong, C., Yu, K. W. and Paek, K. Y. 1999. *Korean Journal of Horticulture Science and Technology*, **17**(3), 352-354.

Cortés-Palomec, A.C., McCauley, R.A. and Ballard, H.E. Jr. 2006. *International Journal of Plant Sciences*, **167**(3): 503-5012.

Crespel, L., Gudin, S., Meynet, J., and Zhang, D. 2002. *Theoretical and Applied Genetics*, **104**: 451-456.

Cvjetan, S., Kajba, D., Pavicic, N. and Pejic, I.I., Genetska Analiza Starog Hvarskog cempresa.

Delaporte, K.L., Conran, J.G., Sedgley, M. 2001. *International Journal of Plant Sciences*, **162**(6): 1317-1326.

Feng, S., Zhao, H., Lu, J., Liu, J., Shen, B. And Wang, H. 2013. *Journal of Genetics*, **92**(2): 205-211.

Guofu, H., Qiang, Y., Fenglan, L., Qi, W. And Baozhong, H. 2007. *Journal of Northeast Agricultural University*, **14**(2): 97-102.

Gustafsson. S., 2000. *Molecular Ecology*, **9**:1863–1872.

Haruki, K., Hosoki, T., Nako, Y. and Ohta, K. 1997. *Journal of Japanese Society of Horticultural Sciences*, **66**(1): 189-192.

Hasan, N., Choudhary, S., Naaz, N., Sharma, N. and Laskar, R.A. 2021. *Journal of Genetic Engineering and Biotechnology*, **19**(1): 128-

Huang, S.C., Tsai, C.C. and Sheu, C.S., 2000. Botanical Bulletin of Academia Sinica, 41.

Hildebrandt, U., Hoef-Emden, K., Backhausen, S., Bothe, H., Bozek, M., Siuta, A. and Kuta, E. 2006. *Plant Systematics and Evolution*, **257**: 205-222.

Hu, J., Pan, L., Liu, H., Wang, S., Wu, Z., Ke, W. and Ding, Y. 2011. *Molecular Biology Report*, **11**: 1-11.

Jinenez, J.F., Sanchez-Gomez, P., Guemes, J., Werner, O. and Rossells, J.A. 2002. *Heredity*, **89**: 387-393.

Joshi, R.K., Kar, B. and Nayak, S. 2011. *Bioinformation*, **5**(9): 378-381.

Joy, P. and Raja, D.P. 2010. *Genetic Resources and Crop Evolution*, **52**: 239-244.

Kadve, S., Yadav, M. and Tiwari, A. 2012. *International Journal of Biomedical and Advance Research*, **3**(6): 473-485.

Kameswari, P.L., Girwani, A. and Radha Rani, K. 2014. *Electronic Journal of Plant Breeding*, **5**(1): 52-57.

Kuanprasert, N., Kuehnle, A.R. and Tang, C.S. 1998. *Phytochemistry*, **49**: 521-528.

Kuligowska, K., Lutken, H., Christensen, B. and Muller, R. 2016. *Breeding Science*, **66**: 300-308.

Kumar, M., Chaudhary, V., Sirohi, U., Singh, M.K., Malik, S. and Naresh, R.K., 2017. *Journal of Pharmacognosy and Phytochemistry*, **7**(5): 2641-2652.

Kumar, P.P., Janakiram, T., Bhat, K.V., Jain, R., Prasad, K.V. and Prabhu, K.V., 2014. *Indian Journal of Agricultural Sciences*, **84**(8): 1024-30.

Kuta, E., Bohdanowicz, J., Slomka, A., Pilarska, A. and Bothe, H. 2012. *Plant Systematics and Evolution*, **298**(2): 445-455.

Lal, S., Mistry, K.N., Shah, S.D., Thaker, R. and Vaidya, P.B. 2011. *Journal of Research in Biology*, **1**(8):667-675.

Lee, J., Lee, P., Lim, Y., Shin, E. and Roh, M. S. 1996. *Acta Horticulturae*, **414**: 137-144.

Li, C., Zheng, Y. and Huang, P., 2020. *Scientific Reports*, **10**(1): 12188.

Mahmood, M.A., Hafiz, I.H., Abbasi, N.A. and Faheem, M. 2013. *International Journal of Agriculture and Biology*, **15**(3): 505-510.

Mostafavi, A.S., Omidi, M., Azizinezhad, R. and Etminan, A. 2021. *Journal of Genetic Engineering and Biotechnology*, **19**(1): 144.

Mukundan, S., Sathyanarayana, B.N., Simon, L., Sendur, S.N. 2008. *Floriculture and Ornamental Biotechnology*, **1**(2): 136-141.

Nowbuth, P., Khitto, G., Bahorun, T. and Vankatasamy, S. 2005. *African Journal of Biotechnology*, **4**(10): 1189-1194.

Obara-Okeyo, P. and Kako, S., 1997. Journal of Horticultural Sciences, 72(2): 263–270.

Patial, H., Shirkot, P., Bhalla, R. and Zinta, G. 2007. *Journal of Ornamental Horticulture,* **10**(2): 78-84.

Person, H.A., Lundquist, K. and Nybom, H. 1998. *Heriditas Landskrona*, **128**(3): 213-220.

Rajaseger, G., Tan, H.T.W., Turner, I.W. and Kumar, P.P., 1997. Annals of Botany, 80: 355-361.

Ranjan, P., Bhat, K.V., Misra, R.L, Singh, S.K. and Ranjan, J.K. 2010. *Scientia Horticulturae*, **123**(4): 562-567.

Roein, Z., Asil, M.H., Sabouri, A. and Dadras, A.R. 2014. *Plant Systematics and Evolution*, **300:** 493-503.

Ruzicka, J., Lukas, B., Merza, L., Göhler, I., Abel, G., Popp, M. and Novak.J. 2009. *Planta Medica*, **75**(11): 1271-1276.

Sarkar, J., Misra, R.L., Bhat, K.V., Singh, A., Singh, S.K. 2010. *Indian Journal of Genetics and Plant Breeding*, **70**(2): 182-188.

Sasikumar, K., Singh, S.K., Mohapatra, T. and Prasad, K.V. 2007. *Indian Journal of Horticulture*, **64**(4): 435-438.

Spiller, M., Linde, M., Hibrand-Saint Oyant, L., Tsai, C.J., Byrne, D.H., Smulders, M.J., Foucher, F. and Debener, T. 2011. *Theoretical and Applied Genetics*, **122**: 489-500.

Srivastava, R.,Shukla, S., Soni, A. and Kumar, A. 2009. *Crop Breeding and Applied Biotechnology,* **9:** 154-163.

Takatsu, Y., Kasumi, M., Manabe, T., Hayashi, M., Inoue, E., Marubashi, W. and Niwa, M. 2001. *HortScience*, **36**(2): 341-343.

Tang, S., Bin, X., Wang, L. and Zhong, Y., 2006. *Biochemical Genetics*, **44**: 444-456.

Tang, T., Zhong, Y., Jian, S. and Shi, S. 2003. *Annals of Botany*, **92**: 409-414.

Takatsu, Y., Miyamoto, M., Inoue, E., Yamada, T., Manabe, T., Kasumi, M., Hayashi, M., Sakuma, F., Marubashi, W. and Niwa, M., 2001. *Scientia Horticulturae*, **91**(3-4): 339-348.

Torres, E., Iriondo, J.M. and Pérez, C. 2003. *American Journal of Botany*, **90**(1): 85-92.

Vainstein, A., Hille, J., Lavi, U. and Tzuri, G. 1991. *Euphytica*, **56**: 225-229.

Vainstein, A., Ben-Meir, H. and Zucker, A. 1993. Proceedings of the XVIIth Eucarpia Symposium 'Creating Genetic Variation in Ornamentals', San Remo, pp. 63-68.

Van Heusden, A.W., Jongerius M.C., Van Tuyl, J.M., Straathof, T.P. and Mes, J.J. 2002. *Acta Horticulturae*, **572**: 131-138.

Van Huylenbroeck, J.M., De Riek, J. and De Loose, M., 2000. *Genetic Resources and Crop Evolution*, **47**(3):.335-343.

Visalakshi, M., Muthulakshmi, R., Ganga, M., Boopathi, N.M., Vasanth, S. and Ganesh, S., 2023. International Journal of Environment and Climate Change, 13(12): 1305-1311.

Wachira, F., Tanaka, J. and Takeda, Y., 2001. *The Journal of Horticultural Science and Biotechnology*, **76**(5):557-563.

Wang, J.Y. and Chuang, K.C. 2013. *Euphytica*, **189**: 421-431.

Wang, T., Wang, Z., Xia, F. and Su, Y., 2016. *Scientific Reports*, **6**(1): 25031.

Wei, Z., Luo, L., Zhang, H., Xiong, M., Wang, X. and Zhou, D. 2012. *American Journal of Botany*, **10**: e1-e5.

Wen, C.S. and Hsiao, J.Y., 2001. *International Journal of Plant Sciences*, **162**: 287-295.

Wolff, K., Zietkiewicz, E. and Hofstra, H. 1995. *Theoretical and Applied Genetics*, **91**: 439-441.

Xiang, N., Hong, Y. and Lam-Chan, L.T. 2003. *Journal of the American Society for Horticultural Science*, **128**(5):731-735.

Yockteng, R., Ballard, H.E., Mansion, Jr. G., Dajoz, I. And Nadot, S. 2003. *Plant Systematics and Evolution*, **241**: 153-170.

Zybartaite, L., Zukauskiene, J., Jodinskiene, M., Janssens, S.B., Paulauskas, A., Kupcinskiene, E. 2011. Žemdirbystė= *Agriculture*, **98**(4): 391-398.

18

Terms used in Plant Taxonomy

The science that includes classification, nomenclature and identification of plants is called plant taxonomy.

Taxa (Sing. Taxon):A general term for taxonomic classification, irrespective of rank.

18.1 Plant Types (habit and habitat)

1. Annual: A plant that completes its entire life cycle, from seed to reproduction to death, in one single year.
2. Biennial: A plant that completes its entire life cycle in two years.
3. Perennial: A plant that grows and reproduces for many years.
4. Monocarpic: plants that flower and set seeds once during its lifetime. They flower only after attaining maturity. They flower gregariously in a single season e.g. Bamboo and Kurinjis.
5. Polycarpic: Plants that flower and set seeds many times during its lifetime.
6. Aquatic: Organisms growing in water.
7. Shrub: A much branched, small, woody perennial plant with several branches from ground level upwards.
8. Tree: A tall, woody, perennial plant with a single trunk which usually bears branches.
9. Climber: A weak-stemmed plant with roots in the ground and climbing the other plants, etc., to support itself with the help of its tendrils and adventitious roots, etc. Usually climbers twist around their support.
10. Creeper: A plant which is unable to support itself, and spreads along the ground.
11. Vine: A creeper of family Vitaceae.
12. Xerophyte: A plant that lives in dry habitats or in a desert.
13. Epiphyte: A plant, growing on stems or branches of other plants, with no roots in the ground, but not taking nutrients on which it is growing.

14. Habit: Appearance of an organism, e.g. herb, shrub, etc.
15. Habitat: The place in which an organism or a community is found.
16. Herb: A small, usually annual plant, with no wood in its stems or roots.
17. Liana: An annual or perennial, elongate plant with weak stem, which is often climbing.
18. Parasite: An organism which takes all its nutrients from the tissues of another organism, usually with harmful effects.

18.2 Root

18.2.1 Root Types

1. Tap root: The main, persistent, primary root of the plant which shows apical dominance and develops from the radicle.
2. Adventitious root: Root arising from any part of the plant body other than the radical. It arises from a organ other than the root e.g. rose, sugarcane, China rose, marigold, tapioca, etc.

18.2.2 Modified Roots

1. Aerial: A root arising from a part of a plant which is above the ground, e.g. betel.
2. Annulated: Root having a series of ring-like swellings e.g. Ipecac.
3. Assimilatory: Green, chlorophyll-containing roots e.g. *Trapanatans* (Fig. 18.8)..
4. Buttress: Roots having plank-like or board-like growth on the upper side. These are supportive structures e.g. *Terminalia bellirica*, Silk cotton tree,
5. Clasping or climbing: These arise on the nodes of weak-stemmed plants, and help them to climb or fix on the support e.g. *Pothos*, Long pepper, Indian ivy.
6. Conical: Cone like, i.e. broad at the base and tapers gradually towards apex, e.g. *Daucuscarota*.
7. Contractile: Roots capable of shortening and contain a wrinkled surface, e.g. *Crocus*.
8. Fasciculated: Fleshy or tuberous roots in a cluster e.g. *Asparagus*, *Dahlia*, *Ruellia*.
9. Fibrous: Fine, thread-like roots e.g. monocotyledons.
10. Fleshy: Succulent roots, e.g. *Dahlia*.

11. Fusiform: Root thickened in the middle and tapering at the ends e.g. Radish.
12. Haustorial or sucking: These are absorbing roots, present within the host of some parasitic plants, e.g. *Balanophora*, *Vanda* orchid, Dodder, broomrape, *Loranthus*.
13. Knee or Aerating: These are vertical or horizontal above ground roots e.g. *Heritiera*.
14. Leaf: Roots arising on the margin of the leaf from adventitious buds, e.g. *Bryophyllum*.
15. Moniliform: Root thickened at certain intervals giving a beaded appearance e.g. *Momordica*, *Basella*, *Portulaca*, etc.
16. Mycorrhizal: Root in symbiotic relationship with fungal hyphae e.g. *Monotropa*.
17. Napiform: Root almost spherical at one end and tapering sharply at the lower part e.g. Turnip, *Beta vulgaris*.
18. Nodulose: Root bearing a small knot at or near the apex e.g. *Curcuma domestica*, *Asparagus sprengeri*, *Maranta*.
19. Pneumatophorous or respiratory: Spongy, aerating roots of marshy plants e.g. *Rhizophora Jussiaea repens*.
20. Prop or stilt: Supportive roots which grow out from the bottom of the trunk into the ground e.g. *Ficusbenghalensis*, screwpine, *Rhizophora*, etc.
21. Tuberous or Tubercular: Fleshy roots appearing like stem tubers e.g. *Ipomoeabatatas*, *Mirabilis jalapa*.

18.3 STEM (Types, Surface, Forms and Modifications

1. Acaulescent: Stemless or with inconspicuous stem.
2. Aerial: Stem which remains above the ground, e.g. *Sesbania*.
3. Angular: Stem showing many angles in a transverse section e.g. *Asparagus*.
4. Arborescent: Woody and tree like e.g. *Mangifera*.
5. Ascending: Inclined upward.
6. Bulb: An underground, short, erect stem covered by fleshy leaves e.g. Onion.
7. Bulbil: A bulblike body developed on the above ground parts e.g. *Agave*.
8. Caudex: A short, hard, overwintering base of a perennial herb e.g. Palms.

9. Caulescent: Possessing a distinct stem.
10. Cladode: A phylloclade in which a branch of a single internode becomes flat or leaf like, e.g. *Asparagus*, *Ruscus*.
11. Cladophyll: A leaf-like, flattened, green stem.
12. Columnar: Erect with a strong trunk.
13. Corm: Solid, round, bulblike, fleshy, underground stem, usually surrounded by membranous scales e.g. *Gladiolus, Amorphophallus*.
14. Culm: Flowering and fruiting stem of sedges (Cyperaceae) and grasses (Gramineae) e.g. *Cyperus*, *Cynodon*.
15. Decumbent: Stem bending in one direction e.g. *Tridex*.
16. Dichotomous: Stem dividing equally into two at tip e.g. *Pandanus*.
17. Diffuse: Branches of stem spreading in all directions e.g. *Boerhaavia*.
18. Eramous: With unbranched stems.
19. Fastigiate: Strictly erect and parallel.
20. Fistular: Stem hollow from inner side e.g. Bamboo.
21. Fruticose: Woody and shrub like.
22. Geniculate: Zigzag stem, or bending abruptly at node e.g. *Vitis*.
23. Glabrous: Smooth or without hairs e.g. Lemon.
24. Glaucous: Shining and smooth.
25. Hairy: Covered with hairs e.g. *Calotropis*.
26. Herbaceous: Soft, non-woody, dying to the ground at the end of growing season e.g. *Eclipta*.
27. Laticiferous: Containing latex e.g. *Calotropis*.
28. Leaf scar: A mark on the stem from which arise branches, leaves; or a leaf, e.g. Sugarcane.
29. Lenticel: Lens-shaped or wartlike pore in the bark.
30. Offset: A short, thick, creeping stem bearing a tuft of leaves above and a cluster of leaves below at the apex e.g. *Eichhornia*, *Pistia*.
31. Pachycauly: Thick, short, frequently succulent stems e.g. Cacti.
32. Phylloclade: A flat or round, green, succulent stem with leaves either ill-developed or modified into spines e.g. *Opuntia*, *Coccoloba*.
33. Phyllode: A flat petiole which has the appearance of a leaf, e.g. Australian *Acacia*.
34. Prickle: A sharp, pointed extension of epidermis or cortex e.g. *Pisonia*.

35. Primocane: The first-year, non flowering stem.
36. Procumbent or Prostrate: Growing flat or parallel on the ground e.g. *Portulaca.*
37. Pterocaulous: Winged stem.
38. Pubescent: Covered with fine silky hairs.
39. Ramose: Branched stem.
40. Rhizome: A horizontal, prostrate, or underground stem bearing scale-like leaves e.g. Ginger, Canna.
41. Runner: A horizontal stem, creeping above ground, usually rooting and producing plants at the nodes e.g. *Cynodon.*
42. Sarcocauly: Fleshy stems.
43. Scandent: Climbing stem.
44. Scape: A leafless flowering stem arising from an underground stem e.g. *Canna.*
45. Scapose: Possessing a scape.
46. Scars: Remains of a point of attachment of leaf, stipule, scale, bud, etc.
47. Stolon: A runner, or a horizontal stem rooting at the nodes e.g. Strawberry.
48. Subterranean: Underground.
49. Supine: Prostrate stem with parts oriented upward.
50. Sucker: Occasionally stem grows horizontally outwards only to a certain extent, but soon it turns up e.g. Chrysanthemum.
51. Tendril: Long, twisting appendage, adapted for climbing e.g. *Vitis*, Passion flower., Antigonon leptopus.
52. Tiller: A grass shoot developed from the base of the stem.
53. Tuber: Thick, enlarged fleshy tip of an underground stem e.g. Potato (Fig. 18.10).
54. Turion: An overwintering bud.
55. Twiner: Stem ascending by coiling on the support without any special device e.g. *Abrus*.

18.4 Leaf

18.4.1Apex of Leaf

1. Acuminate: Drawn out in the form of a long slender tail, or tapers to a protracted point.

2. Acute: Ending into a sharp point in the form of an acute angle but not drawn out (Fig. 18.17B).
3. Apiculate: Ending into a short, sharp, flexible point, or an apicula.
4. Aristate: Tapering to a very narrow, much elongated apex, or bearing a stiff awn.
5. Caudate: Containing a tail-like appendage.
6. Cirrhose: Apex ending into a tendril like structure.
7. Cuspidate: Abruptly and sharply concavely constricted into a sharp and elongated pointed tip.
8. Emarginate: Containing a deep or shallow notch at the apex.
9. Mucronate: Terminating abruptly into a short and sharp point.
10. Mucronulate: Diminutive or smaller form of mucronate.
11. Obcordate: Deeply lobed at the apex.
12. Obtuse or rounded: Blunt or rounded.
13. Retuse: Slightly notched obtuse apex

18.4.2 Base of Leaf

14. Attenuate: Base showing a long gradual taper.
15. Auriculate: Bearing an appendage or ear-shaped part at the base.
16. Connate-perfoliate: Base of opposite sessile leaves, the stem appears passing through them.
17. Cordate: Heart-shaped.
18. Cuneate: Wedge-shaped, triangular with the narrow end at the attaching point.
19. Hastate: Resembling with an arrowhead; halberd-shaped.
20. Oblique: Two sides of base being unequal-sized.
21. Obtuse: Rounded or blunt.
22. Peltate: Attached to its stalk inside the margin.
23. Perfoliate: Sessile leaf whose base completely surrounds the stem.
24. Sagittate: Base like an arrowhead, or triangular.
25. Truncate: Blunt or appearing as if cut off at the end.

18.4.3 Kinds, Modifications, Parts of Leaf

1. Blade: Expanded, flat part of leaf.
2. Bract: Reduced and modified leaf found in inflorescence. These are special leaves from the axil of which one or more flowers arise. Bracts

vary in shape, size, colour and duration e.g. Foliaceous (or leafy), petaloid bracts, involucres, epicalyx, spathe, scaly bracteole, and glume.

3. Bracteole: Small leaf found on the pedicel of flower. When a small leafy or scaly structure is present on any part of flower-stalk (pedicel), it goes by the name of bracteole.
4. Cauline: Leaf arising only on the main axis.
5. Cauline and Ramal: Leaves arising on the main axis as well as side branches.
6. Chaff: Bract present at the base of tubular flower in Asteraceae.
7. Epetiolate: Sessile leaves.
8. Epetiolulate: Sessile leaflets.
9. Epicalyx: Leaves present below the true calyx e.g. Malvaceae.
10. Exstipulate: Without stipules.
11. Floral leaves: Calyx, corolla, stamens and carpels.
12. Fly trap: Hinged leaves of insectivorous plants.
13. Fugacious: Falling soon.
14. Glume: Bract of the spikelets of sedges and grasses.
15. Incomplete: Leaf without one or more parts e.g. without petiole.
16. Involucral bracts: Bracts present below the inflorescence in Asteraceae.
17. Leaflet: A distinct or separate segment of leaf.
18. Lemma: Outer bract of subtending the grass floret.
19. Ligule: Finger-like small part present at the upper end of leaf sheath.
20. Petiolate: Leaf with a distinct stalk.
21. Petiolulate: Leaflets with a stalk.
22. Phyllary: A bract subtending the inflorescence in Asteraceae.
23. Pulvinus: Swollen base of petiole or petiolule.
24. Rachilla: Secondary axis of a compound leaf.
25. Rachis: Main axis of pinnately compound leaf.
26. Radical: Leaves arising from bulb or other underground stem.
27. Scaly leaf: Small, non-green leaf.
28. Sessile: Without petiole.
29. Spathe: An enlarged bract enclosing an inflorescence.
30. Stipule: Scale, gland or blade-like structure at the base of petiole of leaf.
31. Tendril: Coiled, modified leaflet.

18.4.4 Margin of Leaf

1. Biserrate: When each tooth of the serrated margin is again serrated.
2. Ciliate: Margin bearing hairs.
3. Crenate: With blunt, low rounded teeth.
4. Crenulate: Finely crenate e.g. *Bryophyllum* and *Centella.*
5. Dentate: With sharp, marginal teeth pointing outward.
6. Dissected: Cut or deeply divided into many segments.
7. Divided: Cut into distinct segments or sections, extending to the midrib or base of leaf.
8. Entire: With a continuous even and smooth margin; lacking any teeth, lobes or indentations e.g. mango, jack and banyan.
9. Incised: Provided with sharp and irregular incisions.
10. Lacerate: Torn or irregular cut.
11. Lobed: Provided with many lobes extending one third to one half the distance between the midrib and margin.
12. Palmatifid: Cut about halfway down in a palmate form.
13. Parted: Dissected or cut almost to the midrib.
14. Pinnatifid: Parted in a pinnate way, or divided almost to the midrib.
15. Repand: Wavy or undulating in appearance e.g. Mango.
16. Revolute: With the margin rolling inward i.e. towards the underside of the leaf.
17. Runcinate: When the teeth of serrated maring are pointing backward.
18. Serrate: With the marginal teeth pointing towards the apex i.e. cut like teeth of a saw e.g. China rose.
19. Sinuate: With a deeply wavy margin.
20. Spinous: Provided with projecting spines.
21. Undulate: With a slightly wavy margin.

18.4.5 Phyllotaxy of Leaf

1. Alternate: Bearing one leaf at each node (Fig. 18.20A).
2. Distichous: When alternate leaves appear on just two sides of the stem.
3. Opposite: Bearing leaves paired at each node on opposite sides.
4. Opposite decussate: When two successive opposite pairs of leaves occur at right angle to each other.

5. Opposite superposed: When all the successive opposite pairs of leaves occur at the same plane.
6. Radical or basal: When leaves often form a cluster at the ground level.
7. Tristichous, pentastichous and octosticchous: In the alternate arrangement of leaves, if the fourth leaf comes over the first one, the arrangement is called tristichous; if the sixth leaf comes over the first one after completing two revolutions of spiral, it is called pentastichous; and if the ninth leaf comes over the first one after completing three revolutions of the spiral, it is called octostichous.
8. 8.Whorled or verticillate: Bearing three or more leaves at each node.

18.4.6 Shape of Leaf

1. Acicular: Needle shaped.
2. Auriculate: Ear shaped.
3. Cordate: Heart shaped.
4. Cuneate: Wedge-shaped with lower end narrow.
5. Deltoid: Delta-like or triangular.
6. Elliptical: Oval in outline, being narrowed to form rounded ends and widest at the middle.
7. Falcate: Sickle-shaped.
8. Filiform: Long, slender and thread-like.
9. Hastate: Like an arrow-head.
10. Lanceolate: Much longer than broad; or lance-shaped or widening above the base and tapering towards the tip.
11. Linear: Long, flat, and narrow with almost parallel sides
12. Lorate: Strap-shaped, or like a narrow strip of leather.
13. Lyrate: Like a lyre, i.e. having a big terminal lobe and several smaller lateral lobes.
14. Obcordate: Inversely cordate.
15. Oblanceolate: Broad at the middle and tapering towards both the ends.
16. Oblique: When two halves of the lamina are unequal.
17. Oblong: Long, wide with the parallel margins.
18. Obovate: Upper terminal half broader than the lower basal half; opposite to ovate.
19. Ovate: Egg shaped; broad at the base and narrowing towards apex.

20. Pandurate: Resembling obovate with a distinct concavity along each basal side, fiddle-shaped.
22. Pedate: Like the claw of a bird; or palmately lobed.
23. Reniform: Kidney-shaped.
24. Rhomboid: Rhombic-shaped.
25. Rotund: Nearly circular.
26. Runcinate: Irregularly serrate or sharply incised with the teeth pointing towards the base.
27. Sagittate: Like an arrowhead; or triangular.
28. Spathulate: Spoon shaped.
29. Subulate: Tapering from base to apex; awl-shaped.

18.4.7 Types of Stipule

1. Adnate: Attached fully to petiole as wings.
2. Basal: Stipules attached near the base of petiole.
3. Foliaceous: Stipule resembling with a large leaf.
4. Interpetiolar: Present in between the petioles of opposite leaves (Fig. 18.22B).
5. Intrapetiolar: Situated in the axil of leaves, and their margins fuse above the petiole.
6. Lateral: Present laterally on the petiole.
7. Ochreate: Fused to form an ochrea or a fused tubular nodal sheath round the internode.
8. Photosynthetic: Large, green and leaflike or foliaceous.
9. Scaly: Dry, small and membranous stipule.
10. Sheathing or protective: Enclosing a bud or flower.
11. Spinous: Stipule in the form of a hard spine e.g. *Ziziphus*.
12. Tendrillar: Stipule modifying into a tendril.
13. Ventral: Present on the ventral side of the petiole.
14. Vestigial: Very small, minute or remnant.

18.4.8 Types of Leaf

1. Compound leaf: A leaf, in which the leaf blade or lamina remains divided into smaller, blade like parts of leaflets is called as compound leaf. A compound leaf may be palmately compound or pinnately compound.

2. Palmately compound leaf: If the leaflets diverge from a common point at the end of the petiole, in the same way as the fingers from the palm of the hand, the leaf is called palmately compound. In such leaves, if a single leaflet is articulated to the petiole, it is called 'unifoliate', if two or three or four leaflets are articulated to the petiole, it is called bifoliate, trifoliate or quadrifoliate, respectively; and if five or more leaflets are articulated to the petiole, it is called 'multifoliate'.
3. Pinnately compound leaf: If the leaflets are attached to both sides of one central rachis, the leaf is called pinnately compound. In such leaves, if the leaflets are attached directly on the midrib, the leaf is called 'paripinnate', while that which contains an odd terminal leaflet is called 'imparipinnate'. Such a pinnately compound leaf, in which the midrib produces secondary axis, and on the latter are present the leaflets, is called 'bipinnate'; if the midrib of pinnately compound leaf produces secondary axis, and the latter produces the tertiary axis which bears the leaflets, it is called 'tripinnate'. If the leaf is more than thrice pinnate, it is called 'decompound'.
4. Simple leaf: A leaf with the blade in a single part is called simple leaf. The simple blade may, however, be variously divided.

18.4.9 Venation of Leaf

1. Venation: The pattern of veins on the surface of a leaf.
2. Parallel (Fig. 18.25B): When the veins run parallel to each other in the lamina of the leaf, the venation is called parallel e.g. monocots. In this type of venation, if only one principal vein is present, it is called 'unicostate', and if several principal veins are present, it is called 'multicostate'.
3. Reticulate: When the pattern of veins in the lamina of the leaf is like a network, e.g. dicots.
4. Unicostate and multicostate: In the multicostate types, if the veins run in a curved manner from the base of blade to its apex, it is called 'convergent' and if the veins arise at the base of leaf blade and then diverge from one another towards the leaf margin, it is called 'divergent'.

18.5 Inflorescence

The reproductive shoot bearing commonly a number of flowers, or sometimes only a single flower, is called the inflorescence. It may be terminal or axillary and may be branched in various ways. Thus, depending on the mode of branching, different kinds of inflorescence have come into existence. These may primarily be classified into two distinct groups, viz., racemose or indefinite and cymose or definite.

Three theories have been advanced to explain the origin of inflorescence.

1. 1st theory: Nageli (1883) and Pilger (1922) were of the opinion that the primitive type of inflorescence was a panicle and other forms were derived from it. This view was also supported by Goebel in 1931.
2. 2nd theory: Perkin (1914) believed that the primitive type was indicated by a solitary flower and other types with many flowers derived from it as a result of lateral growth of additional branches.
3. 3rd theory: Ricket (1944) considered the dichasium as the primitive type. According to him, a 3-flowered dichasium was the most primitive and a many-flowered dichasium was formed from it as a result of repeated branching.

18.5.1 Basic categories

Cymose, racemose and special types are three basic categories of inflorescences (mode of arrangement of flowers on the floral shoot) of angiosperms.

1. Cymose: In this type the growth of the main axis is checked soon by the development of a flower at the apex, and the lateral axis below the terminal flower also ends in a flower, and thus its growth is also checked. In cymose inflorescence, the terminal flower is the oldest and the young flowers are present on the lower side. Helicoid, circinnus, rhipidium, monochasium/ uniparous, dichasium/ biparous and polychasium/ multiparous are some of the examples of cymose.
2. Racemose: In this type, the main axis does not terminate into a flower, but it grows continuously and gives off flowers laterally in acropetal succession. Here, the youngest flower is present at the apex and the older flowers towards the base. Raceme, spike, spikelet, panicle, catkin, spadix, corymb, umbel and capitulum or head are some of the examples of racemose.
3. Special types: Cyathium (e.g. *Euphorbia, Poinsettia, Pedilanthus*), verticillaster (e.g. *Coleus*, *Ocimum*, *Salvia*), and hypanthodium (e.g. *Ficus* spp.) are some special types of inflorescences.

18.5.2 Types of Inflorescence

1. Catkin: A spike-like, deciduous, elongate, inflorescence, with scaly bracts and unisexual, apetalous, sessile flowers e.g. *Morus*, *Acalypha*, *Betula*, *Quercu*.

2. Circinnus: A modified helicoids cyme having the short pedicels on the developed side.
3. Compound corymb: A branched corymb having pedicellate flowers.
4. Compound cyme: A branched cyme having pedicellate flowers.
5. Compound umbel: A branched umbel having primary rays arising at a common point with a secondary umbel arising from the tip of the primary rays.
6. Corymb: A raceme whose lower stalks are longer than the upper ones, so the inflorescence has a flat top e.g. Candytuft (*Iberis*), wallflower (*Cheianthus*).
7. Cyathium: A cup shaped involucres containing nectar-secreting glands, a centrally placed large female flower, and many male flowers e.g. *Euphorbia pulcherrima*.
8. Dichasium or simple cyme: Determinate, dichotomos inflorescence of pedicellate flowers having pedicels of equal length e.g. Teak, night jasmine, *Ixora*, *Bougainvillea*.
9. Glomerule: An indeterminate inflorescence having dense cluster of sessile or sub-sessile flowers.
10. Head or capitulum: A dense cluster of several sessile or sub-sessile flowers on a compound receptacle or torus. Here, the main axis or receptacle is suppressed, becoming almost flat, and the flowers are also without any stalk so that they become crowded together on the flat surface of receptacle e.g. *Anthocephalus*, Sunflower, marigold, chrysanthemum, China aster, *Mimosa*, Safflower, Cosmos, etc.
11. Helicoid cyme: Curved and unbranched inflorescence of pedicellate flowers having the branches only on one side e.g. *Hemerocallis*, Begonia.
12. Hypanthodium: An inflorescence having sessile flowers on a wall of a concave capitulum, opening by a small ostiole. Male flowers are situated near the periphery and female flowers in the centre e.g. *Ficus*.
13. Panicle: Branched inflorescence with pedicellate flowers arranged in the form of number of racemes e.g. *Delonix*, *Brassica*.
14. Pleiochasium: It is a compound dichasium in which each dichasium has three lateral branches. It is also called polychasium or multiparous cyme.
15. Raceme: An indeterminate, unbranched inflorescence with a single axis and the flowers arranged along the main axis on pedicels e.g. *Hamelia patens, Cassia, Delonix regia, Gladiolus*.

16. Scorpioid cyme: A zig-zagcymose inflorescence that appears to coil like a scorpion's tail. In this type, the branches develop alternately on opposite side of the rachis e.g. *Strelitzia*, *Crassula*, *Geranium*, *Hyoscyanus* and *Heliotropium*.
17. Secund: Flowers arranged only on one side of the rachis.
18. Solitary axillary: Single-flowered, flower attached in the axis.
19. Solitary terminal: Single-flowered, flower attached at the apex.
20. Spadix: A fleshy or thick, spikelike inflorescence with very small flowers usually enclosed in a spathe e.g. *Anthurium*, *Musa*, and palms.
21. Spike: Elongate, unbranched, indeterminate inflorescence with sessile flowers e.g. *Polianthus*, *Adhatoda*, *Amaranthus*, *Achyranthes*.
22. Spikelet: A small spike, or the basic inflorescence unit of Cyperaceae and Gramineae e.g. Grasses, bamboo, paddy, wheat, and sugarcane.
23. Thyrse: A compact and compound panicle having an indeterminate main axis and laterally determinate axes.
24. Umbel: An inflorescence in which all the pedicels are of the same length and arise from the same point e.g. Fennel, carrot, *Cumin*, etc.
25. Umbellet: A secondary umbel in a compound umbel.
26. Verticillaster: Whorled dichasial cymes arranged at the nodes of an elongate axis e.g. Labiatae.

18.6 Flower (parts and types)

The flower is a specialized shoot of limited growth, bearing reproductive organs- microsporophylls (or stamens) and megasporophylls (or carpels) or only one, often with two accessory whorls-calyx and corolla, sometimes only one or even none at all. A typical or complete flower consists of four whorls- two lower accessory whorls- calyx and corolla, and two upper essential or reproductive whorls- androecium and gynoecium or pistil.

1. Accessory whorls: Calyx and corolla.
2. Achlamydous: Flower without calyx and corolla e.g. Betel.
3. Actinomorphic or regular: Flowers with radial symmetry; or flowers which can be divided into exactly equal halves by any vertical section passing through centre e.g. Brinjal, chilli.
4. Amorphic: Flowers without symmetry.
5. Androecium: All stamens of a flower.
6. Androgynophore: Stripe bearing both androphore and gynophores together.

7. Androphore: Elongated internodal part between corolla and androecium.
8. Anthophore: Elongated internodal part between calyx and corolla.
9. Bisexual or hermaphrodite: Flowers having both the sex organs.
10. Bract: Modified leaf which develops in the axil of the flower.
11. Bracteate: With bracts.
12. Bracteolate: With bracteole.
13. Bracteole: Small leaf borne on pedicel of flowers.
14. Calyx: Whorl or sepals.
15. Carpel: A unit of gynoecium, or floral organ that bears ovules.
16. Complete: Flowers having all the four floral whorls, i.e. sepals, petals, androecium and gynoecium.
17. Corolla: Whorl of petals.
18. Ebracteate: Without bracts.
19. Epicalyx: Leaves resembling sepals below the true calyx.
20. Epigynous: Flowers with inferior ovary; or those having the floral parts situated above the ovary.
21. Essential whorls: Androecium and gynoecium.
22. Gynoecium or pistil: Group of carpels of the flower.
23. Gynophore: Elongated axis between androecium and gynoecium.
24. Haplomorphic: Petals or tepals colored.
25. Hypanthium: Fused basal portion of sepals, petals or stamens around the ovary.
26. Hypogynous: Flowers with superior ovary; or those having the floral parts situated below the ovary.
27. Monochlamydous: Flowers with only one whorl.
28. Pedicel: The stalk of the flower.
29. Pedicellate: With pedicel.
30. Perianth: When there is no differentiation of calyx and corolla.
31. Perigynous: Flowers with half-inferior ovary; or those having floral parts situated around the ovary.
32. Petal: Individual unit of corolla.
33. Pistillate flower:Flower in which only the pistils are present, there are no stamens.
34. Pleomorphic: Actinomorphic flowers with number of reduced parts.

35. Polyphore: A receptacle having several distinct carpels.
36. Receptacle or Thalamus or Torus: Tip of the axis bearing floral appendages.It is swollen end of the flower axis on which floral leaves, viz., sepals, petals, stamens and carpels are inserted.
37. Sepal: Individual unit of calyx.
38. Sessile: Without stalk.
39. Stamen: Individual unit of androecium.
40. Tepal: Individual unit of perianth.
41. Truss: Cluster of flowers growing on one stalk.
42. Unisexual flower: Flower with only one sex, i.e. either male or female.
43. Whorl: Circular arrangement of foliage or flowers around an axis or at a node.
44. Zygomorphic or monosymmetrical: Flowers with bilateral symmetry; or flowers which can be bisected into similar halves in only one plane; or flowers with parts usually irregular and reduced in number, e.g. Gulmohor, *Cassia*, etc.

18.7 Calyx

The calyx is usually green (sepaloid), and sometimes coloured (petaloid). In its symmetry, the calyx may be regular, zygomorphic or irregular. It may again be polysepalous (sepals free) or gamosepalous (sepals united). The calyx may be modified into pappus, as in Compositae. If the calyx falls off as soon as the floral bud opens, it is said to caduceus, as in poppy. The calyx is said to be deciduous, if it falls off when the flower withers. But sometimes, when it remains adherent to the fruit, it is known as persistent.

1. Anterior lobes: Abaxial lobes; or lobes that remain away from axis.
2. Aposepalous: With separate sepals.
3. Aspalous: Without sepals.
4. Bilabiate: Consisting of two lips.
5. Caducous: Sepals which wither soon.
6. Campanulate: Bell-shaped.
7. Chorisepalous: With separate sepals.
8. Cupulate: Cup-like.
9. Dorsal side: Abaxial, or back, or lower side of the sepals.
10. Fringed: Modified margin of sepals or petals.
11. Gamosepalous: Fused sepals.

12. Infundibuliform: Funnel shaped.
13. Pappus: Reduced, scaly or hairy calyx of Asteraceae.
14. Persistent: Sepals which persist even in the fruit.
15. Petalloid: Coloured sepals, except green.
16. Polysepalous: When sepals are free.
17. Posterior lobe: Adaxial lobe; or lobe that remains next to axis.
18. Sepal: A unit of calyx.
19. Spurred: When one or more sepals are produced into spur.
20. Synsepalous: With fused sepals.
21. Tubular: Like a tube; cylindrical.
22. Urceolate: Urn-shaped.
23. Ventral side: Upper or top side of sepal or petal.

18.8 Aestivation

The mode of arrangement of the sepals or petals, more particularly the latter, in a floral bud with respect to the members of the same whorl is known as aestivation.

1. Imbricate aestivation: Out of the five sepals or petals, one external, one internal, and other three are partly external and partly internal e.g. *Cassia*, Gulmohor, *Caesalpinia*, etc.
2. Quincuncial aestivation: Out of the five sepals or petals, two external, two internal, and remaining one is partly external and partly internal.
3. Twisted or contorted aestivation: When margins of each part are overlapped regularly, i.e. one edge of the sepal or petal are overlapped by the preceding part e.g. China rose, cotton, lady's finger, etc.
4. Valvate aestivation: When the sepals or petals meet edge to edge without overlapping each other e.g. Custard apple, *Calotropis*, *Artabotrys*, etc.
5. Vexillary aestivation: Out of the five sepals or petals the posterior one is the largest and covers the two lateral sepals or petals, and the latter in turn overlap the two smallest and anterior sepals or petals, e.g. Papilionaceae.

18.9 Corolla

The corolla may also be regular or radially symmetrical, zygomorphic or bilaterally symmetrical, or irregular. Like the calyx again, the corolla may be gamopetalous or polypetalous, according as the petals are united or free. In the former case, the petals may be united partially or wholly.

18.9.1 General terms

1. Apetalous: Without petals.
2. Apopetalous: Having separate petals.
3. Carina: Keel or two fused petals of a papilionaceous flower.
4. Choripetalous: Apopetalous.
5. Gamopetalous: With fused petals.
6. Lodicule: Scale-like perianth of Gramineae.
7. Petal: An individual unit of corolla.
8. Polypetalous: With free petals.
9. Standard or vexillum: Wider petal in a papilionaceous flower.
10. Sympetalous: With fused petals.
11. Wing: Lateral petals e.g. Fabaceae.

18.9.2 Types of Corolla

1. Bilabiate or bilipped: Zygomorphic gamopetalous corolla with two unequal lips and divisions.
2. Calcarate: Spurred corolla.
3. Campanulate or bell shaped: Shape resemble to that of bell e.g. *Phasalis*, *Campanula*, *Diospyros*.
4. Carinate: Keeled.
5. Caryophyllaceous: When all the five petals of the corolla are clawed and have their limbs spreading out e.g. *Dianthus*.
6. Cruciate or cruciform: All the four petals are arranged in the form of a cross e.g. Cruciferae viz. mustard, cabbage, cauliflower, etc.
7. Hypocrateriform or salver-shaped corolla: Basal portion of corolla is narrow and tubular having abruptly expanding flat apical portion e.g. *Ixora*, *Vinca*, *Ipomoeaquamoclit*.
8. Infundibuliform or funnel-shaped: Shape like a funnel e.g. *Datura*, *Ipomoea*, railway creeper, morning glory, etc.
9. Ligulate or strap shaped: Zygomorphic, gamopetalous, strap-shaped corolla with a short narrow tube and strap-like upper portion, e.g. Sunflower.
10. Papilionaceous or butterfly-like: When five petals of the corolla are arranged in a butterfly like shape. It consists of a posterior largest *vexillum*, two lateral *alae*, and two posterior fused petals called *keel*, e.g. Papilionaceae, *Clitoria*, *Crotalaria*, etc.

11. Personate or masked: It is also a bilabiate corolla but here the corolla mouth is closed because two lips are very close to one another, e.g. *Antirrhinum, Linaria.*
12. Rosaceous: When five or more petals are spreading like those of rose e.g. Rose, tea, *Prunus*, etc.
13. Rotate or wheel-shaped: Narrow corolla tube, and the limbs of petals are at right angle to the lobe e.g. Jasmine, Night jasmine, brinjal, etc.
14. Salverform or trumpet-shaped
15. Tubular or cylindrical:Petals fused to form a tube e.g. Sunflower.
16. Urceolate or Urn-or pitcher-shaped:

Appendages of the corolla:

The corolla or the perianth is sometimes provided with outgrowths or appendages of various kinds. For instance, in snapdragon, the tube of the corolla is slightly diluted on one side like a pouch or sac. It is then said to be saccate or gibbous. In some cases, as in balsam, garden nasturtium, larkspur, orchids, etc., the perianth is prolonged into a tube known as the spur and it is then said to spurred. The contains nectar. In many flowers of Ranunculaceae, Compositae, Labiatae, Rubiaceae, etc., a special gland known as the nectar develops, containing nectar.

Sometimes, by a transverse splitting of corolla, an additional whorl may be formed at its throat. This additional whorl may be made up of lobes, scales or hairs, free or united, and is known as the corona (crown). The corona may be seen in passion flower, oleander. A beautiful cup-shaped corolla is seen in daffodil. The corona adds to the beauty of the flower and is thus an adaptation to attract insects for pollination.

18.10 Perianth

1. Gamotepalous: With fused tepals.
2. Lodicule: Scale like perianth part of Gramineae.
3. Perianth: Collective term applied for calyx and corolla.
4. Petalloid: Resembling with petals.
5. Polypetalous: With free tepals.
6. Sepalloid: Resembling with sepals.
7. Tepal: Individual unit of perianth.

18.11 Androecium

The androecium (*andros*, male) is composed of a number of stamens or microsporophylls. Each stamen consists of a filament, anther and connective . Each of the two anther-lobes has two chambers or loculi, called the pollen-sacs or microsporangia. Thus, there are altogether four loculi in each anther. But in many cases, there are only two or even one. Each chamber is filled with pollen grains. A sterile stamen (without pollen grains) is known as a staminode, as in *Dianthus*, *Pterospermum*, *Mimusops*, etc.

1. Adnate: When filament runs throughout the entire length of the anther from the base to the top.
2. Allagostemonous: Condition in which stamens are attached to petals and torus alternately.
3. Alternipetalous: Stamens present alternately with the petals.
4. Anther: Pollen-bearing portion of stamen.
5. Antipetalous: Stamens opposite to petals.
6. Antiphyllous: Stamens opposite to tepals.
7. Antisepalous: Stamens opposite to sepals.
8. Apostemonous: With separate stamens.
9. Appendicular: Stamen with a modified or protruding connective.
10. Basifixed: Filament attached with the base of the anther lobe.
11. Connective: Filament extension of the anther between thecae.
12. Diadelphous: With the filaments of stamens united or connate in two groups with their anthers being free, e.g. Pea, beans, etc.
13. Diandrous: Flowers with two stamens.
14. Didymous: With stamens in two equal parts.
15. Didynamous: When two out of the four stamens are larger and remaining two are smaller in size e.g. *Ocimum*, *Leucas*, etc..
16. Diplostemonous: When stamens are arranged in two whorls, outer whorl opposite to sepals, inner whorl opposite to petals.
17. Dorsifixed: When the filament is attached on the dorsal side of the anther lobes.
18. Epipetalous: When the stamens remain attached to the petals e.g. *Datura*, *Ixora*, potato, sunflower, etc.
19. Epiphyllous: When the stamens remain attached with the petals e.g. Liliaceae.

20. Episepalous: When the stamens remain attached with the sepals.
21. Extrorse: Stamen which dehisces longitudinally outward.
22. Filament: Stalk of the stamen.
23. Gynandrous: When stamens are united with gynoecium either wholly or by their anthers only e.g. *Calotropis*, orchids, etc.
24. Introse: Stamen which dehisces longitudinally inward.
25. Laminar: When stamens are leaflike with no distinct anther and filament.
26. Laterose: Stamen which dehisces longitudinally and laterally.
27. Locule: Chamber of an anther.
28. Monadelphous: When filaments of one group of stamens are fused but their anthers are free e.g. China rose, cotton, lady's finger, etc..
29. Monandrous: Flower with one stamen.
30. Obdiplostemonous: When the stamens are arranged in two whorls, of which the outer whorl of stamens being opposite to petals.
31. Petalloid: Stamens resembling with petals, showing no distinction between anther and filament.
32. Petalostemonous: With filaments of stamens fused to petals with free anthers.
33. Phaenantherous: With exerted stamens.
34. Polyadelphous: When filaments are united in several groups with their anthers remaining free e.g. *Bombax*, Lemon, Castor, etc.
35. Polyandrous: With several free stamens.
36. Poricidal: Dehiscing through a pore at the theca apex.
37. Stamen: Pollen-bearing floral organ of angiosperms.
38. Staminode: Sterile stamen with reduced anthers.
39. Synandrous: With anthers as well as their filaments being united in the form of one group e.g. Cucumber, ash gourd, *Colocasia*, etc.
40. Syngenesious: With fused anthers and free filaments e.g. Sunflower, marigold, etc.
41. Tetrandrous: Flowers with four stamens.
42. Tetradynamous: Two out of a total of six stamens are short while the remaining four are long e.g. mustard, rape, radish, etc.
43. Theca: Half of the portion of anther containing two pollen sacs.
44. Triandrous: Flowers with three stamens.

45. Tridynamous: Stamens arranged in two equal groups of three.
46. Valvular: When the stamen dehisces through a pore covered by a flat of tissue.
47. Versatile: Dorsifixed but it appears as if the anther is swinging freely on the filament.

18.12 Gynoecium

The gynoecium or pistil is composed of one or more carpels or megasporophylls which bear female spores or megaspores. The pistil may be simple or compound. In a compound pistil the carpels may be free, as in lotus, *Michelia*, rose, *Magnolia*, *Artabotrys*, etc.

18.12.1 Parts, Fusion Types, Carpel Parts And Types, Style Types and Ovary Position

1. Apocarpous: With free or separate carpels.
2. Astylocarpellous: Without a stipe and a style.
3. Astylocarpepodic: With a stipe and without a style.
4. Astylous: Without any distinct style.
5. Carpel: Organ of the flower that bears ovules.
6. Epigynous: The condition in which the sepals, petals and androecium are attached to the floral tube above the ovary e.g. ovary is inferior and rest of floral members, superior e.g. sunflower, guava, gourds, apple, pear, etc.
7. Funiculus: The stalk with which the ovule is attached to the placenta.
8. Geniculate style: The style which bents abruptly.
9. Gynobasic style: The style which is attached at the base of the ovary in central depression.
10. Heterostylous: With styles of different shapes or lengths.
11. Homostylous: With styles of same shapes or lengths.
12. Hypogynous: The condition in which sepals, petals and androecium are attached to the floral tube below the ovary i.e. ovary occupies the highest position on the thalamus. Ovary is said to be superior and the rest of the floral members inferior e.g. China rose, Magnolia, brinjal, mustard, etc.
13. Inferior ovary: The ovary in the epigynous condition.
14. Locule: The cavity of the ovary.

15. Monocarpellary to polycarpellary: With one to several carpels; bicarpellary if two carpels are present, tricarpellary if three carpels are present, and tetra and pentacarpellary, if five carpels are present.
16. Ovary: Ovule-bearing part of the gynoecium.
17. Ovule: Young seed made up of nucellus and integuments.
18. Parts of a carpel: Funiculus, locule, ovary, ovule, placenta, stigma, style.
19. Parts of gynoecium: Carpel, locule, ovary, ovule, placenta, stigma, style.
20. Parts of ovule: Chalaza, embryo sac, integuments, micropyle, nucellus, raphe.
21. Perigynous: The condition in which the sepals, petals and androecium are attached to the floral tube around the ovary i.e. ovary is said to be half-inferior. In some perigynous flowers, the ovary may be partially sunken in the thalamus e.g. rose, primrose, *Prunus*, *Lagerstroemia*, etc.
22. Pistillode: A sterile carpel.
23. Placenta: Ovule-bearing region of the wall of ovary.
24. Raphe: Part of an ovule present in the form of a longitudinal ridge on its outer integument.
25. Replum: False septum.
26. Semicarpous: With fused ovaries of adjacent carpels and their free styles and stigmas.
27. Semi-inferior: The position of ovary in the perigynous condition.
28. Stigma: Uppermost, pollen-receptive organ of gynoecium.
29. Stipe: Stalk of the gynoecium.
30. Style: Long portion of gynoecium between ovary and stigma.
31. Stylocarpellous: Carpel without a stipe and with a style.
32. Stylocarpepodic: With a stipe and a style.
33. Stylodious: With single free carpel.
34. Superior: The position of ovary in the hypogynous condition.
35. Syncarpous: With united or fused carpels throughout their whole length e.g. *Dianthus*, *Plumbago*, *Vinca*, *Nerium*, etc.
36. Synovarious: The condition when the ovaries of adjacent carpels are fused while their styles and stigmas separate.
37. Synstylovarious: The condition when the ovaries and styles of adjacent carpels are fused while their stigmas separate.
38. Terete: Cylindrical and elongate.

39. Unicarpellous: Stylodious.
40. Unilocular to multilocular: Ovary with one chamber or locule is called unilocular, with two chambers is called 'bilocular', with three and four chambers are called trilocular and tetralocular, and with many locules is called 'multilocular'.

18.12.2. Types of stigma

1. Capitate: Like a cap or head.
2. Clavate: Club-shaped.
3. Crested: Possessing a terminal tuft or ridge.
4. Decurrent: Long and extending downward.
5. Diffuse: Spread over a wide surface.
6. Discoid: Like a disc.
7. Fimbriate: Fringed.
8. Lineate: In the form of small lines.
9. Lobed: Divided into some lobes.
10. Plumose: Like a feather.
11. Terete: Elongate and cylindrical.

18.12.3. Ovules

Each ovule is attached to the placenta by a slender stalk known as funicle. The point of attachment of the body of ovule to its stalk or funicle is known as the hilum.

1. Raphe: The funicle continues beyond the hilum alongside the body of ovule, forms a sort of ridge.
2. Chalaza: The upper end of raphe, which is the junction of integuments and nucellus.
3. Nucellus: Main body of the ovule.
4. Integuments: Two coats surrounding the ovule body.
5. Micropyle: A small opening present at the apex of integuments.
6. Embryo sac: Ovary cell lying embedded in the nucellus towards the micropylar end. This sac bears embryo and is most important part of ovule.

18.12.4. Forms of ovules

1. Orthotropous or straight: When the ovule is erect or straight so that funicle, chalaza and micropyle lie on one and the same vertical line e.g. Polygonaceae, Piperaceae, Casuarinaceae.
2. Anatropous or inverted: When the ovule bends along the funicle so that the micropyle lies close to the hilum. The micropyle, chalaza, but not the funicle, lie on the same straight line. This is commonest form of ovule.
3. Amphitropous or transverse: When the ovule is placed transversely at a right angle to its stalk or funicle e.g. Lemna.
4. Campylotropous or curved: When the transverse ovule is bent round like a horse-shoe so that the micropyle and chalaza do not lie on the same straight line e.g. Cruciferae, Graminae.

18.13 Placentation

The placenta is a ridge of tissue- a parenchymatous outgrowth – in the inner wall of the ovary to which the ovule or ovules remain attached. The placentae most frequently develop on the margins of carpels, either along their whole line of union, called the suture, or at their base or apex. The manner in which the placentae are distributed in the cavity of ovary is known as placentation.

1. Axile: The placenta develops along the central axis in a compound ovary with septa. It occurs in a bilocular to multilocular ovary e.g. China rose, lady's finger, potato, tomato, etc.
2. Basal: The placenta develops at the base of the ovary, i.e. directly on the thalamus. It occurs in a unilocular ovary e.g. Compositae.
3. Free-central: The placenta develops along the central axis in a compound ovary without septa. It occurs in unilocular ovary e.g. *Dianthus*, soapwort, etc.
4. Laminar or Laminate or Superficial: The placenta develops over the inner surface of the ovary wall. It occurs in a multicarpellary ovary e.g. water lily.
5. Marginal: The placenta develops along the margin of the simple ovary. It occurs in monocarpellary and unilocular ovary e.g. Leguminosae family.
6. Parietal: The placenta develops on the wall or intruding portions of a compound ovary. It occurs in bicarpellary to multicarpellarybut unilocular ovary e.g. Papaya, poppy, orchids, Cruciferae, etc.
7. Pendulous or Apical: The placenta develops at the top of the ovary and the ovules remain suspended in a pendulous manner.

18.14 Fruit

1. Achene: Developing from a monocarpellary superior ovary, the achene is a one-seeded, dry, indehiscent, simple fruit having its seed attached to fruit wall only at one point, e.g. *Mirabilis*, *Clematis*.
2. Amphisarca: A fleshy simple, berry-like fruit with a woody rind, e.g. *Lagenaria*.
3. Balusta: A simple, dry, many-seeded and many-loculed indehiscent fruit with a tough pericarp, e.g. *Punica*.
4. Berry: Fleshy simple fruit with succulent pericarp, e.g. banana, grape.
5. Bibacca: A multiple type of fruit formed as a fused double berry, e.g. *Lonicera*.
6. Carcervulus: A dry schizocarp fruit which develops from bi to multicarpellary, syncarpous, superior ovary with many locules, e.g. Labiatae.
7. Calybium: A hard, unichambered dry indehiscent fruit derived from an inferior ovary, e.g. *Quercus*.
8. Capsule: A dry dehiscent fruit derived from a bi-to multicarpellary superior or inferior ovary, e.g. *Papaver*.
9. Caryopsis: A dry indehiscent fruit derived from a monocarpellary ovary in which pericarp is inseparably fused with testa, e.g. Graminae.
10. Cremocarp: A dry schizocarpic fruit derived from bicarpellary, syncarpous, inferior ovary with one ovule in each locule. It dehisces from two indehiscent, single-seeded maricarps attached with carpophore, e.g. Umbelliferae.
11. Cypsela: A dry indehiscent achene type of fruit derived from unilocular, inferior ovary, e.g. *Helianthus*.
12. Drupe: A fleshy fruit with a stony endocarp and edible mesocarp, e.g. Mango, *Prunus*.
13. Druplet: A small drupe, e.g. *Rubus*.
14. Etaerio of achenes: Also called achenecetum. It is an aggregation of achenes, e.g. Ranunculus.
15. Etaerio of berries: Also called baccacetum, it is an aggregation of berries, e.g. *Ariabotrys*.
16. Etaerio of drupes: Also called drupecetum, it is an aggregation of drupes, e.g. *Rubus*.

17. Etaerio of follicles: Also called follicetum, it is an aggregation of follicles, e.g. *Delphinium*.
18. Etaerio of samaras: Also called samaracetum, it is an aggregation of samaras e.g. *Liriodendron*.
19. Follicle: A dry, dehiscent fruit derived from a monocarpellary, superior ovary, and dehisces only along one suture e.g. *Delphinium* (Fig.18.57B).
20. Hesperidium: A fleshy, thick-skinned berry with hard and leathery pericarp. It is derived from a polycarpellary, multilocular, superior ovary. Bulk of this fruit is derived from glandular hairs, e.g. *Citrus* (Fig.18.60F).
21. Lomentum: A dry dehiscent legume that separates transversely between seed sections, e.g. *Acacia*.
22. Nut: A dry, indehiscent, one-seeded fruit with a hard pericarp. It is derived from unilocular ovary, e.g. *Anacardium*.
23. Pepo: A berry with a leathery nonseptate rind, developing from tricarpellary, syncarpous, inferior ovary with parietal placentation, e.g. Cucurbitaceae.
24. Pome: A fleshy fruit surrounded by the fleshy thalamus and developing from a two-or-more-celled, syncarpous, inferior ovary, e.g. Apple.
25. Pyrene: Fleshy fruit with each seed covered by a bony endocarp, e.g. *Ilex*.
26. Regma: Dry, shizocarpic fruit derived from tricarpellary, syncarpous, superior, trilocular ovary and bears spinous tubercles, e.g. *Ricinus*.
27. Samara: Dry, indehiscent simple fruit derived from bicarpellary, syncarpous ovary. Its pericarp is winged, e.g. *Elm*.
28. Silicula: Dry, dehiscent fruit derived from two or more carpels. It dehisces along two sutures and leaves a persistent partition after dehiscence. It is as broad as, or even broader, than long, e.g. *Iberis*.
29. Siliqua: Dry, dehiscent fruit derived from bicarpellary, syncarpous ovary with parietal placentation and a false septum. It is longer than board, e.g. *Brassica*.
30. Sorosis: A multiple fruit developing from the spadix or spike in which the flowers usually fuse by their succulent sepals, and axis bearing them becomes woody or fleshy forming a compact mass, e.g. Pineapple.
31. Syconus: A multiple fruit derived from hypanthodium type of inflorescence. Here the achenes develop on the inside of a hollowed-out fleshy receptacle e.g. *Ficus*.
32. Utricle: Dry, indehiscent, small, bladder-like, one-seeded fruit e.g. *Chenopodium*.

References

Nageli, C.W. Von. 1883. Mechanisch-physiologischeTheorie der Abstammungslehre, A Felix, Munchen. Pilger, R.K.F.1922. *Bibliotheca Botanica*, **23**: 1-38.

Goebel K 1931. *Blütenbildung und Sproßgestaltung* (*Anthokladien und Infloreszenzen*). 2 Suppl Jena Fischer Organographie der Pflanzen.

Parkin, J. 1914. *Botanical Journal of The Linnaean Society,* **42** :511-563.

Rickett, H.W. 1944. *Botanical Review*, **10**: 187-231.

19

Intellectual Property Rights Related Issues In Ornamental Plants

Agriculture including forestry, range management and sustenance of plant biodiversity have been the core of Indian civilization that dates back to several millenniums. The enormous plant diversity arising due to the varied soil, environment and altitude variations that India enjoys has resulted in the evolution of various crop species and varieties suiting to different domestic and commercial needs. The Gene Fund created under the law is designed to encourage, protect, promote, and utilize the plant biodiversity for plant breeding activity in perpetuity. The Gene Fund in turn gets nourished by the flow of annual fee arising out of the sale of seed of the registered varieties. The keepers (services of the men and women; individuals and communities) of plant variation in the agrobiodiversity hotspots are to be beneficiaries of the gene fund. National Gene Fund shall be used for supporting and rewarding farmers and farming communities, who are engaged in conservation and improvement of genetic resources of economic plants and their wild relatives, particularly from the areas identified as hotspots of agrobiodiversity.

An understanding of intellectual property rights for plant breeders has become increasingly important over the last few decades. It is also a confusing topic to many in the industry. Confusion is understandable when we are presented with various types of protection, many acronyms, terms used interchangeably and incorrectly, regulations that vary internationally and a lack of understanding or consistent interpretations of laws that govern these rights.

Plants are unique from other marketed products due to their complex nomenclature, ease of propagation and the many levels of distribution and promotion involved in their marketing. The protection is required for a plant cultivar needs to be considered in relation to all of these factors in order to be correctly implemented and maintained. Breeding and commercialization of new ornamental plant varieties is an economically important activity both for the industry and the well being of our society. Breeding or developing a new variety requires a long time. In horticultural crops, it takes 3-10 years from initial crossing of parent cultivars to develop a commercial variety. It is important to encourage development of new variety by providing incentives such as Plant Breeders' Rights (PBR) over exclusive production and marketing of seed of an improved variety. When the new cultivar is introduced and commercialized, plant breeders finally begin to get paid for their years of hard work in the form of royalties. Legal protection for new plant varieties provides the means by which companies, institutions, and independent plant breeders can receive a return on their investments of time and money as well as justify their involvement in research and new plant development. In addition, legal protection enables plant breeders to be recognized for their craft and skill and encourages plant breeding as a profession. Plant variety protection provides an incentive for continued breeding and to ensure the availability of new varieties that meet future demands and keep the industry thriving.

19.1 Global initiatives

Important initiatives taken by inter-governmental organizations, conglomerates of breeders resulted in the establishment of dedicated organizations to address the issues related to plant variety protection.

The USA had passed the "Plant Patent Act" 1930 (US) to offer protection to plants. Plant patents were limited to vegetatively propagated plants (except potato and Jerusalem artichoke) and thus benefited only breeders of fruit and ornamental plants. Forty years later, in 1970, the USA passed the Plant Variety Protection Act 1970 (US) for sexually produced plants and tuber-propagated plants (Aguirre, 2007).

Europe, struggled for years to find a method to protect plant breeders' rights. Attempts were made to modify laws and use trade organizations, however, these systems failed to offer any secure, satisfactory results (Heitz, 1999). To overcome these difficulties, ASSINSEL, the Association of Plant Breeders for the Protection of Plant Varieties convened between 1957 and 1961 to develop a unified system for plant breeders' rights. Negotiations led to the adoption of the International Convention for the Protection of New Varieties of Plants, "The UPOV Convention" on December 2, 1961. Minor revisions were made in 1972 (the 1972 Act) and major revisions were made in 1978 (the 1978 Act) and 1991 (the 1991 Act).

Globally organizations like Community Plant Variety Office (CPOV), International Association of Breeders of Asexually Reproduced Ornamental and Fruit Varieties (CIOPORA), International Union for the Protection of New Plant Varieties (UPOV) and Plant Patent Office offer protection to new varieties evolved in all crops. It is a fact that ornamental crops dominate in number of plant breeder's rights awarded in different IPR regimes like UPOV, CPVO and USPTO.

19.1.1 Community Plant Variety Office (CPOV)

The CPOV is an agency of the European Union, located in Angers, France. Its function is to administer a system of plant variety rights. It works like the office for harmonization in internal market, it grants plant breeders' rights in more than fifteen European countries and validity of rights is for 25-30 years. Ornamental crops remains the most important in terms of applications filed each year. Of the ornamental crops, rose, *Chrysanthemum*, *Petunia*, *Pelargonium*, *Lilium*, *Impatiens*, *Gerbera*, *Phalaenopsis*, *Dianthus* and *Osteospermum* are top 10 crops that are protected over the years.

19.1.2 International Association of Breeders of Asexually Reproduced Ornamental and Fruit Varieties (CIOPORA)

The CIOPORA is an international non-governmental organization, representing the interests of breeders of asexually-reproduced ornamental and fruit-tree varieties worldwide. Top priority of CIOPORA is constant development of systems of protection which both international state organizations and single states to provide protection of intellectual property concerning ornamental and fruit-tree plants. It has 115 members of plant breeding companies and individuals involved in breeding.

19.1.3 International Union for the Protection of New Plant Varieties (UPOV)

The UPOV is an intergovernmental organization with headquarters in Geneva, Switzerland. The acronym, UPOV, is derived from the French name of the organization, *Union internationale pour la protection des obtentions végétales*, as the convention was signed in Paris in 1961. UPOV, although a separate intergovernmental organization, works closely with the World Intellectual Property organization (WIPO). The UPOV is undoubtedly the most influential organization in determining future directions and regulations for plant protection worldwide. At present, there are 71 member countries of UPOV. The purpose of International Convention for the Protection of New Varieties of Plants is to ensure that the member states party to the Convention acknowledges the achievements of breeders of new plant varieties by making available to them an exclusive property right, on the basis of a set of uniform and clearly defined principles. The UPOV system advocates the testing of a candidate variety in terms of novelty, distinctness, uniformity and stability in order to offer the plant breeders' rights. The candidate variety is entitled to get protection if it can be distinguished by at least one essential character from any other existing variety. In order to conduct the DUS tests, UPOV has developed guidelines in 246 crops out of which 46 belong to fruits, 51 to vegetables, 74 to ornamental crops (Table 19.1), 8 to medicinal and aromatic crops and 4 to spices (Prasad and Janakiram, 2013).

Table 19.1: DUS test guidelines available with UPOV

Category	Test guidelines available for
Ornamental crops	Rose, African violet, Elatior Begonia, Poplar, Poinsettia, Carnation (vegetatively propagated varieties), Chrysanthemum, Freesia (vegetatively propagated varieties), Zonal pelargonium, Alstromeria, Kentucky Bluegrass, Meadow Fescue, Tall Fescue, Rhododendron, Streptocarpus, Lily, Lupins, Willow (tree varieties only), Gerbera, Kalanchoe, White Cedar, Sunflower, Anthurium, Lagerstroemeria, Norway Spruce (ornamental varieites), Juniper, *Euphorbia fulgens*, Narcissi (including daffodil), Christmas Cactus, Tuberous begonia hybrids, Lachenalia, Gladiolus, Easter Cactus, Regal Pelargonium, Exacum, Tulip, Protea, Leucospermum, Red Fescue, Sheep's Fescue, Hair Fescue, Reliant Hard Fescue, Shade Fescue, Pseudovina, Statice, Dieffenbachia, Hydrangea, Spathiphyllum, Pot Azalea, Aster, Evening primrose, Zantedeschia, Firelily, Rescue Grass, Alaska Bromegrass, Bromus Auleticus, Cymbidium, Iris (bulbous), Kangaroo Paw, Amarayllis, Celosia, Ornamental apple, Lotus, Trefoil, Lavender, New Guinea Impatiens, Strawflower, Everlasting Daisy, Dendrobium, Petunia, Phalaenopsis, Catharanthus, Clematis, Verbena, Antirrhinum, Argyranthemum, Brachycome, Wax flower, Dahlia, Nemesia, Portulaca, Garden Rocket, Marigold.

Source: http://www.upov.int/en/publications/tg_rom/tg_idex.html

19.1.3.1 Plant Patents

The USA offers protection for new cultivars of plants, plant phenotypes and plant propagation methods. The US patent system offers protection usually for a period of 17 or 20 years. After the expiry of the period, the variety becomes free variety and can be asexually reproduced, used and/or sold without permission from the inventor(s). To be eligible for the protection, the inventor(s) must demonstrate that-

- The cultivar is new and distinctly different from its antecedents and known related cultivars in at least one distinguishing characteristics. The distinguishing characteristics might be plant size, flower colour, flower form or leaf colour, etc.
- The inventor(s) must declare that they are the original and sole inventor(s) of new cultivar.
- The new cultivar has been discovered in a cultivation area, like greenhouse or nursery except wild species/ relatives or species that freely occur "in nature".

- The inventor(s) must declare that new cultivar is reproduced asexually and that unique characteristics of new cultivar are uniform and stable over generations of asexual reproduction.
- Must include a visual details with botanical description, coloured photographs, DNA fingerprinting, differentiating from its antecedents and known related cultivars; location and method by which new cultivar has been reproduced.

Propagation of a patented plant for any purpose without permission from the owner is illegal and an infringement of the plant patent owner's rights.

19.1.3.2 Examination of variety

Protection of a new variety can only be granted after examination of the variety shows that it complies with the requirements for protection. An examination of a new plant variety should establish that it is distinct, unique and stable and novel. A acronym used to describe is 'DUS'. Globally, guidelines have been developed for 160 genera and species and are published in separate versions in English, French, German and Spanish. A criterion for granting protection to a new variety has been summarized as-

Novelty	The criteria of 'novelty' replaces that of 'non-obvious' which the requirement for protection of new invention. It is used in the sense that new variety should not have been commercially exploited for more than a year before granting protection.
Distinctiveness	The new variety must possess a trait that makes it distinguishable from other known varieties by one or more identifiable morphological, physiological, or other characteristics.
Uniformity	The new variety must be uniform in appearance under specified environment of its adaptation or stable in its essential characteristics.
Stability	The new variety must be stable in appearance and its claimed characteristics over successive generations under specified environment.
Denomination	It is the name of new variety that is chosen by the breeder. The variety must be given a denominations for identification; the denomination must not be liable to mislead or to cause confusion as to the characteristics, value or identity of new variety and/or identity of the breeder

19.1.3.3 Plant breeders' rights vs. U.S. plant patents

Plant breeder's rights (PBR), plant variety rights (PVR), plant variety protection (PVP) are all titles used by various States for plant variety protection. They are, however, commonly referred to as PBR. PBR is not used to describe any protection in the U.S. however, when PVP is used to refer to protection of seed and tuber propagated plants and plant patents indicate protection of asexually reproduced plants. PBR requires an examination of the variety whereas a plant patent application requires a detailed botanical description and the USPTO relies on the information provided by the applicant and the knowledge of the examiners and/or the United State Department of Agriculture (USDA) for determination of DUS and novelty. In this regard, PBR can be more difficult to obtain, especially if any reversion or instability is observed. Another difference is that PBR applications are typically published, or at least a notice of the application is posted, shortly after filing, whereas a plant patent is not published until 18 months after filing. The cost of PBR varies greatly depending on the State, but typically an annual maintenance fee is required. No such fee is required for a plant patent.

19.1.3.4 Trademarks

A plant may be produced by many different growers, marketed through a complex chain of distribution and then promoted, written about, and labeled by many different parties with many different agendas. The most unique feature of plants is their nomenclature. The fact that plant has a cultivar name (fancy or nonsensical), a common name, a scientific name, and possibly a selling name or trademark associated with it, adds a dimension that does not exist in other industries. Trademarks can be of great benefit to a plant breeder. They may be used to heighten the reputation and exposure of a breeder and enhance the marketing of a breeder's plants. A trade mark is linked to the name of the product. It is possible to obtain this right per country. The term trademark is defined as any word, phrase, symbol, or device, or any combination thereof that is used to identify and distinguish one's goods or products from those of others and indicates the *source* of such goods or products. Quite simply, *a trademark is a brand.* Examples of effective trademarks in the industry include: Flower Fields®, Proven Winners®, and My Favorite®, etc.

19.1.3.5 Licensing and royalties

License agreements are used to extend rights of use for patents, trademarks, or plant breeders' rights to another party. The construction and content of a license agreement is dependent on the laws governing each. A licensor may grant an exclusive license to one party or license many parties on a non--exclusive basis. In all cases, the owner of the rights retains all ownership. If ownership is to be transferred, then an assignment of rights is needed. In many countries, licenses as well as assignments require registration of user agreements. A license may require a lump--sum payment, periodic royalty payments based on sales, or both. The owner of a variety issues license agreements to specify the terms and conditions for use of the protected variety.

Royalty payments are typically accessed on a per-plant-sold or other unit basis (unrooted, rooted propagules) basis. A patent owner will need to determine a fair royalty amount or analyze an offer from a prospective licensee.

There are several factors that should be considered when setting a royalty rate: (i) typical royalty rate for the class of plant (e.g. annual vs. an herbaceous or woody perennial plant), (ii) the ease of propagation (e.g. tissue culture vs. stem cuttings), (iii) the prospective number of plants that can be produced and sold, (iv) market price, (v) the type of market (e.g. specialty retail vs. large outlet), (vi) uniqueness of plant (vii) services provided (e.g. marketing support provided by licensee), and (viii) geographical market (e.g. Europe vs. U.S.)

19.1.3.6 Life time

In the past ornamental varieties has protection period of 20 years or more, like for example 'Baccara, Sonia and Motrea' in roses, 'Sim's ' in carnation, 'Spider' in chrysanthemums. Now-a-days, the life time of varieties is becoming much shorter. In general, growers do not replant the old varieties, but will change to new ones. The main reason is the progress with new varieties and their better price compared to old varieties. Consumers are prepared to pay extra for these new varieties.

19.2 Indian Initiatives

The Indian initiatives to offer protection to the intellectual property mainly pertains to the Protection of Plant Varieties and Farmers' Rights Act 2001 and Geographical Indications which was acted in October and PPV& FRA rules were notified in 2003. Consequently, the PPV & FR was established on 11, November, 2005 by Government of India, with the following objectives:

1. To provide an effective system for protection of plant varieties and rights of farmers and plant breeders.
2. To recognize the farmers in respect of their contribution made in conserving, improving and making available plant genetic resources for development of new plant varieties.
3. To protect plant breeders' rights to stimulate investment for R&D and develop new varieties.
4. To facilitate the growth of seed industry to ensure production and availability of high quality seed/ planting material.

19.2.1 Protection of Plant Varieties and Farmers' Rights Act 2001

After India became signatory to the Trade Related Aspects of Intellectual Property Rights Agreement (TRIPS) in 1994, a legislation was required to be formulated to address the issue of plant breeders rights. The provision of this agreement requires member countries to provide for protection of plant varieties either by a patent or by an effective *sui generis* i.e. system of its own (drawing its essence from Union for Protection of (New) Plant Varieties, UPOV) system or by any combination thereof. Thus member countries have the choice to frame legislation that suits their own system. The existing Indian Patent Act, 1970 excluded agriculture and horticultural methods of production from patentability. The *sui generis* system for protection of plant varieties has integrated the rights of breeders, farmers and village communities and taking care of the concerns for equitable sharing of benefits. The act covers all categories of plants, except microorganisms. The variety which is to be protected must be new, distinct, uniform and stable. The protection offers for a period of 18 years for trees and vines; 15 years for other plants and extant varieties. The Government of India identified 35 crops to take up DUS testing in order to develop national test guidelines. Of the 35 crops, 8 belong to vegetables and two belong to ornamental crops. The PPV& FRA has initiated the 2nd phase of developing the test guidelines for other important crops (Prasad and Janakiram, 2013).

19.2.1.1 DUS Test Centres

Authority is maintaining and funding 42 DUS test Centres for different crops with a mandate of maintaining and multiplication of reference varieties/ example varieties and generation of database for DUS descriptors as per DUS guidelines of respective crops. The list of DUS test Centres for ornamental crops is available on the official website of the authority (Rana, 1995).

DUS testing centers in India

Crop	Centres
Chrysanthemum & rose	IARI, New Delhi, IIHR, Bengaluru, NBRI, Lucknow
Damask rose & periwinkle	CIMAP, Lucknow
Orchids (*Cymbidium*, *Dendrobium* and *Vanda*)	NRC, Orchids, Sikkim; SFRI, Itanagar & KAU, Thrissur
Marigold	IARI, New Delhi
Bougainvillea	IARI, New Delhi, IIHR, Bengaluru, NBRI, Lucknow
Tuberose	DFR, Pune & IIHR, Bengaluru
Gladiolus	DFR, Pune & IIHR, Bengaluru
Jasmine	IIHR, Bengaluru & TNAU, Coimbatore
China aster	MPKV, Rahuri & IIHR, Bengaluru

19.2.1.2 Registration

Section 29.2 of the Act provides that Genera and species for the purpose of registration of varieties will be notified by Central Government in official Gazettes. So far, 45 crop varieties have been notified for the purpose of registration. For these crop species, PPV&FR Authority has developed "Guidelines for the Conduct of Species Specific Distinctiveness, Uniformity and Stability," tests or "Specific Guidelines". The purpose of these Specific Guidelines is to provide detailed practical guidance for the harmonized examination of DUS and, in particular, to identify appropriate characteristics for the examination of DUS and production of harmonized variety descriptions. The notified flower species include rose and chrysanthemum. The DUS test guidelines of crop species *viz.*, Bougainvillea, China aster, Chrysanthemum, Rose, Damask Rose, Gladiolus, Jasmine, Periwinkle, Tuberose have been finalized.

19.2.1.3 Definitions related to registration

As per the PPV&FR Act, 2001, the registration can be granted for different kinds of varieties like New, Extant, EDVs etc. The relevant definitions related to registration are as under:

a. **Variety**: "Variety" means a plant grouping except micro-organism within a single botanical taxon of the lowest known rank, which can be-
   (i) defined by the expression of the characteristics resulting from a given genotype of that plant grouping;
   (ii) distinguished from any other plant grouping by expression of at least one of the said characteristics; and
   (iii) considered as a unit with regard to its suitability for being propagated, which remains unchanged after such propagation, and includes propagating material of such variety, extant variety, transgenic variety, farmers' variety and essentially derived variety.

b. **Farmer**: "farmer" means any person who-
   (i) cultivate crops by cultivating the land himself; or
   (ii) cultivates crops by directly supervising the cultivation or land through any other person; Or
   (iii) conserves and preserves, severally or jointly, with any other person any wild species or traditional varieties or adds value to such wild species or traditional varieties through selection and identification of their useful properties;

c. **Farmers' Variety**: "farmers' variety" means a variety which -
   (i) has been traditionally cultivated and evolved by the farmers in their fields; or
   (ii) is a wild relative or land race or a variety about which the farmers possess the common knowledge.

d. **Cultivar**: means a variety of plant variety of plant that has originated and persisted under cultivation or was specially bred for the purpose of cultivation.

e. **Folk variety**: means a cultivated variety of plant that was developed, grown and exchanged informally among farmers.

f. **Landrace:** means primitive cultivar that was grown by ancient farmers and their successors.

g. **Extant Variety**: means a variety available in India which is-

 * notified under section 5 of the Seeds Act, 1966 (54 of 1966); or

 * farmers' variety; or

 * a variety about which there is common knowledge; or

 * any other variety which is in public domain.

h. **EDV**: "essentially derived variety", in respect of a variety (the initial variety), shall be said to be essentially derived from such initial variety when it –

 (i) is predominantly derived from such initial variety, or from a variety that itself is predominantly derived from initial variety, while retaining the expression of the essential characteristics that results from the genotype or combination of genotypes of such initial variety;

 (ii) is clearly distinguishable from such initial variety; and

 (iii) conforms (except for the differences which result from the act of derivation) to such initial variety in the expression of the essential characteristics that result from the genotype or combination of genotype of such initial variety.

 The breeder of the essentially derived variety shall have the same rights as the plant breeder of other new varieties. Activities that exempt from Breeders' Rights include: (i) acts done privately and for non-commercial purposes (ii) acts done for experimental purposes and (iii) acts done for purpose of breeding and exploiting other varieties, provided they are not essentially derived.

i. **Variety of Common Knowledge**: Under section 20(1) read with section 2(j)(iii) of PPV & FR Act, 2001 that the Registry will process the applications for registration of an extant variety about which there is common knowledge (Section 2(j) (iii) of PPV & FR Act, 2001) if the following conditions are met:

 A1. If a variety which is not released and notified under the Seeds Act, 1966 but is well documented through publications and is capable of satisfying the definition of 'variety', or

A2. The candidate variety should either have an entry in any official register of varieties or in the course of being made, or

A3. The candidate variety should find inclusion in a reference collection or is having a precise description in a publication, or

A4. By any other means a variety has become a matter of common knowledge and

B. The variety is under cultivation or marketing during the time of filing of application for registration.

C. The true representative seed of the variety should be available at the time of filing of application.

D. A candidate variety should have been sold or otherwise disposed of in India one year prior to the date of filing of the application and it should not have been sold or otherwise disposed of 13 years prior to the date of filing of application and in case of trees and vines it should not have been sold or otherwise disposed of 16 years prior to the date of filing of application.

j. **Essential Characteristics**: "essential characteristics" means such heritable traits of a plant variety which are determined by the expression of one or more genes of other heritable determinants that contribute to the principal features, performance or value of the plant variety.

19.2.1.4 Conditions of protection by PBR

i. Production for commercial purposes, offering for sale and selling propagating material becomes the exclusive right of the holder of the PBR title.

ii. Production, sale etc. of material for consumption or industrial use does not fall under the protection.

iii. The protection is limited to the propagating material produced for commercial purposes, which means that a grower is allowed to reserve a portion of his harvest for use as seed for his own next crop without the permission of the holder of the right. This is called farmers' exemption but a farmer who does this is not allowed to sell such material to any one else.

iv. Exchange of similar material of different cultivars between farmers is illegal.

v. The minimum period of protection prescribed by the UPOV convention is 15 years, with 18 years for grapes, forest and ornamental trees.

vi. Protection is not granted for an unlimited period.

vii. The use of propagating material from a protected cultivar for scientific purposes is not dependent on permission of the holder of the right. Breeders are, however, unhappy with this.

viii. The use of protected cultivar for the creation of genetic variability for plant breeding purposes is not dependent on permission of the holder of the right.

This free accessibility of the protected cultivars as under (VIII) above is generally appreciated by the breeders. It has worked as stimulus to plant breeding. It has allowed movement of genes from breeder to breeder. The breeding community as a whole has benefited from free gene traffic. There is one exception to the rule of free accessibility of protected varieties for breeding purposes. That relates to parental cultivars being used in crosses to produce propagating material of a hybrid variety.

19.2.1.4.1 Rights of farmers

Under section 39 of the PPV&FR Act, 2001 farmers' rights have been given. These are:

* A new variety bred/developed by farmer will be registered in like manner as a breeder of a variety.

* Farmer engaged in conservation of genetic resources of land races and wild relatives of economic plants, improvement through selection and preservation shall be entitled for recognition and reward from Gene Fund provided that the said material has been used as donors of genes in varieties registrable under this Act.

* Farmers can save, use, sow, re-sow, exchange, share or sell farm produce including seed of a protected variety but cannot sell branded seed of a variety protected under this Act.

* Breeder shall disclose expected performance of the protected variety under given conditions and farmers can claim compensation if variety fails.

* Sec 42 refers to protection against innocent infringement

* Sec 43 refers to authorization required for EDV developed from farmers' variety

* Sec 44 provides that farmers are exempted for payment of any fee for any proceeding/ application/ DUS Test.

19.2.1.4.2 Researchers' rights

Section 30 states that nothing contained in this Act shall prevent-

(a) The use of any variety registered under this Act by any person using such variety for conducting experiment or research i.e. bona fide research purposes.

(b) The use of a variety as an initial source of variety for the purpose of creating of other varieties:

Provided that the authorization of the breeder of a registered variety is required where the repeated use of such variety as a parental line is necessary for commercial production of such other newly developed variety.

19.2.1.4.3 Breeders' rights

(a) Breeder shall be required to deposit seeds or propagating material including parental line seeds of registered varieties in the National Gene Bank (Sec 27).

(b) A certificate of registration for a variety issued under this Act shall confer an exclusive right on the breeder or his successor, his agent or licensee, to produce, sell, market, distribute, import or export the variety [Sec 28(1)].

(c) Breeders can appoint agent/licensee and can opt for civil remedy in case of infringement of his rights.

19.2.1.4.4 Communities' rights

(a) Section 41 provides for settlement of any claim attributable to the contribution of the people of that village or local community, in the evolution of any variety registered under this Act.

(b) Compensation to be determined by PPV&FRA and deposited in Gene Fund.

19.2.2 Compulsory Licensing (Sec 47)

The authority can grant compulsory license, in case of any complaints about the availability of the seeds of any registered variety to pubic at a reasonable price. The license can be granted to any person interested to take up such activities after the expiry of a period of 3 years from the date of issue of certificate of registration to undertake production, distribution and sale of seed or other propagating material of the variety.

(a) Authority will hear both parties and in public interest, may order breeder to grant a license to the applicant on payment of a fee.
(b) Period of compulsory license will be maximum up to period of protection.
(d) Authority can settle terms and conditions, revoke or modify compulsory license.

19.2.3 Benefit sharing

The benefit sharing mechanism denotes that any person/ group of persons or government organizations can file claim for benefit sharing from their contribution in evolution of variety registered under the act.

(a) The claims under benefit sharing can be submitted by Citizens of India or Firms/ NGOs formed or established in India.
(b) Such claims after due examination and after receiving counter-claims will be disposed by the Authority.
(c) The amount thus determined shall be deposited by the breeder in the manner referred to in clause (a) of sub-section (1) of section 45 in the National Gene Fund.

19.2.4 National Gene Bank

Section 19 of the PPV&FR Act provides that DUS test should be conducted once proper seed and DUS test fee has been deposited by the applicant. Rule 29(10) prescribes that seeds or propagules and parental lines of varieties under registration submitted for DUS test and special tests should be deposited at the National Gene Bank of PPV&FRA. Section 27 provides that the applicant/ breeder should deposit seeds/propagating material including parental lines of hybrids of registered variety in the National Gene Bank of PPV&FRA. Thus, the PPV&FR Authority established its National Gene Bank at Delhi to handle the orthodox or true seeds of the varieties of crop species notified for registration. Further, to take care of the tree/vines/ other vegetatively propagated crop species, five field gene banks are under establishment at different locations depending upon the agro-climatic regions. These gene banks have following major purposes:

i. To act as repository of the seed or propagules during the process of registration
ii. To provide the true reference sample of the registered varieties for any future use

19.2.5 National Gene Fund

The National Gene Fund to be constituted under the Act shall be credited thereto-

(a) The benefit sharing from the breeder.

(b) The annual fee payable to the authority by way of royalties.

(c) By the compensation provided to the communities.

(d) Contribution from any national and international organization and other sources.

The fund will be applied for disbursing shares to benefit claimers, either individuals or organizations, and for compensation to village communities. The fund will also be used for supporting conservation and sustainable use of genetic resources, including *in situ* and *ex situ* collection and for strengthening the capabilities of the *panchayat* in carrying out such conservation and sustainable use.

DUS testing centers

Crop	Main centers and sub-centers
Chrysanthemum &Rose	IARI, New Delhi, IIHR, Bengaluru & NBRI, Lucknow
Jasmine	IIHR, Bengaluru & TNAU, Coimbatore
Damask rose, Periwinkle	CIMAP, Lucknow
Bougainvillea	IARI, New Delhi & NBRI, Lucknow
Orchids (*Cymbidium*, *Dendrobium*, *Vanda*)	NRCO, Sikkim; SFRI, Itanagar & KAU, Thrissur
Marigold	IARI, New Delhi
Tuberose, Gladiolus	DFR, Pune & IIHR, Bengaluru
China aster	MPKV, Rahuri & IIHR, Bengaluru
Turmeric, Ginger	IISR, Calicut & ICAR-NEH region, Barapani

19.2.6 Geographical Indications

Geographical indication (GI) is the 3rd important IPR law covered under the TRIPS and concerned to agriculture. Indian law on GI of goods (Registration and Protection) Act was enacted in 1999 and enforced from 15th September, 2003 with the GI Registry located at the Patent Office, Chennai. Unlike the other IPRs, which are created and owned by individuals or institutions, the GI is a collective IPR of many people belonging to a territory of a country or the country itself. For this reason GI is not transferable or licensable. GI refers to an essentially attributable association between a given quality, reputation or other characteristics of any goods such as agricultural, natural or manufactured, which is valued in commerce, with its either origin or production or manufacture or any act of processing or preparation in a definable territory, or a region or a country. For this exclusive relationship between the region/ territory/ country and the quality and distinctiveness of given goods, it commands a commercial reputation conferring premium market space in demand and price. Examples of horticultural products having distinctiveness for GI are as follows:

Geographical indications/ Quality tag/ Branding	**State**
Darjeeling-Green Tea and White Tea; Laxman Bhog Mango (2008); Khirsapati (Himsagar) Mango (2008); Fazli mango grown in the district of Malda (2009); Kohitur-Mango	West Bengal
Mysore-Agarbatti, Sandalwood oil & soap, Betel leaf (2005) & Mallige (2007, jasmine); Udupi-Mallige (jasmine); Hadagali & Shankarapura Mallige (jasmine); Nanjanagud-Banana (2005); Coorg- Arabica coffee; Coorg-Green cardamom (syn. Coorg Clipped Cardamom, 2007) & Coorg-Mandarin oranges (2004); Devanahalli-Pummelo (2009); Appemidi-Mango (2009); Kamalapur-Red banana (2009); Byadagi-Chilli, Udupi-Mattu Gulla brinjal; Bengaluru- Bangalore Blue (grapes, 2012); Bangaluru- Pink (Rose) onion (25th March, 2015); Chikmagalur- Arabica coffee; Bababudangiri-Arabica coffee, Monsooned- Malabar Robusta coffee (2007); Monsooned Malabar Arabica Coffee (2007); Sirsi-Supari (arecanut);	Karnataka
Kangra-Tea	Himachal Pradesh
Malabar-Pepper (2007), Alleppey-Green cardamom(2007); Vazhakkulam-Pineapple (2009); Wayanad-Robusta coffee; Chengalikodan-Nendran Banana; Central Travancore-Jaggery; Tirur-Betel leaf	Kerala
Allahabad Surkha-Guava (2007); Malihabadi-Dusseheri mango (2008)	Uttar Pradesh
Uttarakhand-Tejpatta or Tamalpatra (31st May, 2016)	Uttarakhand
Eathomozhy-Tall coconut (2008); Nilgiri (Orthodox) logo; Virupakshi-Hill banana (2008); Sirumalai- Hill banana (2008); Madurai-Mallige (jasmine); Erode-Turmeric;Kodaikanal-Malai Poondu (i.e. Hill Garlic);	Tamil Nadu
Naga-Chilli; Naga-Tree Tomato; Naga-Cucumber	Nagaland
Arunachal-Wakro Orange (2014)	Arunachal Pradesh
Sikkim-Large cardamom; Dalle Khursani- Chilli (September, 2021)	Sikkim
Mizo-Chilli (cv. Bird's Eye, 2004); Mizo-Ginger ()	Mizoram
Tripura-Queen pineapple (2015)	Tripura
Assam-Orthodox Logo;Karbi Anglong-Ginger (25th March, 2015); Tejpur-Litchi (27th March, 2015); Kaji Nemu i.e. Assam Lemon (2022)	Assam
Kachai-Lemon (2014); Tamenglong-Orange (); Elephant chilli ();	Manipur
Memong Narang (mandarin orange); Khasi Mandarin (2014);	Meghalaya
Tezpur litchi (2014); Shahi-Litchi (2018); Fazli-Mango; Bhagalpuri Jardalu-Mango (2018); Magahi Paan (i.e. Betel leaf; 25th February, 2018)	Bihar
Feni-Wine; Khola-Chilli; Moira-Banana; Harmal-Chilli	Goa
Guntur- Sannam Chilli; Krishna- Banginpalli (Mango) (2017); Araku valley-Arabica coffee;	Andhra Pradesh

Geographical indications/ Quality tag/ Branding	State
Mahabaleshwar-Strawberry (2010); Alibagh-White onion (15th July, 2022); Nashik-Grapes (2010); Jalna-Sweet orange (2015); Jalgaon-Brinjal (for *bharta* making), Basrai-Banana (30th November, 2016) ; Marathwada-Kesar Mango; Solapur-Pomegranate (2016); Sangli-Turmeric & Raisins; Devgad- *Hapus* (i.e. Alphonso mango); Ratnagiri-Alphonso mango; Lasalgaon-Onion; Purandar-Fig (2016); Beed-Custard apple (2016); Vengurla-Cashew nut; Dahano-Gholvad Chikoo (Sapota); Bhivapuri (Nagpur)-Chilli; Vidharbha-Waygao turmeric; Nagpur-Mandarin orange (2013); Satara- *Waghya ghevada* (i.e. French bean, 2016); Sindhudurg & Ratnagiri Kokum; Junner-Shivneri *hapus*; Nandurbar-Amchur; Latur-Coriander & slender tamarind, Nanded-tambul Prasad (2022); Vasmat-turmeric (2024)	Maharashtra
Gir- Kesar Mango (2011); Kandhamal-Haldi	Gujarat
Ganjam- Kewda flower & Kewda rooh; Kandhamal-Haladi (2020);	Odisha
Kashmir-Saffron (2022); Gucchi-Mushroom (2021); Raktsey Karpo (2022) from Ladakh	Jammu & Kashmir

According to Indian GI Act, any association of persons or producers or any organization or authority established under any law representing the interest of producers of the concerned goods are eligible to apply for GI. Using the flexibility of this provision, efforts are underway in the country to appropriate the collective IP of farmers associated with some reputed agricultural commodities by the organized processors and traders.

19.2.7 Awareness Generation

The PPV& FR Act, 2001 is one of the unique and innovative Acts of India. It involves diversified stakeholders ranging from tribals, farmers, research institutions, seed industry, government, legal systems and so on. The Act, being enacted merely a decade ago and made functional only five years back, requires massive awareness campaign to be undertaken for creating desired awareness about its provisions. Thus, there is a need that all the stakeholders should work in tandem and close collaboration for transmitting the true essence of the Act up to the grass roots.

19.2.8 Salient achievements

- The Plant Patent Act was enacted in 1930 in USA.
- The Plant Variety Protection Act was enacted in USA in 1970 to provide IPR for sexually propagated plants and tuber multiplied crops.
- TRIPS includes copy rights, trade marks, geographical indications, layout designs, trade secrets and patents.
- Since 1930, the US Plant Patent Act has provided incentives for the breeders of new varieties of ornamental plants including roses.

- In 1994, the US Patent and Trade Mark office granted 498 plant patents, of which 127 were directed to distinctive new rose varieties and chrysanthemum varieties took 2nd place.
- The 1st patent was awarded for the rose cv. New Dawn.
- India's KSG Farm and Nursery received US Patents for two aglaonema hybrids of distinctive shape, colour and form, named 'Jewel of India' and 'Emerald Star'.
- The Indian Patent Act of 1970 allows process patent and no product parent for foods, chemicals, drugs and pharmaceuticals.
- There are 3 types of patents viz. plant patent, design patent and utility patents.

1) Plant patent is granted for newly invented asexually propagated plants, which provides protection for 20 years.
2) Design patent protects the ornamental characteristics and life span of this patent is only for 14 years.
3) Utility patent is most common patent used by universities and companies to protect their results of research and development.
   a) Utility patents are granted for protection of inbred lines and hybrids and their seeds, variants, mutants and trivial modification of hybrid.
   b) Utility patent also covers breeding methodologies and engineered gene constructs.
   c) In ornamental plants, utility patents have been granted for bicoloured petal patterns in *impatiens*, scented geranium by genetic transformation, flower form and method of breeding of New Guinea *impatiens* and plant hybrids germplasm of My Favorite® chrysanthemums.

- In 1930, the Plant Patent Act was enacted for protection of vegetatively propagated plants covering fruits and ornamental plants and US was the only country to issue plant patents.
- In 1970, the Plant Variety Protection Act was enacted to provide IPR for sexually produced and tuber propagated plants.
- The Association of Plant Breeders for Protection of Plant Varieties (ASSINSEL) held for International Convention for protection of New Varieties of Plants in Paris during 1961 for protection of rights of Plant Breeders and Farmers and to encourage development of new varieties of plants.

- The International Union for the Protection of New Varieties of Plant (UPOV), an inter-governmental organization with headquarter in Geneva was constituted in Paris convention in 1961, to set basic rules and principles for plant variety protection.
- The UPOV has standardized descriptors for following ornamental crops- Alstroemeria, amaryllis, anthurium, aster, carnation, *Celosia*, chrysanthemum, *Cymbidium*, narcissus, *Dendrobium*, *Dieffenbachia*, everlasting daisy, freesia, gerbera, gladiolus, hydrangea, iris, *Pelargonium peltatum*, *Pelargonium domesticum*, *Pelargonium zonale*, kalanchoe, *lilium*, New Guinea impatiens, petunia, *Phalaenopsis*, *Euphoria pulcherrima*, potted azalea, rhododendron, *Helianthus annus*, *Begonia tuberhybrida*, tulip, verbena, zantedeschia, saintpaulea, Calluna, *Trapa natans*, *Catharanthus*, firelily, Exacum, Ifafa lily, gentiana, *Epiphyllopsis*, *Oenothera*, *Limonium*, *Clematis*, Calluna, *Ornithogalum*, *Euphorbia milli*, *Schlumbergera*, Epiphyllopsis, *Begonia elatior*, *Euphorbia fulgens*, *Eustoma*, *Picrantha*, *Forsythia*, *Gentiana*, *Guzmania*, *Hypericum*, *Rosa scorzonerea*, kangaroo paw, Scots heather, *Surruria*, *Streptocarpus*, Nerine, *Osteospermum*, and *Pentas*.
- During 2001, the Protection of Plant Varieties and Farmers Right Act was passed by Indian Government in tune with WTO resolution.
- New Varieties are subjected to DUS tests for each crop and registration valifity is for 18 years for trees and vines.
- New Plant variety protected under faster development of seed industry, improvement in seed quality, enrichment of genetic resources and incentives to plant breeders.
- FAO commission developed concept of Farmers Rights in 1979 on conserving, improving and making available plant genetic resources in centres of origin or diversity.

References

Heitz, A. 1999. Plant Variety Protection and Cultivar Names under UPOV Convention. *In*: A. Andrews, A.C. Leslie and C. Alexander (eds.), *Taxonomy of Cultivated Plants: Third International Symposium*, Royal Botanic Gardens, Kew, pp. 59-65.

Prasad, K.V. and Janakiram, T. 2013. *Indian Horticulture*, **58**(2): 29-35.

Rana, R.S. 1995. Intellectual Property Rights: Protection of Plant Varieties. *In*: Ed. R.S. Rana, Gupta, P.N., Rai, M. and Kochher, S. Genetic Resources of Vegetable Crops: Management, Conservation and Utilization, NBPGR, pp: 421-427.

Singh, P.K. 2013. *Indian Journal of Genetics*, **71**(2): 190-198.

20

Acanthaceae (Acanthus Family)

A family of over 250 genera and 2,500 species, Acanthaceae are distributed mainly in tropics, but also in Mediterranean region, Australia and USA. Central America, Brazil, Africa and Indo-Malaysia are the 4 chief centres of distribution. About 70 genera and over 340 species have been reported from India. Some of the largely represented genera along with the approximate number of their worldover reported species are *Justicia* (300), *Strobilanthes* (250), *Barleria* (230), *Thunbergia* (200), *Aphelandra* (200), *Decliptera* (150), *Blepharis* (100), *Staurogyne* (80), *Mendoncia* (60), *Crossandra* (60) and *Acanthus* (50).

Selected Indian Genera

Acanthus, Adhatoda, Barleria, Eranthemum, Justicia, Peristrophe, Ruellia, Thunbergia.

Economic Importance

1. Ornamental plants include *Aphelandra aurantiaca*, *Barleria cristata*, *B. gibsoni*, *Beloperone guttata* (shrimp plant), *Crossandra infundibuliformis*, *Eranthemum bicolor*, *E. nervosum*, *Fittonia gigantea*, *Jacobinia carnea*, *Justicia gendarussa*, *Ruellia brittoniana*, *Thunbergia coccinia*, *T. erecta*, *T. fragrans* and *T. grandiflora*.
2. Plants of medicinal value include (i) *Adhatoda vasica*- leaves are useful in cough, asthma and bronchitis, (ii) *Asteracantha longifolia*- seeds and roots are used against cough, (iii) *Barleria prionitis*- leaves and and roots are used against bronchitis & cough, (iv) *Barleria cristata*- seeds are used as an antidote for snake bite, (v) *Hygrophila spinosa*- leaves and roots are used against jaundice and rheumatism, (vi) *Peristrophe bicalyculata*- leaves and roots are used against snake bite, (vii) *Ruellia prostrata*- leaf juice is used in earache.
3. A blue dye is obtained from the leaves of *Strobilanthes custa*.

Description of the Family

1. General habit: Mostly perennial herbs or shrubs, rarely trees; some are twiners (*Thunbergia, Mendonica*), or xerophytes (*Barleria, Blepharis*), or halophytes (*Acanthus ilicofolius*), or spiny shrubs (*Acanthus montanus*).
2. Leaf: Simple, opposite decussate, exstipulate, entire, rarely pinnately lobed with small spines on the margins (*Acanthus mollis*); cystoliths, appearing as protuberances or streaks, are usually common on stems and leaves.
3. Inflorescence: Usually a dichasial cyme, frequently condensed in the leaf axils; sometimes racemose spike (*Adhatoda vasica*), or flowers solitary axillary (*Thunbergia*).
4. Flower: Usually bracteate, bracteolate, pedicellate or subsessile, complete, hermaphrodite, zygomorphic, hypogynous; bracteoles are often large, coloured and more or less enclosing the flower; in *Peristrophe* out of the 2 bracts the posterior one is larger than the anterior one, and 4 laterally placed bracteoles cover the calyx.
5. Calyx: Usually 5 united (*Justicia*) or free (*Adhatoda*) sepals; valvate (*Justicia*,) or quincuncial (*Adhatoda*); sometimes sepals are 4 or rarely 3; rarely the calyx is much reduced and divided into teeth-like structures (*Thunbergia*).
6. Corolla: Usually 5 lobed, gamopetalous, imbricate or twisted; bilipped (2/3) or bilabiate personate, of which the upper lip is bifid or bilobed and the lower lip is trolobed; sometimes the upper lip is not developed (*Acanthus*); corolla is tubular in *Thunbergia* and *Ruellia*.
7. Androecium: Usually 4 didynamous stamens (*Ruellia, Thunbergia*), or only 2 stamens (*Adhatoda, Justicia, Peristrophe*), epipetalous, usually exserted; rarely 5 stamens (*Pentstemonacanthus*); frequently 1-3 staminodes present; anthers bilobed with often one smaller lobe than the other, connective often long introrse, longitudinal dehiscence.
8. Gynoecium: Bicarpellary, syncarpous, carpels median; ovary superior, bilocular, one to many anatropous ovules in each locule, axile placentation; style narrow and long; stigmas 2, of which the anterior one is often larger; usually with a hypogynous, nectar-secreting disc.
9. Fruit and seeds: Fruit usually a capsule, loculicidal to the very base; rarely a drupe (*Mendoncia*). Seeds 1 to many, of which the funiculus develops into a hook-like retinaculum or 'jaculator', often non-endospermic.

10. Pollination: Flowers are suited for insect pollination because of coloured bilabiate corolla and abundant nectar in hypogynous disc. Protandry favours cross pollination.
11. General floral formula: Br, Brl, ·|·, ⊠, $K_{(5)\text{ or }5}$, $C_{(2/3)}$, $A_{2\text{ or }4}$, $G_{\underline{(2)}}$.

Description of a Common Plant

A) *Justicia gendarussa* Burm

Habit: A perennial herb. Leaf: Simple, stipulate, opposite decussate, subsessile, lanceolate, obtuse, unicostate reticulate. Inflorescence: Dichasial cyme arranged in racemose manner. Flower: Bracteate, bracteolate, pedicellate, complete, zygomorphic, pentamerous, hypogynous. Calyx: 5 sepals, united, valvate. Corolla: 5 petals, united, 2/3 bilabiate personate, of which posterior lip is bilobed while anterior lip trilobed, valvate. Androecium: 2 stamens, polyandrous, epipetalous, dithecous, anther lobes situated at unequal heights, dithecous, basifixed, introrse. Gynoecium: Same as in description of family. Fruit: Capsule. Floral formula: Br, Brl, ·|·, ⊠, $K_{(5)}$, $C_{(2/3)}$, A_2, $G_{\underline{(2)}}$.

21

Amaranthaceae (Amaranths or Pigweed Family)

A family of about 175 genera and 2,500 species, Amaranthaceae are mostly distributed in tropical but also in temperate regions. About 18 genera and over 50 species have been reported from India. Some of the larger genera with the number of their approximately reported species and common names are *Alternanthera* (200, alligator weed), *Gomphrena* (100, globe amaranth), *Iresine* (80, gizzard plant), *Amaranths* (60, pigweed) and *Celosia* (60, cock's comb).

Gomphrenas are 30-40 cm tall, bushy with erect stems, hairy, light green with oblong-ovate leaves tapering towards the base. Round button like or spherical flower head is a cylindrical spike 1-2 cm long and 1 cm in diameter and may be terminal and/or axillary, and are born on long, thin stems above the foliage. Each flower head has tiny, conspicuous, papery, colourful bracts which enclose the inconspicuous true flowers. Bracts ovate, three mm long, apex acuminate, bracteoles 4.5-5.5 mm long. Ovary is sub-globose with two stigmas. Seed is brown, lustrous, 1.5 mm long (Wagner *et al.*, 1999). It has both C_3 and C_4 pathway.

Selected Indian Genera

Achyranthes, *Alternanthera*, *Amaranths*, *Celosia*, *Digera*, *Gomphrena* and *Pupalia*.

Economic Importance

1. Ornamental plants of the family known for their muticoloured foliage and beautiful inflorescence are *Amaranthus caudatus* (Foxtail), *A. tricolor*, *A. salicifolius*, *A. paniculatus, A. spinosus*; *Celosia cristata C. argentea* (cocks comb; white, purple or yellowish-red panicles), *Deeringia amaranthoides*, *Gomphrena globosa* (globe amaranth), *Iresine herbisii* (purplish-red foliage) and *I. lindenii* (red foliage).
2. **Edible plants:** Leaves and young shoots are used as vegetable of *Amaranths blitum* (Amaranth or chaulai), *A. caudatus*, *A. hybridus*, *A.*

spinosus (kantili chaulai), *A. tricolor* (bari chaulai), *A. viridis*, *Celosia argentea* and *Digera muricata* (Lehsua). Grains of *Amaranths caudatus* are roasted and eaten as 'Ramdana'.

3. Inflorescence paste of *Achyranthes* aspera (Latjeera or chirchita) is used as an antidote against snake and scorpion bites.
4. Widespread troublesome weeds of this family belong to several genera including *Acnidia*, *Amaranths*, *Gomphrena* and *Iresine*.

Description of the Family

1. **General habit:** Mostly annual or perennial weedy herbs, rarely woody shrubs or climbing shrub (*Deeringia amaranthoides*).
2. **Stem:** Anomalous secondary growth occurs by the formation of collateral vascular bundles from cambium in the pericycle. In the scattered vascular bundles, the fascicular cambium is functionless.
3. **Leaf:** Simple, alternate (*Amaranths*) or opposite decussate (*Achyranthes*), exstipulate, usually entire or nearly so, often thickly covered with hairs.
4. **Inflorescence:** Dense or congested axillary or terminal spikes (*Amaranths*), heads, racemes or flowers solitary; often spiny because of the firm spiny tips on the inflorescence bracts; usually the basic unit is either a 3-flowered dichasium unit (*Celosia*) or 7 to 11-flowered dichasium units (*Pupalia*).
5. **Flower:** Bracteate, bracts scaly, membranous, dry or scarious; bracteolate, bracteoles 2, dry, membranous, large and lateral; usually bisexual or when unisexual (Amaranthus) the species mostly dioecious or polygamodioecious; actinomorphic, pentamerous, hypogynous, incomplete, apetalous, very small.
6. **Perianth:** Uniseriate consisting of 5, free or basally connate tepals, green or sepalloid, dry or membranous; imbricate; only 3 tepals in *Amaranthus viridis*, but 5 free tepals in *A. spinosus*.
7. **Androecium:** Usually 5 stamens present opposite the tepals, often monadelphous (*Celosia*, *Achyranthes*), separate in *Amaranthus*, often with equal number of staminodes alternating with the anthers (*Celosia*, *Achyranthes*); staminodes are fimbriated in *Achyranthes*; filaments united at the base into a short tube, but in *Amaranthus*, the filaments are free; anthers dithecous, dorsifixed (*Achyranthes*) or versatile (*Amaranthus*); introrse, longitudinally dehiscent; only 1 or 2 stamens in *Nothosaerva*.
8. **Gynoecium:** Bi to tricarpellary, syncarpous, superior, unilocular, with a single pendulous or campylotropous ovule in the locule, basal

placentation; rarely several ovules (*Celosia*.) are present; styles 1 to 3; stigma 1 to 3, capitate.

9. **Fruit and seeds:** Fruit usually a utricle (*Achyranthes*) or nutlet (*Digera*), or a capsule (*Celosia*), or rarely a berry or drupe. Seeds with curved embryo and mealy endosperm.
10. **Pollination and dispersal:** Pollination is entomophilous because numerous small flowers are aggregated in dense or congested inflorescence. Dispersal tales place either by wind or by animals.
11. **General floral formula:** Br, Brl, ⊠, P_5, $A_{(5-5\ \text{staminodes})}$, $G_{\underline{(2\ \text{or}\ 3)}}$.

References

Wagner. W. L., Herbst, D.R. and Sohmer, S.H. 1999. Manual cf the Flowering Plants of Hawaii, Revised edition, Bishop Museum, USA, pp. 192.

22

Amaryllidaceae (Earlier Agavaceae) (Daffodil Family)

A family of about 73 genera and over 1,600 species (Hickey and King, 1988), Amaryllidaceae are usually distributed in tropical and sub-tropical regions. About 9 genera and over 60 species have been reported from India. Some of the larger genera along with the number of their world over reported species are *Crinum* (110), *Hippeastrum* (75), *Narcissus* (60), *Hymenocallis* (50), *Haemanthus* (50), *Cyrtanthus* (47), *Zephyranthes* (40), *Galanthus* (20), *Polianthes* (15), and *Pancratium* (15).

Hippeastrum (popular name *Amaryllis*) is one of the most beautiful bulbous ornamental plants grown in the world. It is very suitable for planting in pots, greenhouses, window gardens, flowering beds, rockeries, popular as cut flower, and useful in floral arrangements. *Hippeastrum* and *Amaryllis* are synonyms for 'Trumpet lily' or 'Night Star lily' in Indian gardens. In USA, *A. belladonna* is known as *Callicore rosea* sometimes as *Hippeastrum equestre*, and in South Africa as 'March lily', being a variable species. Plants of the genus *A. belladonna* are known as 'Belladonna lily', 'Naked Lady' or 'Amarillo'. *Hippeastrum hybridum* is popularly known as 'Royal Dutch' amaryllis or 'Dutch' amaryllis hybrids. Hippeastrum takes its name from the *Hippeus* meaning 'a knight rider' or horse and '*astron'* meaning a star, due to some fancied resemblance in *H. equestre*, perhaps of the equitant leaves and the star shaped corolla opening.

Tuberose is a bulbous ornamental plant valued for its beauty, sweet fragrant flowers and excellent keeping quality. It is a night blooming plant, commonly known as '*Rajanigandha* or *Nishigandha*'. The plant has many medicinal uses in *Ayurveda* form of medicine. Flowers are used for artistic garlands, bouquets, floral ornaments and buttonholes. It is grown commercially for cut flower and essential oil; so popular in India. The variegated types are best suited in landscape gardening. The lack of diversity of flower types, flower size, and flower colour in tuberose has been attributed to non cultivation on wider scale in the world.

Selected Indian Genera

Agave, Amaryllis, Crinum, Hippeastrum, Pancratium, Narcissus, Zephyranthes.

Economic Importance

1. Common ornamental plants of Amaryllidaceae are *Agave americana* (century plant), *A. mexicana, A. vera-cruz* (Kuwarbuti), *Alstroemeria aurantiaca, Amaryllis belladona* (Amaryllis), *Clivia miniata* (kafir lily), *Crinum asiaticum* (Sukhdarsan), *Curculigo capitulata, Eucharis grandiflora* (Amazon lily), *Haemanthus coccineus* (blood lily), *Hippeastrum vittatum, Hymenocallis americana* (spider lily), *Hypoxis hirsuta* (stargrass), *Leucojum vernum* (snowflake), *Narcissus pseudo-narcissus* (daffodil), *Nerine filifolia, Polianthes tuberosa* (tuberose), *Zephyranthes grandiflora* (Zephyr lily), *Z. candida* and *Z. tubispatha.*
2. The oil obtained from the flowers of several species of *Narcissus* and *Polianthes* are used in high grade perfumery.
3. The fibres obtained from *Agave cantala, A. sisalana* (sisal hemp), *A. vera-cruz* and *Furcraea foetida* (Mauritius hemp) are used for making cordage, mats, socks, shoe soles, ropes, twines, etc. The fibres obtained from the leaves of *Curculigo latifolia* are used for making fishing nets.
4. Fruits of *Curculigo latifolia* and roots of *C. orchioides* and *Alstroemeria* are ground and eaten as flour.
5. Tubers of *Curculigo orchioides* are used in the treatment of asthma and jaundice.

Description of the Family

1. General habit: Usually perennial, bulbous (*Crinum*) or rhizomatous (*Hypoxis*), xerophytic herbs leafing only in the rainy or spring seasons, some are arborescent (*Agave, Furcraea*).
2. Leaf: Mostly basal, simple, alternate, often distichous, exstipulate, sessile with sheathing leaf base (*Crinum*); very large, more or less linear, strap-shaped or ensiform; entire, parallel venation.
3. Inflorescence: Usually on a long leafless scape, with one or more spathes or bracts, cymose, but generally condensed in the form of an umbel or head; sometimes solitary or paired flowers are present (e.g. *Narcissus*, Fig. 22.1; *Zephyranthus*, *Pancratium gloriosa.*); inflorescence appears after several years in *Agave*.
4. Flower: Bracteate, complete, bisexual, actinomorphic (*Crinum*, Fig. 22.2) or zygomorphic (*Amaryllis*), trimerous, epigynous; large, showy and usually fragrant.

5. Perianth: Tepals 6, arranged in 2 whorls of 3 each, not differentiated into calyx and corolla; petioled; free or often united basally to form a long tube; salver-shaped or funnel shaped (*Crinum*); large and showy, valvate or twisted or imbricate; sometimes corona present (*Narcissus*) and looks like an extra whorl of perianth-segments in between the normal whorl of perianth and stamens.
6. Androecium: Stamens 6, arranged in 2 whorls of 3 each, epiphylous; present opposite the tepals; inserted at the throat of perianth tube; filaments usually long and coloured (*Crinum*); anthers dithecous, basifixed or versatile (*Crinum*), introrse, dehiscence by longitudinal slits or rarely by apical pore (*Galanthus*); Corona is prominent in some cases, as *Pancratium*, *Narcissus*, *Eucharis*, etc., sometimes some stamens become staminodial.
7. Gynoecium: Tricarpellary, syncarpous; ovary inferior, rarely half-inferior, trilocular or rarely unilocular (*Alstroemeria*); numerous anatropous ovules in each locule; axile placentation; style 1; stigma capitates or trilobed.
8. Fruit and seeds: Fruit a loculicidal capsule or berry (*Haemanthus*). Seeds with small and straight embryo, and endospermic.
9. Pollination: It is entomophilous. Insects are attracted because of gaudy-coloured, often scented perianth and presence of nectar. Nectar is secreted in the inner tepals in *Galanthus* and in corona in *Narcissus*, *Hippeastrum*, etc.
10. General floral formula: Br, □ or ·|·, ⊠, P_{3+3}, A_{3+3}, $G_{(3)}$ OR , ⊠, P_{3+3}, A_{3+3}, $G_{(a)}$.
11. Bulb: It is simplest type. Food storage occurs only in leaf bases as there are no scales. Each bulb is made up of several bulb units. Bulb unit is in the axil of next oldest leaf. Each bulb unit comprises 4 leaves and terminal inflorescence. Three bulb units are initiated each year i.e. 12 leaves and 3 inflorescences. In twin-scales, thickness and length of outer scale affect the rate of propagation e.g. *Hippeastrum* (Fig. 22.3).

Taxonomy of Amaryllis

Narain and Khoshoo (1977) have made brief review of important events about the origin and evolution of garden amaryllis. Confusion exists in the nomenclature of scattered hybrids produced by various sources. Linnaeus (1753) described both under *Amaryllis* L. whereas Herbert (1819) separated *Hippeastrum* Herb., from *Amaryllis* L. Traub (1958) mentioned many species under *Amaryllis* L. During the International Botanical Congress held in Paris in 1959, it was decided to opt only for *A. belladonna* L. under the genus *Amaryllis* L. and others under the genus *Hippeastrum* Herb. Moreover, through intergeneric hybridization, both are so merged with each other that both the genera are taken as synonyms, and the modern cultivars are referred to as *Hippeastrum* hybrids. Many countries prefer using only *Hippeastrum* for the varieties of the both genera.

Separation of genus Hippeastrum from Amaryllis

Uphoff (1938) insisted that *Amaryllis* stand for American genus that has been named *Hippeastrum* Herb. Goldblatt (1984) opined that *A. belladonna* applies to species from South Africa known as 'Cape belladonna'. Approximately 42 species have been transferred from *Amaryllis* to *Hippeastrum* in accord with the decision at 14th International Botanical Congress in 1987, on application of former name to the South African genus. The genus has been intensely bred and cultivated since early 19th century to produce large colourful showy flowers and thus *Amaryllis* had become a polymorphic (diverse) genus with about 50 species.

Taxonomy of Tuberose (*Polianthes tuberosa*)

Originally, Linnaeus had used the name *Polianthes floribus alternis* in *Hortus Cliffortianus* (1738). Later in *Species Plantarum* he grouped this and hyacinthus tuberosus of Casper Bauhin and Carlos Clusius under the genus *Polianthes* and the species *tuberosa*.

Polianthes is a monotypic genus belonging to family Agavaceae and is closely related to Bravoa from which it differs in having a perianth tube, which is widened upwards. There are some 12-15 species (*P. tuberosa*, *P. playphylla*, *P. palustris*, *P. montana*, *P. longiflora*, *P. geminiflora*, *P. gracilis*, *P. pringleri*, *P. sessiliflora*, *P. nelsonii*) under the genus *Polianthes*. Nine among them have white flowers, one white and red and two were having red flowers. All were found growing wild with the exception of *P. tuberosa* which has never been found anywhere except under cultivation. The only close relative of *P. tuberose* is believed to be *P. gracilis*. Except *P. tuberosa*; other seems to be absolete, single forms are fertile whereas doubles are sterile; therefore in a breeding program double form is normally not used as male parent. There is self incompatibility (SI) whether it is a single or double form. Crop is highly cross pollinated due to gametophytic SI, sterility and dichogamy. The exact cause of sterility is unknown.

It is a monotypic genus closely related to *Bravoa*. The systematic position of genus *Polianthes* had long been a subject of discussion. Baker (1888) classified this genus under the family Amaryllidaceae; sub-family-Agavoideae, sub-order Agaveae and genus *Polianthes*. Hutchinson (1934) ultimately placed the genus *Polianthes* under the family Agavaceae in tribe-6 of *Poliantheae*. Systematic position of this genus has been discussed on the basis of karyotypes. Sato (1938) observed that karyotypes of five genera in Agavaceae, which includes *Polianthes* genus too, are similar. These are different from other taxa in Amaryllidaceae and rather resemble those of Liliaceae. On the basis of cytological studies, Joshi and Pantulu (1941) and Sharma and Ghosh (1956) suggested that it would be better to divide *Agavoideae* into two groups, one including only Australian genus *Doryanthes* and other comprising remaining six genera i.e. *Agave, Furcraea, Beschorneria, Polianthes, Prochnyanthes* and *Pseudobravoa* which are all American. Bailey and Bailey (1977) placed the genus under family Agavaceae. There is enormous scope to develop cultivars through breeding. Some coloured forms of popular scented tuberose would of great asset. Bigeneric crosses with allied should be tried. In this connection, studies on the genus *Manfreda* would be worthy. Most flowers in this genus are greenish or brownish, but two known species have pinkish flowers. Some species are fragrant and some are colourless. *Manfreda longiflora* and *M. maculosa* are delightfully fragrant. *Manfreda virginia* though not colourful is also a sweet scented one. *Manfreda brunnea* and *M. potosina* are like miniature form of *M. virginia*. Also these might be useful in crosses with *Polianthes*. There are many other Mexican *Manfreda* species that might likewise be of interest in a breeding program (Bhattacharjee, 1995).

Floral Biology of Amaryllis

All amaryllis flowers are strongly protandrous. Flowers remain open for 4-6 days. Once flower is open, the anthers dehisce through longitudinal splitting. Anther dehiscence is between 8 and 10 AM. Stigma becomes receptive 12-14 hours after anther dehiscence. At maturity, the style and stigma point upwards. The best medium for pollen viability test through *in vitro* is 2 per cent sucrose supplemented with 0.5 per cent Mg, 1 per cent ATP and 10 ppm GA_3. *In vitro* germination of *A. vittata* pollen with 5 and 10 µg per ml ABA was recorded (Sindhu *et al.*, 1986). Freshly shed pollen can be stored in household refrigerator for a month or longer at 4-10^{0}C (Pfeiffer, 1936). Pollen grains stored for more than five months at 10^{0}C and less than 50 per cent humidity gives 50-75 per cent germination.

Floral Biology of *Polianthes*

Fourteen stages of floral bud development were enumerated by Kosugi and Kimura (1961). Stage I- Differentiation, Stage II and III- Spike Differentiation and Stage IV to IVX- Tepal formation stages including stamen, pistils and pollen formation stage.

The reproductive phase is marked by differentiation of floral parts, elongation of floral stalk and flowering. The single type takes 55-65 days for appearance of spike from planting. The spike takes 26-29 days for opening of florets. The anther formation is reported to be variable in cv. Pearl Double (9.7 % buds had anthers) and in cv. Suvasini (90.7 % buds had anthers). Florets of single usually have a functional pistil and floret length varies from 5-7 cm; spike lasts on plant for 15-18 days. In case of double type, plant produces 40-50 florets per spike and spike length varies from 75 to 100 cm. Most florets have rudimentary pistil with 0.1 per cent florets have functional pistil. Spike lasts on plant for 18-22 days (Bhattacharjee, 1984). Stigma remains receptive for 3-5 days after anthesis.

Studies by Jayasree *et al.* (2002) indicate that in cvs. Mexican Single and Shringar, the size of fertile pollen grain ranges from 45.4 to 75.6 μm and seed are fertile while in cvs. Pearl Double and Suvasini, pollen fertility is 86.8 per cent and mean pollen size is 60.5 μm and seed is sterile because of absence of female reproductive organ. The cv. Shringar has 76 per cent pollen fertility and size of fertile pollen grain range from 49.12 to 73.4 μm and seed is fertile. In cv. Suvasini, all six anthers are not present in all buds, but most had 3-5 anthers. Pollen fertility is 21.6 per cent and size of fertile pollen grain ranges from 71.3 to 88.6 μm and this cultivar is seed sterile.

There was no pollen tube production, if stigmas were pollinated when florets were at bud stage, the day before flowering, the day of flowering or one day after opening. Considerable pollen tube growth was seen when florets were pollinated 2-3 days after anthesis. In another study by Ranchana *et al.* (2014), pollen germination was observed on stigmas of florets collected at 4 to 72 hours after cross-or-self pollination. After 72 hours of pollination, most of the self pollinated pollen tubes had penetrated the top third of the style only, but cross-pollen tubes had grown up to the base of the style and reached 40 to 60 per cent of the ovules. Stigma remains receptive for 3-5 days after anthesis. On selfing neither 'Double' varieties nor 'Single' varieties produce any seed due to incompatibility. Neither selfed double varieties nor selfed single varieties produce any seed due to self incompatibility (Shen *et al.*, 1987). Seeds from singles exhibit better germination than those from doubles (Srivastava and Sridhara, 1999). Crossability between 'Single' and 'Double' types and reciprocals reveals great variation in fruit set and seed set.

References

Bailey, L.H. and Bailey, E.Z. 1977. Polianthes. In: Hortus Third, Macmillan Publishing Co. Inc. New York, pp. 893.

Baker, P.1888. Tuberosa gracilis. In: Handbook of Amaryllideae including the Alstroemerieae and Agaveae, G. Bell and Sons, pp. 159.

Bhattacharjee, S.K. 1995. Tuberose: In: Advances in Horticulture (Vol. XI), Medicinal and Aromatic Plants. Malhotra Publishing House, New Delhi, pp. 1-20.

Goldblatt, P . 1984. Taxon, 33: 511-516.

Herbert, W. 1819. Botanical Magazine, London, pp. 25-26.

Hickey, M. and King, C. 1988. 100 families of flowering plants, (2nd edition), Cambridge University Press, Cambridge.

Hutchinson, J. 1934. The Families of Flowering Plants. Vol. 2. Clarendon Press, Oxford, 3rd ed., pp. 816-819.

Jayasree, P.L., Srinivas, M., Krishnan, R. 2002. The Nucleus, 44(3): 196-201.

Joshi, A.C. and Pantulu, J.V. 1941. Journal of Indian Botanical Society, 20: 37-71.

Kosugi, K. and Kimura, Y.1961. Technical bulletin of the Faculty of Agriculture, Kagawa University, 12: 230-234.

Linnaeus, C. 1753. Species Plantarum, 1: 292-293.

Narain, P. and Khoshoo, T.N. 1977. Indian Journal of Horticulture, 34(1): 80-85.

Pfeiffer, N.E. 1936. Herbertia, 3: 103-104.

Ranchana, P., Kannan, M. and Jawaharlal, M. 2014. 6th Indian Horticulture Congress, November, 6-9, TNAU, Coimbatore. Abst. No. 88, pp. 81.

Sato, D. 1938. Cytologia, 9: 203-242.

Sharma, A.K. and Ghosh, C. 1956. Genetica, 28: 99-111.

Shen, T.M., Huang, K.L. and Huang, T.S. 1987. Acta Horticulturae, 205: 71-72.

Sindhu, R.K., Basra, A.S. and Malik, C.P. 1986. Journal of the Plant Growth Regulation, 4(3): 293-298.

Srivastava, H.C. and Sridhara, C.J.1999. Journal of Medicinal and Aromatic Plant Sciences, 21(4): 934-936.

Traub, H.P. 1958. The Amaryllis Manual. The Macmillan Publishing Co., New York, pp. 158-172.

Uphoff, J. C. T. 1938. Herbertia, 5: 100-111.

23

Apoycynaceae (Dogbane Family)

A family of about 400 genera and 4,555 species of mostly tropical and sub-tropical distribution, but a few species distributed also in temperature regions. About 30 genera and over 60 species of Apocynaceae have been reported from India. Some of the larger genera along with their approximately reported species in bracket are *Rauvolfia* (100), *Tabernaemontana* (100), *Parsonsia* (100), *Aspidosperma* (80), *Prestonia* (65), *Strophanthus* (60), *Apocynum* (7), *Plumeria* (6) and *Catharanthus* (5). Some plants are of medicinal value and some are highly ornamental.

Periwinkle (syn. Vinca) is grown as an ornamental bedding plant in gardens and parks and is known for its colourful flowers and ever-blooming nature. In India, it is commonly known as *sada bahar*, *sada phuli*, and *barmasi*. It is one of the top ten best selling plant in USA. Its importance has increased owing to medicinal properties and as a model species for study of plant's secondary metabolism. The plant contains more than 70 different alkaloids in leaves and roots which contribute to anti-cancerous (vincristine and vinblastine) and anti-hypertensive (ajamalicine) properties, respectively.

Selected Indian Genera

Apocynum, *Catharanthus*, *Nerium*, *Plumeria*, *Tabernaemontana*, *Thevetia*, *Rauvolfia*,

Economic Importance

1. Ornamental plants include *Allamanda nerrifolia* (golden yellow flowers), *Alstonia scholaris* (pink flowers), *Beaumontia grandiflora* (large white flowers), *Carrisa carandas* (white-pink flowers), *Catharanthus roseus* (sadabahar), *Mandevilla taxa* (pink flowers), *Nerium indicum* (kaner), *Plumeria* (pagoda tree), *Tabernaemontana divaricata* (Chandni tree), *Thevetia peruviana* (Peeli kaner), and *Vallaris solanacea* (Dudhi bel).
2. Plants of medicinal value include (i) *Rauvolfia serpentina* (Sarpagandha)

used in the treatment of hypertension and mental disorders; drug is obtained from dried roots; the alkaloid reserpine can lower the blood pressure and tranquilize mental patients suffering from schizophrenia, (ii) *Thevetia peruviana* (seeds are used in rheumatism), (iii) *Vallaris solanacea* (latex is useful in toothache & inflated gums), (iv) *Alstonia scholaris* (bark is used in malaria and dysentry), (vi) *Cerbera manghas* (latex is used as an emetic and purgative), (vii) *Strophanthus hispidus* (seeds yield the drug strophanthin) (viii) *Wrightia tomentosa* (bark and roots are used as antidote to snake bite).

3. Rubber is obtained from the coagulated latex of several species of *Carpodinus*, *Funtunia*, *Hancornia*, *Landolphia* and *Rhynchodia*.
4. Poisonous plants of the family belong to *Acokanthera*, *Nerium*, *Strophanthus* and *Thevetia*. If one leaf of *N. oleander* or *T. peruviana* is ingested, it may be fatal to an adult human. Poisoned arrows are made from the bark extract of *Acokanthera abyssinica* and seeds of several species of *Strophanthus*.
5. Fruits of *Carissa carandas* (karonda) are edible and used in pickles.
6. An indigo-like dye is obtained from the seeds of *Wrightia tinctoria*.

Description of Family

1. General habit: Trees (*Alstonia*, *Holarrhena*), shrubs (*Thevetia*, *Nerium*), herbs (*Catharanthus*), woody twiners (*Allamanda*) or even succulent (*Adenium*), usually twining shrubs with milky sap in laticiferous vessels.
2. Leaves: Simple, exstipulate, opposite decussate or sometimes alternate (*Thevetia*) or in whorls of 3, entire, rarely stipulate (*Tabernaemontana*).
3. Inflorescence: A panicle, cyme or raceme or flowers solitary (*Catharanthus*), usually dichasial cyme.
4. Flower: Bracteate, bracteolate, complete, hermaphrodite, actinomorphic, pentamerous; hypogynous but rarely perigynous or epigynous (*Plumeria*).
5. Calyx: 5 sepals, usually gamosepalous, sometimes free (*Thevetia*, *Catharanthus*) or deeply lobed; valvate (*Catharanthus*), quincuncial (*Thevetia*); odd sepal posterior; rarely only 4 sepals; sometimes glandular appendages (*Squalellae*) present on the adaxial side.
6. Corolla: 5 petals, united into a tube, usually salver-shaped or funnel-shaped; twisted or rarely valvate; rarely only 4 petals; often hairy within or contain some corona-like outgrowths at the mouth of corolla tube.
7. Androecium: 5 stamens, epipetalous, alternate with the corolla lobes,

filament short; attached at the mouth or deeper in the corolla tube; anthers are dithecous, basifixed, often sagittate; introrse, dehiscence longitudinal; anthers sometimes basally awned; sometimes bear hairy appendage (*Nerium*) over the lobes.

8. Gynoecium: Usually bicarpellary, carpels apically united, superior or rarely half-inferior (*Plumeria*), 1-or 2-locular, with 2 to many anatropous and pendulous ovules in each locule; marginal placentation; style one and simple; stigma is characteristically thickened, massive and bilobed; receptive surface of stigma is situated below; rarely the carpels are 3-5.

   A nectariferous disc is often present around or the base of the gynoecium in several genera, e.g. *Thevetia*, *Catharanthus*, *Allamanda* and *Rauvolfia*.
9. Fruits and seeds: Fruit usually consisting of 2 follicles, or a berry (*Landolphia*), drupe (*Cerbera*), capsule (*Allamanda*), or 2 indehiscent mericarps. Seeds often flat with straight embryo and a crown of hairs.
10. Pollination and dispersal: Pollination is entomophilous. Insects are attracted mainly because of often large and conspicuous petals and the presence of nectar. Because the receptive stigmatic surface is located on the underside of the stigma, the cross pollination is the rule of the family. Self pollination is absent.

    Dispersal is affected mainly by wind because of winged seeds and presence of crown of hairs.
11. General floral formula: Ebr, $\oplus$, $\boxtimes$, K_5, $C_{(5)}$, $A_{(5)}$, $\underline{G_{(2)}}$.

Description of a Common Plant

A) *Catharanthus roseus* (Periwinkle or *Sadabahar*)

Habit: Erect, everblooming ornamental plant with milky latex. Leaf: Simple, opposite decussate, exstipulate, subsessile or petiolate, elliptic-obovate, entire, mucronate, unicostate reticulate. Inflorescence: Dichasial cyme or solitary axillary. Flower: Ebracteate, pedicellate, complete, hermaphrodite, actinomorphic, pentamerous, hypogynous, rosy-purple or pink. Calyx: 5 sepals, polysepalous, valvate. Corolla: 5 petals, gamopetalous, twisted, hypocrateriform. Androecium: 5 stamens, epipetalous, inserted at the mouth of corolla tube, filaments short; anthers sagittate, dithecous, basifixed or dorsifixed, introrse. Gynoecium: Bicarpellary, syncarpous; ovaries free and superior; many ovules, marginal placentation; two hypogynous nectarines are present, one on anterior and another on posterior side of the ovary. Fruit: Follicle. Floral formula: Ebr, $\oplus$, $\boxtimes$, K_5, $C_{(5)}$, A_5, $\underline{G_{(2)}}$.

Floral Biology

Periwinkle produces fairly large number of flowers per plant per day. An anther dehiscence occurs almost simultaneously with anthesis in non-self pollinating strains. Anther dehiscence in self pollinating strains occurs one day after anthesis. Intraflower self pollination in self pollinating strains is due to continued growth of gynoecium beyond the base of anthers and even after anthesis in majority of flowers. Stigma remains below the base of anthers even two days after anthesis did not come into contact with pollen up to flower drop. Self pollinating stains produce 18 to 24 seeds per fruit on natural intraflower self pollination. Non-self pollinating strains are fully self fertile and produce 22-25 seeds per fruit. Pollen fertility has been fairly high and ranges from 71 to 92 per cent. The percentage of fruit set in absence of pollinators is 75-85 per cent.

24

Araceae (Arum Family)

A family of 140 genera and over 4,075 species, Araceae are distributed mostly in tropical and some also in temperate regions of the world. About 25 genera and over 140 species have been reported from India, chiefly from western and southern parts. Some larger genera along with the number of their worldover reported species are *Anthurium* (500), *Philodendron* (275), *Arisaema* (150), *Amorphophallus* (100), *Rhaphidophora* (100), *Pothos* (75), *Alocasia* (70), *Monstera* (50), *Arum* (15), *Colocasia* (8), *Acorus* (2) and *Pistia* (1). *Colocasia* (*Colocasia antiquorum*) and 'money plant' (*Pothos aureus*) are two famous plants of Araceae.

Anthurium is a beautiful group of tropical aroids. It is cultivated globally for its attractive flowers and foliage. The flowering stalks have a long vase life and add an ethnic beauty to any floral decoration. Anthurium also called as 'Flamingo flower' or 'Boy flower', both referring to the structure of the spathe, and spadix. 'Anthra' is a worldwide knowledge center for the cultivation of anthurium, based on Holland and Germany. 'Anthura' performs activities in the field of research and development breeding, tissue culture technology, training and publishing cultivation know- how. 'Anthura' collaborates with research stations and test locations in Thailand, China, USA, Japan and India.

Selected Indian Genera

Acorus, *Alocasia*, *Anthurium*, *Amorphophallus*, *Colocasia*, *Pistia*, *Pothos*.

Economic Importance

1. Widely used vegetables are obtained from (i) the stem and root-stocks of *Alocasia indica* (mankanda), (ii) tuberous corms and leaves of *Amorphophallus campanulatus* (elephant foot yam), (iii) tubers of *Colocasia esculenta* syn. *C. antiquorum* (Arvi, Colocasia), and (iv) corms and leaves of *Remusatia vivipara* and *Xanthosoma niger*.
2. Industrial alcohol and starchy baby foods are also prepared from the tubers of *Colocasia esculenta*.

3. Certain plants are of medicinal value. These include (i) *Acorus calamus*, of which the rhizomes are beneficial in chronic diarrhea and dyspepsia, (ii) *Alocasia macrorrhiza*, of which the stem juice is used to relieve pain from scorpion bite, (iii) *Amorphophallus campanulatus*, of which the corms are used in treating piles and dysentery and (iv) *Scindapsus officinalis*, of which the dried fruits have anthelmintic properties.
4. Plants of ornamental value, grown commonly in gardens for their often variegated and strikingly handsome leaves are *Acorus calamus* var. variegatus, *Alocasia indica* var. metallica, *Caladium bicolour*, *C. picturatum*, *Colocasia antiquorum*, *Dieffenbachia picta*, *Monstera deliciosa*, *Pothos aureus*, *Scindapsus officinalis* and *S. pictus*.
5. Ornamental plants cultivated mainly for their flowers and spathes include several species of *Anthurium*, *Arisaema* (Indian turnip), *Dracunculus* (dragon arum), *Lysichiton* (western skunk cabbage), *Sauromatum* (Voondoo lily) and *Spathiphyllum*.
6. *Pistia stratioites* is cultivated for being used in aquaria.

Description of the Family

1. General habit: Highly variable; usually perennial, rhizomatous or tuberous herbs (*Arum*), climbers (*Pothos*), or tree like (*Philodendron*, *Dracontium*); rarely free-floating aquatic herbs (*Pistia*); some are epiphytes (*Anthurium*) or occur in marshy conditions (*Lasta*); often raphides or calcium crystals present in the sap of the plants.
2. Root: Fibrous, adventitious and usually of two types i.e. (i) climbing or clasping roots which cling to the support, and (ii) absorbing roots which enter into the soil and absorb nutrients; velamen develops in the aerial roots of epiphytic species.
3. Stem: Subterranean or underground in the form of tubers (*Arum*), corms (*Amorphophallus*), rhizome (*Acorus*) or root stocks (*Colocasia*); or aerial (*Pothos*) showing monopodial branching or sympodial; accessory buds often develop in leaf-axils.
4. Leaf: Variable in shape and types; single (*Arisaema*) or a few; basal (*Arisaema*), or cauline and alternate (*Pothos*); often hastate (*Typhonium*), sagittate or round; pinnately or palmately divided; petiolate (*Philodendron*) or sessile (*Pistia*); venation is usually parallel but rarely reticulate (*Arum*).
5. Inflorescence: Flowers usually grouped together to form a more or less cylindrical spadix subtended by a large coloured spathe; the upper

portion of spadix is usually naked and the lower portion bears whorls of small unisexual flowers, the male flowers above and the female below (Fig. 24.1); usually some sterile male and female flowers are present close to fertile whorls of male and female flowers; male and female flowers develop on different spadix in some genera, e.g. *Arisaema*.

6. Flower: Ebracteate, ebracteolate, usually unisexual but sometimes bisexual (*Acorus*, *Pothos*), actinomorphic, hypogynous or epigynous, dimerous or trimerous, small, often bad-smelling.
7. Perianth: Usually absent in unisexual flowers (*Arum*, *Pistia*) and present in bisexual flowers (*Acorus*, Fig. 24.2); if present, the perianth lobes are small, scale-like, 4 to 6 in number, present generally in 2 whorls of 2+2 or 3+3, free or rarely connate (*Spathiphyllum*).
8. Androecium: Usually 6 or lesser stamens present in one or 2 whorls, located opposite the perianth lobes, free or united into a synandrium (*Colocasia*, *Spathicarpa*); anthers dithecous, introrse, dehiscing by slits or pores; female flowers usually bear staminodes.
9. Gynoecium: Varied in structure but often reduced to a single carpel; sometimes 2 to 9 united carpels; superior or rarely inferior and embedded in the spadix; one to many locules and one to numerous ovules in each locule; placentation is basal (*Arisaema*, *Typhonium*), axile (*Pothos*) or parietal (*Arum*); style various, sometimes greatly reduced, stigma variable.
10. Fruit and seeds: Fruit usually a berry, clustered closely on the spadix and appearing as a multiple fruit. Seeds with or without endosperm.
11. Pollination: Flowers are protogynous and insect-pollinated. Insects are attracted because of the coloured spathe and characteristic bad smell emitted by inflorescences. Insects come into contact with the fertile female flowers and transfer to them any pollen they may be carrying from the male flowers of another plant. Some species of *Arisaema* show self-pollination.
12. General floral formula:
    a) Male flowers: Ebr, ⊠, ♂, P_0, $A_{3+3 \text{ or } 2+2}$, G_0.
    b) Female flowers: Ebr, ⊠, ♀, P_0, A_0, $G_{\underline{1} \text{ or } (\underline{2-9})}$.
    c) Bisexual flowers: Ebr, ⊠, ⊠, $P_{3+3 \text{ or } 2+2}$, $A_{3+3 \text{ or } 2+2}$, $G_{\underline{1} \text{ or } (\underline{2-9})}$.

Floral Biology Of Anthurium

Croat (1980) studied floral biology in some species of *Anthurium*. It produces flowers throughout the year. One flower emerges from each leaf axil. More flowers are produced during rainy season than in winter. The spathe and spadix are borne on a leafless stem or peduncle. Depending on the variety, the spadix contains 50-300 conspicuous flowers; arranged in a series of spirals. The flowers are protogynous, bisexual, actinomorphic, and hypogynous. The female phase extends up to 28th day from opening of spathe. After completion of female phase, there is an interphase of 7-15 days; after which male phase commences. The duration of male phase is 14-20 days. The bisexual flowers are embedded within the spadix, rhomboidal in shape, with four triangular tepals, which enclose the elliptical stigma.

Flowers develop acropetally; an individual flower has four perianth segments (petals) arranged in a four sided configuration which envelops four stamens with four loculed anthers. Pistil is cylindrical; ovary is two celled with 1-2 ovules. As the pistil develops, a stout style exerts to expose the receptive stigma. New flowers become receptive each day, because of uneven maturation of individual female flowers; beginning at the spadix base and advance gradually towards the top over a period of two weeks. Anthesis and dehiscence occur between 8 and 10 am. Pollens are starchy and non-starchy in nature and are available for several days on anthurium flowers. Stigma becomes receptive about 6-7 days before anther dehiscence. Seed is engulfed in juicy mucilaginous pulp and seed germinate within 10-40 days.

Protogyny is mainly responsible for allogamy in this crop and variety of pollinators *viz.*, ants, bees, flies, humming bird and weevils have been recorded. Plant produces large number of pollen grains per anther depending upon the variety. Pollen production is positively correlated with plant height, flower size and anther size and it ranges from 1213 to 1725 and 9.7 to 17.9 per cent, respectively in cvs. Lady Jane and Pink. Pollen grains are more or less uniform in size, round in shape with a single germ pore. During cooler period from December to April most anthurium varieties shed their pollen in an environment-controlled greenhouse.

References

Croat, T.B. 1980. *American Journal of Botany*, **67** (6): 888-904.

25

Asteraceae (Sunflower Family)

Asteraceae or Compositae is one of the largest families of flowering plants, represented by about 1,707 genera and over 23,600 species which are worldwide in distribution. Radford (1986) has mentioned the number of genera to be 'One thousand to 2000'. About 140 of its genera and over 700 species have been reported from India. Some of the largely represented genera include *Senecio* (2,000+), *Eupatorium* (1,200), *Vernonia* (1,000), *Hieracium* (1,000), *Centaurea* (600), *Aster* (500), *Helichrysum* (500), *Cousinia* (400), *Artemisia* (400), *Baccharis* (400), *Mikania* (250), *Bidens* (230), *Crepis* (200), *Inula* (200), *Erigron* (200), *Chrysanthemum* (160), *Cirsium* (150), *Ligularia* (150), *Helianthus* (110), *Lactuca* (100), *Hypochoeris* (100) and *Carduus* (100).

In the *Chrysanthemum* genus, the modern autumn flowering perennial *Chrysanthemum morifolium* Ramat is most important. The name of *C. morifolium* Ramat has been changed to *Dendranthema grandiflora* Tzvelev (Anderson, 1987; Heywood and Humphries, 1977 and Kitamura, 1978). Chrysanthemum (*D.*x *grandiflora*) is an internationally recognized floriculture crop. It ranks second among commercial flowers in the international market. It is known as 'Queen of East' in Japan while in USA, it is considered as the 'Queen of the fall flowers'. Flowers are in great demand in landascape, for making garland, wreath, *veni* and religious offerings. Pot chrysanthemums are commonly known as 'mums' and large flowered cultivars are used as potted plants.

Dahlias with their most gorgeously coloured flowers are very popular in the Indian gardens and are widely used for garden display exhibition, indoor decoration and flower arrangement. It is one of the most flamboyant flowers of winter season. Medicinal and nutrition value of tubers of dahlia is also mentioned. Diversity of forms and colours makes it the 'King of flowers'. Large decoratives are the most popular type of dahlias in Indian gardens and very popular in Europe as cut flowers. According to present trend, Indian bids fair to be one of the foremost countries producing superb dahlia cultivars.

Selected Indian Genera

Ageratum, *Aster*, *Callistephus*, *Calendula*, *Carthamus*, *Chrysanthemum*, *Cosmos*, *Dahlia*, *Eclipta*, *Eupatorium*, *Gaillardia*, *Helianthus*, *Helichrysum*, *Lactuca*, *Launaea*, *Sonchus*, *Tagetus*, *Vernonia*, *Zinnia*, etc.

Economic Importance

1. Ornamental plants: Well-known ornamental species of the family include *Aster amellus*, *Calendula officinalis*, *Centaurea moschata*, *Chrysanthemum carinatum*, *C. coronarium*, *C. morifolium*, *Coreopsis grandiflora*, *Cosmos bipinnatus*, *Dahlia excelsa*, *D. pinnata*, *Dimorphotheca sinuata*, *Gaillardia pulchella*, *Helianthus annuus*, *Helichrysum petiolatum*, *Tagetus erecta*, *Zinnia elegans* and *Z. linearis*.
2. Disease-causing plants: *Ambrosia artemissifolia* (Rugweed)-pollen causing hay fever; *Eupatorium urticaefolium* causes milk sickness in animals; *Pyrethrum hysterophorus* (carrot grass)- pollen cause skin allergy.
3. Medicinal plants: *Anthemis nobilis* dried capitula are used against dyspepsia; *Artemisia cina* flower heads provide the drug santonin, an antidote against intestinal worms; Leaves of *Blumea balsamifera* are used against excitement and insomnia while *B. lacera* leaves are given to cure cholera; heads of *Grindelia camporum* are used in bronchitis and whooping cough; leaves of *Inula helium* are effective against tuberculosis; Leaves of *Lactuca virosa* have sedative properties; capitula of *Sphaeranthus indicus* are used for curing stomachache and piles; capitula of *Spilanthes paniculata* are chewed to relieve toothache; leaves of *Tanacetum vulgare* are effective in chronic ulcers and rheumatism; roots and rhizomes of *Taraxacum officinale* provide the drug 'taraxacum' used as a mild laxative and leaves of *Tussilago farfara* are used in asthama, cough and colds.
4. Insecticides: Powdered by capitula of *Centrantherum anthelmenticum*, *Chrysanthemum coccineum* and *Pyrethrum cinerarifolium* are used in preparing insecticides.
5. Edible products: Tuberous roots of *Cynara scolymus*, *Dahlia tuberosa*, *Helianthus tuberosus* (Jerusalum artichoke) and *Tragopogon porrifolius* (vegetable oyster) are edible and used as food by man; *Lactuca sativa* (salad lettuce or Garden lettuce) leaves are eaten raw as salad; roasted seeds or *Taraxacum officinale* are used in place of coffee.
6. Oils: Oil obtained from seeds of *Carthamus tinctorius* (Safflower) is used for manufacture of soaps, paints, varnishes, etc.; hair-oil, 'Bhrangraaj',

is obtained from leaves of *Eclipa alba*; fatty oil, obtained from seeds of sunflower, is edible; a strong, aromatic essential oil, used in perfumery, is obtained from seeds of *Tagetus patula*; oil from seeds of *Xanthium strumarium* is used as illuminant.

7. Dye: A red dye safflower, obtained from *Carthamus tinctorius*, is used for colouring candles, butter and several liquors.

Why Asteraceae occupy highest position among Angiopsperms?

1. Cosmopolitan distribution from arctic to tropical zone.
2. Herbaceous growth habit.
3. Inflorescence capitulum.
4. Presence of large number of flowers in a small space which makes it possible to pollinate several flowers easily by a single insect.
5. Presence of both actinomorphic and zygomorphic flowers.
6. Presence of both unisexual as well as bisexual flowers.
7. Presence of calyx in the form of pappus.
8. Gamopetalous corolla.
9. Syngenesious anthers.
10. Protandry of androecium.
11. Reduction of carpels to 2 and number of ovules to only one.
12. Presence of inferior ovary and basal placentation.
13. Cypsela fruit.
14. Elborate mechanism of dispersal.

Description of Family

1. General habit: Usually annual or perennial herbs, but some are shrubs and only a few are trees (*Leucomeris*, *Vernonia arborea*); *Senecio*, the large genus of the family includes herbs (*S. vulgaris*), shrubs (*S. magnificus*), trees (*S. cruentus*) and even climbers (*S. scandens*); majority of the temperate zone Compositae are herbaceous; sap watery or milky; usually with rhizomes, stolons, tubers or fleshy roots; plants often with spiny or various types of vestiture or surface coverings, such as tomentose, lanate, pannose, strigose, etc.; few are xerophytes, e.g. *Hoplophyllum*.
2. Leaves: Usually alternate, sometimes opposite (*Dahlia*, *Helianthus*), rarely whorled (*Eupatorium*); mostly simple, sometimes needle-like or reduced to scales, margins often pinnately or palmately lobed or divided but leaves only rarely truly compound; exstipulate; frequently in basal rosettes; leaves

usually with oil passages; some contain latex (*Lactuca*, *Crepis*).

3. Inflorescence: The primary inflorescence is a head or capitulum with many flowers (called florets) borne on a conical, flat, concave, or convex receptacle (Fig. 25.2); each floret is often subtended by a receptacular bract called 'pale' or 'chaff'; the receptacle is often subtended by an involucres or bracts called 'phyllaries', bracts or phyllaries are of various shapes, sizes and textures, and present in one or more series. Usually the heads or capitula are arranged in various ways, such as racemes, panicles, corymbs or compound heads; in *Echinops*, the heads are compound, but each smaller head contains only one flower. The capitulum may be heterogamous or homogamous.
4. Flower: Flowers are actinomorphic or zygomorphic; bisexual, or unisexual, or even neutral (both stamens and carpels are aborted); pentamerous; epigynous.

   In heterogamous capitulum inflorescence e.g. Daisy, the central florets are disc florets which are bisexual and actinomorphic; and peripheral florets are ray florets which are ligulate, generally female or neutral and zygomorphic.

   In homogamous capitulum inflorescence, all florets are similar, usually bisexual, actinomorphic or zygomorphic, rarely all unisexual.
5. Calyx: Represented by pappus of bristles, awns or scales; or absent; epigynous.
6. Corolla: 5 petals, gamopetalous, valvate; represented by 3 basic types; (i) 5-lobed and tubular, (ii) ligulate with 3 to 5 teeth, (iii) bilabiate with 3 lobes in the upper lip and 2 lobes in the lower tip. In *Mutisia*, all florets are bilabiate while in *Lactuca* and *Cichorium* all florets are ligulate.
7. Androecium: 5 stamens, epipetalous, alternating with corolla lobes; anthers nearly always united into a tube around the style and filaments free i.e. syngenesious; dithenous, connective often prolonged, introrse, longitudinally dehiscent; in *Silybium*, the stamens unite by anthers as well as by filaments.
8. Gynoecium: Bicarpellary, syncarpous; ovary inferior, unilocular; one ovule, anatropous; basal placentation; style 1, usually 2 branched; stigma of diverse forms; an epigynous nectar-secreting disc is also present at the base of the corolla tube surrounding the style.
9. Fruit and seed: Fruit a cypsela, often crowned by the persistent pappus, sometimes enclosed by persistent bracts. Seed is 1 per fruit; with large and straight embryo and no endosperm.

10. Pollination: The pollination mechanism is entomophilous and noteworthy in Asteraceae. A single insect may pollinate numerous flowers of a capitulum in single visit. Insects are attracted by the conspicuous ligulate ray florets of the capitulum. This also benefits the often inconspicuous disc florets. Ray florets also protect the nectar from rain. Cross pollination is promoted also by the protandrous condition of the androecium. In case of failure of cross-pollination, self-pollination may also take place.
11. General floral formula: ·|· or ⊕, ⊠, K_{pappus}, $C_{(5)}$, $A_{(5)}$.

1. Taxonomy of China Aster

Leaves alternate, vary in shape or size. Leaves around middle of the stem are few centimeter long and are borne on winged petioles. Leaf blades have serrated margin. The bloom/ head/ capitulum/ inflorescence holds several hundred individual florets. The peripheral florets (petals) of the head are ray florets which are long, ligulate, generally female and zygomorphic. The central disc florets are bisexual and actinomorphic. The flower head is lined with layers of phyllaries, those in outer layer large and leaf like up to 3 cm long. The flower head contains one or two rings/ rows of ray florets, up to 3.5 cm long, of variable colour (red/pink/ violet/ purple/ white) and yellow disc florets in the centre. It has a pappus composed of one outer layer of reddish scales and two inner layers of white bristles.

Floral biology of China Aster

China aster is self pollinated crop. Strube (1965) studied some aspects of floral biology of China aster. Florets of a head, mature from the outside rows of ray florets inwards to the central disc florets i.e. centripetal. Successive rings of the florets mature each day. Several days may require between the maturity of first and last floret. The ray florets do not produce pollen while in disc florets, stamen and pistil do not mature simultaneously. The stigma of the individual floret unfolds after the pollen is discharged from the flower. The China aster is therefore, mentioned as geitonogamous.

2. Taxonomy of Chrysanthemum

The genus *Chrysanthemum* comprises about 200 species, majority have been sub-divided into 38 satellite genera of the chrysanthemum complex (Anderson, 1987). Taxonomic reclassification of chrysanthemum complex has been done repeatedly. Linnaeus classified species of this complex into 5 major genera as *Argyranthemum* Webb ex Schultz Bip., *Chrysanthemum sensu stricto* L., *Dendranthema* Des Moul., *Leucanthemum* Miller, and *Tanacetum* L. Numerous independent satellite genera fill out the remainder of the complex.

Table 25.1: Revised taxonomic designations of genera and commercial species within the chrysanthemum complex (Boase *et al*., 1997; Anderson, 1987)

Genera	Species	Previous name	Common name	Characters
Primary genera				
Argyranthemum Webb. ex. Schultz Bip.	*frutescens* Schultz Bip.	*C. frutescens* (3n=27)	Paris/ Marguerite Daisy	Popular as pot plant
Chrysanthemum L.	*coronarium* Schousboe.		Rainbow/ Garland chrysanthemum	Winter season annual
	segetum L.		Corn marigold	
Dendranthema Des. Moulins	x *grandiflora* Tzvelv.	*C.* x *morifolium* (6n=54)	Florist's Chrysanthemum/ hardy garden mum	Single daisy flower type, perennial, hybrid species with the composite inflorescence & involved in evolution of Florists' chrysanthemum
	indicum L.	*C. indicum* (4n=36) *C. boreale,* and *C. sinense*	Mother chrysanthemum	
	japonicum Makino	*C. japonicum*		
	weyrichii (Maxim.) Tzvelv.	*C. weyrichii*		
Leucanthemum Miller.	x *superbum* Berg. ex. Kert.,	*C.* x *superbum* *C. maximum*	Shasta daisy (Max daisy or Moon daisy)	Perennial, used as cut flower
	vulgare Lam.	*C. leucanthemum*	Ox-eye daisy	
Tanacetum L.	*balsamita* L.	*C. balsamita*	Costmary	Grown in temperate regions
	cinerariifolium (Trev.) Vis	*C. cinerariifolium*	Pyrethrum	
	coccineum Willd.	*C. coccineum* (2n=18)	Persian chrysanthemum/ Painted daisy/ Colour daisy	Perennial, seed propagated, grown in temperate zones
	macrophyllum Schultz Bip.	*C. macrophyllum*	Fever few	-
	ptarmiciflorum Schultz Bip.	*C. ptarmiciflorum*	Dusty Miller	-
	parthenium Schultz Bip.	*C. partheniume* cv. Santana	Fever few	Day neutral plant
	vulgare L.	*C. vulgare*	Tansy	-

Genera	Species	Previous name	Common name	Characters
Independent Satellite genera				
Ajania Polj.		-	-	-
Arctanthemum Tzvelv.		-	-	-
Balsamita P.Miller	*major* Desf.	*C. balsamita*	Costmary	-
Coleostephus Cass.	*miconis* L. Cass	*C. miconis*	-	-
Glossopappus G. Kunze.	*macrotus* (Durieu) Briq.	-	-	-
Heteranthemis Schott.	*viscidehirta* Schott.	-	Sticky oxeye	-
Hymenostemma Kunze ex. Willkomm	*pseudoanthemis* (G.Kunze) Willk	-	-	-
Ismelia Cass.	*carinatum*	*C. carinatum*	Painted daisy/ Tricolour chrysanthemum	Winter season annual
Leucanthemella Tzvelv.	*serotinum* Tzvelv.	*C. serotinum*	High daisy	-
Leucanthemopsis (Giroux) Heywood	*alpina* (L.) Heywood	*C. alpinum*	-	-
Nipponanthemum Kitamura	*nipponicum* (Franch. ex. Maxim) Kitamura	*C. nipponicum*	Nippon Oxeyed daisy	-
Phalacrocarpum (DC.) Willkomm	*oppositifolium*	-	-	-
Pinardia Cass.	*anisocephala* Cass.	-	-	-
Prolongoa Boiss.	*pectinata* (L.) Boiss.	-	-	-

Engler and Prantl (1926) originally divided the genus *Chrysanthemum* into eight sections of which, four sections contain annual species while remaining sections consists of perennial species. Cultivated chrysanthemum (*D.* x *grandiflora*) and pyrethrum (*Tanacetum cinerariifolium*, *T. coccineum*), are the two most important domesticated species within the genus. These were taxonomically classified into section VI *Pyrethrum*. Long stalked capitulum with a composite inflorescence is the most common characteristic shared by the 50 species found in this section (Anderson, 1989). Taxonomic classification is based on embryo sac development, cypselar anatomy, plant habit, molecular markers, and phytochemical characteristics (Harling, 1951; Heywood, 1958; 1959). Heywood and Humphries (1977) have included one species *Leucathemum frutescens* having anemone flowers as a diagnostic key.

Floral biology of Chrysanthemum

Chrysanthemums are normally self incompatible and entomophilous. Bees (*Apis mellifera*) are the pollinators. The time immediately after flower opening is ideal for effecting crosses. Stigma remains receptive up to 15 days and reaches peak on the 11th day when style elongates distinctly and presents a definite angle (Zhao *et al.*, 2008). Pollen grains are trizonocolporate. Pollen shape is circular and semicircular to angular. Intravarietal variation in pollen size was noted. No relationship between pollen size and chromosome number was observed (Endo and Inada, 1990). Pollen fertility is accelerated by flowering stage, season and flower character. Pollen longevity lasts for about 16 days under room temperature and 35-45 per cent relative humidity. At diploid level disc floret numbers are variable whereas ray floret number is static. Disc florets, in self pollination, are responsible for increasing the variability in ovule number. In polyploidy species ovule populations are dynamic and interactive. Seed set occurs only in disc florets and greatest number of disc florets in the inflorescence is formed under short day conditions and at relatively low temperature in late summer (Dvoryaninova and Beztian, 1984). All varieties do not set seeds. Only a few varieties set seed in large flowered chrysanthemum whereas more varieties set seed in small flowered ones.

Taxonomy of *Zinnia* spp.

The genus *Zinnia* consists of 20 species of annual herbs or perennial shrubs or sub-shrubs (McVaugh, 1984). *Zinnia* species are classified into two sub-genera: *Diplothrix* and *Zinnia*.

Sub-genus *Diplothrix* consists of six species of perennial cespitose shrubs or sub-shrubs distributed in south western USA and north Mexico. The basic chromosome number is n=10. Diploid and polyploid taxa have also been documented. The only species cultivated extensively as an ornamental is Rocky Mountains zinnia (*Z. grandiflora*) or Plains zinnia.

Sub-genus *Zinnia* consists 13 species, distributed in warm temperate and tropical regions of North America with basic chromosome number n=12 (Torres, 1963a, b). It is classified into three sections *viz., Mendezia, Tragoceros,* and *Zinnia.*

Section *Mendezia* found in western coastal Mexico State contains eight species, of which six are herbaceous annual and two are perennial. Chromosome no. n=10, 11 and 12 have been reported. *Zinnia elegans* is the sole species in this section which is cultivated extensively.

Section *Tragoceros* contains several species of herbaceous annuals with basic chromosome no. n=11 (Torres, 1963a). Species have small flowers (ligules <1 cm in length) and none are cultivated as ornamentals.

Section *Zinnia* consists of three species *viz., Z. haageana, Z. peruviana* and *Z. violacea* with chromosome no. 2n=2x=24 for all three species.

Following key should serve to identify members of *Z. acerosa* complex, as currently perceived (Turner, 2012).

1. Rays bright lemon yellow---*Z. citrea*
1. Rays white or pale lemon yellow
    2. Ray florets 2-3; disc florets 2-4 (5)--------------------------*Z. oligantha*
    2. Ray florets 5-8; disc florets 5-20
        3. Leaves elongate, 3 cm long, glabrous----------------*Z. coahuilana*
        3. Leaves variously curvate, 1-2 cm long, clearly pubescent
            4. Stems mostly prostrate----------------------- *Z. guanajuatensis*
            4. Stems mostly erect
            5. Ray florets 5-8, ligules drying pale yellow; disc florets 10-20 ------- --*Z. austrotexana*
            5. Ray florets mostly 5, ligules drying chalky white; disc florets 5-10 --------------------------------------*Z. acerosa.*

Floral Biology of zinnia

Within a capitulum, florets mature sequentially from outermost to innermost whorl. Anthers in disc floret form a tube. Anthers dehisce inside of tube and elongating style pushes the pollen to the apex of disc floret. Disc florets are dichogamous. The stigmatic surface is not exposed for pollination until after the stylar lobes have reflexed. Unpollinated pistil (*Z. violacea*) remains turgid for more than 3 weeks after style lobes reflects. Seed set diminishes as pistils age (Miyajima, 1998). Pollination within five days from pistil unfolding results in high seed set. Pollen of zinnia is trinucleate and short lived less than one day at more than 20^{0}C (Hoekstra, 1983). Pollen grain number per tubular floret is 12,000-18,000. Each zinnia floret produces a single seed (achene). Capitulum mature in about 7-8 weeks from pollination. Poor pollination is major cause of low seed set of open pollinated zinnias. Capitula with more tubular florets produce more seeds than those with fewer tubular florets. A tubular floret produces abundant pollen, but pollen flow to ray stigmas is limited due to infrequent visits by pollinators. Moreover, in double flowered capitulum, newly opened ray florets overlap on the pistils, and cause low seed set. The composition of capitulum is a major cause of these problems.

References

Anderson, N.O. 1987. *HortScience*, **22**(2): 313.

Anderson, N.O. 1989. Rapid generation cycling and inbreeding depression in chrysanthemums. Ph.D. Dissertation, University of Minnesota, St. Paul, U.S.A.

Boase, M.R., Miller, R. and Deroles, S.C. 1997. *Plant Breeding Reviews*, **14**: 321-361.

Dvoryaninova, K.F. and Beztian, T.V. 1984. *Horticulture Abstract*, **55**(8): 6231.

Endo, M. and Inada, I.1990. *Plant Breeding Abstract*, 1992, **62**(6): 5679.

Engler, A. and Prantl, K. 1926. Die Natuerlichen Pflanzenfamilien. W. Engelmann, Leipzig, **4**(5): 277-278.

Harling, G. 1951. *Acta Horticulturae Bergiani*, **16**: 2-54.

Heywood, V.H. 1958. *Botanical Society of British Isles. Proceedings*, **3**: 177-179.

Heywood, V.H. 1959. *Agriculture Lusitana*, **20**: 205-216.

Heywood, V.H. and Humphries, C.J. 1977. Anthmideae systematic review, *In*:V.H. Heywood, J.B. Harborne, and B.L. Turner (Eds.). The biology and chemistry of the Compositae. Academic Press, New York, pp. 851-898.

Hoekstra, F.A. 1983. Physiological evolution in angiosperm pollen: Possible role of pollen vigor, in: D.L. Mulcahy and E. Ottaviano (eds.), *Pollen: Biology and Implications for Plant Breeding*, Elsevier Science Publishers, Amsterdam, pp. 35-41.

Kitamura, S. 1978. *Dendranthema* at Nipponanthemum. *Acta Phytotaxonomica et Geobotanica*, 29: 165-170.

McVaugh, R. 1984. *Flora Novo-Galiciana, Volume 12. Compositae*. The University of Michigan Press, Ann Arbor, Michigan.

Miyajima, D. 1998. *HortScience,* **33**: 696-698.

Radford, A.E. 1986. Fundamentals of Plant Systematics, Haper and Row Publishers,Inc. New York.

Strube, H. 1965. *Dische Gartenb*, **12**: 134-136.

Torres, A.M. 1963a. *Brittonia*, **15**: 1-25.

Torres, A.M. 1963b. *Brittonia*, **15**: 290-302.

Turner, B.L. 2012. *Phytoneuron*, **19**: 1-8.

Zhao, H., Chen, F., Wang, Y., Chen, S., Fang, W. and Guo, W. 2008. *Acta Horticulturae,* **766**: 405-412.

26

Bignoniaceae (Bignonia Family)

Represented by about 120 genera and over 800 species. Bignoniaceae are distributed mainly in tropical and sub-tropical, and a few in temperate regions. About 15 genera and 40 species are found in India. Some of the larger genera along with their approximate number of worldover reported species are *Tabebuia* (100), *Arraabidaea* (70), *Jacaranda* (50), *Anemopaegma* (30), *Tecoma* (16), *Catalpa* (11) and *Crescentia* (5).

Selected Indian Genera

Jacaranda, Kigelia, Millingtonia, Tecoma, Heterophragma, Oroxylum and *Dolichandrone.*

Economic Importance

1. Ornamental plants of the family include handsome trees (*Catalpa bignonioides, Haplophragma adenophyllum, Jacaranda mimosaefolia,, Kigelia pinnata, Millingtonia hortensis, Spathodea campanulata* and *Tabebuia spectabilis*), shrubs (*Bignonia unguis-cati*), beautiful tendril-climbers (*Adenocalymna alliaceum, Bignonia venusta* syn. *Pyrostegia venusta*), aerial rootlet climbers (*Campis grandiflora* and *C. radicans*) and herbaceous genus *Incarvillea.*
2. Some of the timber-yielding plants include *Catalpa, Dolichandrone, Haplophragma, Millingtonia, Oroxylum, Radermachera, Spathodea, Stereospermum, Tabebuia* and *Tecoma.*
3. An inferior quality of cork is prepared from the bark of *Millingtonia hortensis* (Akash neem or Indian cork tree).
4. *Crescentia cujete* (Calabash-tree) is grown for its gourd-like fruits, which, when hollowed out and dried, are used for holding liquids.

Description of the Family

1. General habit: Mostly woody lianas or climbers, climbing with the help of tendrils; or trees or shrubs; only rarely herbs; some climb with the help of aerial roots (*Campsis radicans*), others are twiners (*Pandorea,*

Tecomaria) and most genera are tendril climbers. Anatomically, wedges of phloem develop in stem in between secondary xylem.

2. Leaf: Usually opposite, pinnately compound, exstipulate, terminal leaflet usually modifying into tendril; tendrils may be simple or branched, hooked, or provided with some adhesive discs; sometimes the leaves are simple, and rarely whorled (*Diplanthera*).
3. Inflorescence: Usually cymose (dichasial cyme) with bracts and bracteoles.
4. Flower: Bracteate, bracteolate, complete, hermaphrodite, zygomorphic, hypogynous, often showy.
5. Calyx: Usually 5 united sepals, 5 toothed, often campanulate, valvate.
6. Corolla: Usually 5 united petals, campanulate or funnel-shaped, imbricate; sometimes bilipped (2/3) with the upper lip of 2 petals and lower lip of 3 petals.
7. Androecium: Typically 4 epipetalous stamens, didynamous, alternipetalous, with a posterior staminode; sometimes 2 (*Catalpa*) and rarely 5 (*Oroxylum*) stamens; usually the anther cells one above the other; dithecous, introrse, longitudinal dehiscence.
8. Gynoecium: Bicarpellary, syncarpous, superior, bilocular, numerous anatropous ovules in each locule, axile placentation; sometimes unilocular (*Kigelia*) with parietal placentation; style simple, terminal; stigma bilipped; a hypogynous nectar-secreting disc is present.
9. Fruit and seeds: Fruit usually a capsule, sometimes fleshy and indehiscent. Seeds often winged, non-endopspermic, with straight embryo.
10. General floral formula: Br, Brl, ·|·, ⊠, $K_{(5)}$, $C_{(5)}$, $A_{4+1\ \text{staminode}}$, $G_{\underline{(2)}}$.

27

Cactaceae (Cactus Family)

Cacti are eulogized as 'Plants of 21st century'. They have gained popularity due to many exciting shapes, blooms, ease of culture and long life without least care. Cacti are categorized as fleshy stem succulent with photo-synthetically active stem. Good for indoor decoration in pot, outdoor decoration, landscaping, public garden, rock garden, etc.

The family Cactaceae includes approximately 131 genera and 1,750 species. But different authors mention the number of genera differently between 30 to 200. Members are distributed chiefly in semi-desert or drier regions of tropical and sub-tropical America, and also in Africa, Mauritius, Seychelles, India, Sri Lanka and Australia. Mainly the Cactaceae are the succulents of New World. Main species of Old World is *Rhipsalis*. Some of the larger genera along with their number of reported species are *Mammillaria* (200-300), *Opuntia* (250), *Echinocereus* (75), *Rhipsalis* (60), *Cereus* (50), *Cephalcereus* (48), *Echinopsis* (35), *Epiphyllum* (21) and *Pereskia* (20).

The classification of the family Cactaceae remains uncertain as of August 2023. Since the mid-1990s, the system produced by the International Cactaceae Systematics Group (ICSG) of the International Organization for Succulent Plant Study has been used as the basis of many published classifications. Hunt (2006) recognized Cactaceae family has 4 sub-families viz. Pereskioideae (the leaf cactus, 20 spp.), Opuntioideae (300 spp.), Cactoideae (1,000 spp.) and Maihuenioideae (2 spp.). Three tribes: Pereskieae, Opuntieae, Cereae. Tribe Cereae has 8 sub-tribes: Cereanae, Hylocereanae, Echinocereanae, Echiocactanae, Cactanae, Coryphanthanae, Epiphyelanae, Rhipsolidanae. The cacti of sub-tribe 'Echinocereanae' are commonly known as 'Hedgehogs'.

Selected Genera and Species

Some commonly grown ornamental cacti include species of *Cereus*, *Cephalocereus*, *Echinocereus*, *Echinocactus*, *Mammillaria*, *Opuntia*, *Rebutia*, and *Stenocereus*. Stems of several cacti are used as emergency fodder. Fruits of *Opuntia* are edible; a red dye, cochineal, used in cosmetics, is obtained from several species of *Opuntia*. Some species of *Cereus* an *Opuntia* have medicinal properties. Uses of cacti are:

1. Outdoor growing: The most beautiful flowering and colourful cacti and succulents are grown outdoors. In open-air cultivation the plants body develops faster, the shape also grown to its best and flowers develop beautifully.
2. Bowl culture: Any kind of bowl or dish, glazed or unglazed, china clay, wooden or plastic material is suitable for growing cacti and succulents, provided that depth is enough to hold plant upright e.g. *Opuntia imbricata*, *O. cylindrica*, and *Crassula argentea*, *C. arborescens*.
3. Cold greenhouse: Most of the plants will grow well in the cold frames or greenhouse, besides, many others that are too big or unsuitable in other ways for indoors.
4. As fragrant flowers: Some cacti produce sweet fragrance along with their beauty of plant e.g. *Astrophytum silversrii*, *Echinocereus luteus*, *Echinopsis multiplex*, *Hylocereus extensus*, *Nyctocereus serpentinus*.
5. For desert garden: Many cacti and succulents are suitable for displaying in desert garden e.g. *Mammillaria cephalo*, *Cephalocereus senilis*, *Cleistococtus strausii*.
6. For hanging baskets: E.g. *Aporocactus flagelliformis*, *Rhipsalis cassutha*, *Hatiorasalic ornioedes*, and Kalanchoe, Euphorbia, etc.

Description of Family

1. General habit: Fleshy, succulent or xerophytic, perennial herbs, shrubs or trees of diverse forms, with spines or bristles or both.
2. Roots: Root system generally shallow, with elongated, slender but fleshy roots.
3. Stem: Fleshy, of various shapes, rarely bearing normal leaves, usually provided with spines or barbed bristles (glochidia) which are generally considered to be modified leaves; spines and glochidia develop from small cushion-like structures called 'areoles'; in several genera, the stem assumes leaf-like flattened form (e.g. *Epiphyllum*, *Rhipsalis*).
4. Leaves: Simple, highly reduced or scale like, or even absent in many genera, or flat and fleshy; spines or tuft of bristles develop from areoles.
5. Inflorescence: Flowers mostly solitary, showy and borne upon or near the areoles; only rarely corymbose or paniculate (*Pereskia*).
6. Flower: Exceptionally beautiful, large, showy but only very rare (*Mammilaria*, (Fig. 27.1), *Opuntia*, *Cereus*, *Echinocereus*,

Schlumbergera); (Fig. 27.2); sessile; mostly bisexual and actinomorphic or nearly zygomorphic; epigynous.

7. Perianth: Perianth parts numerous, showing a gradual transition from sepals to petals; spirally arranged; sepals often petalloid; petals epigynous and present in several series; often fuse to form a perianth tube or hypanthium.
8. Androecium: Stamens numerous, in several series or groups, epipetalous or inserted at the base of petals; anthers 2-celled, basifixed, introrse; longitudinally dehiscent.
9. Gynoecium: Two or many carpels, syncarpous; ovary inferior, unilocular, numerous anatropous ovules; parietal placentation, very rarely basal placentation; style 1 and simple; stigmas often as many as the carpels.
10. Fruit and seeds: Fruit usually a many-seeded berry, often spiny, scaly, bristly or glochidiate. Seeds with scanty or no endosperm, often arillate, with curved or straight embryo.
11. General floral formula: ⊠, ⊠, $P_{\infty \text{ or } (\infty)}$, A_{∞}, $G_{(2-\infty)}$.

Reference

Hunt, D.R. 2006. *The New Cactus Lexicon* (two volumes), Milborne Port: dh books.

28

Caryophyllaceae (Pink Family)

It is represented by about 100 genera and over 2,200 species which are either annual or perennial, Caryophyllaceae are cosmopolitan in distribution, and mainly found in the north temperate or warm temperate regions of the world. Over 20 genera and about 100 species have so far been reported from India. Some of the larger genera, along with their common names and number of species in parenthesis include *Silene* (catchfly, 500), *Dianthus* (carnation, 300), *Arenaria* (sandwort, 250), *Gypsophila* (baby's breath, 125), *Stellaria* (chickweed, 100), *Cerastium* (mouse-ear chickweed, 100), *Drymaria* (50), *Spergula* (40), *Saponaria* (30), etc.

Carnation, commonly named as 'Divine flower', 'Clove pink', and 'Gilly flower', is one of the most beautiful and commercially important flower crops of the world. It is excellent for cut flowers, bedding, pots, edging and rock gardens due to its exquisite shape, wide range of colours, light weight and good vase life. In Europe and USA, perpetual carnation is cultivated on large scale in greenhouses for production of cut flowers. In terrain parts of France and Holland, carnation flowers are used for perfume extraction and only light coloured flowers are useful for this purpose.

Selected Genera and Species

Genera	Importance
Arenaria	*A. grandiflora* used as ornamental;
Dianthus	*D. caryophyllus* (carnation) used as cut flower & yields perfume, *D. barbatus* (sweet William), *D. plumarius* (pink), *D. sinensis* (rainbow pink) are ornamental, used as diuretic & against gonorrhea; *D. anatolicus* is used to cure intermittent fever;
Drymaria	*D. cordata* is used to check soil erosion
Gypsophila	*G. paniculata* (baby's breath) used as ornamental;
Saponaria	*S. vaccaria* (cowcockle) used as ornamental, against urinary bladder diseases; *S. officinalis* (soap wort) roots contain saponin & used for washing wood and silk
Silene	*S. pendula* (catchfly) used as ornamental; juice of *S. apetala* is useful in eye troubles

Genera	Importance
Spergula	*S. arvensis* seeds are used in lung tuberculosis, diuretic; fodder for cattles is obtained from *S. arvensis*
Stellaria	*S. media* leaves are used against inflammations of digestive and respiratory tracts; entire plant of *S. semivestita* has anti-cancerous properties.

Description of Family

1. General habit: Mostly annual or perennial herbs, rarely suffrutescent shrubs; arctic forms are reduced and possess caespitose habit.
2. Stem: Herbaceous; rarely with a basal woody part; nodes swollen; branching usually dichotomous but sometimes monopodial.
3. Leaf: Simple, usually opposite decussate, rarely alternate (*Dysphania*), connected at the base by a transverse line; linear to lanceolate, entire; usually exstipulate, and if stipulate then the stipules are scarious (*Paronychia*); leaves highly reduced in arctic forms (*Lychnis*).
4. Inflorescence: Flowers either solitary terminal (*Arenaria*) or arranged in dichasial cymes (cincinus-type).
5. Flower: Bracteate, bracteolate (*Spergula*), complete, hermaphrodite but rarely unisexual (*Lychnis*), actinomorphic; hypogynous but perigynous in *Arenaria*; pentamerous but tetramerous in *Sagina*; many bracts appear like epicalyx in *Dianthus*.
6. Calyx: 5 sepals, polysepalous or united into a tube at the base; rarely 4 sepals (*Sagina*); quincuncial or imbricate and persistent.
7. Corolla: 5 petals, free; sometimes 4, rarely minute or none (*Sagina*); often notched at the tip; differentiated into a claw and limb; caryophyllaceous; petals deeply bifid in *Stellaria media*.
8. Androecium: Stamens typically 5 or 10, or as many as petals, or twice as many as petals, arranged in 1 or 2 whorls; obdiplostemonous i.e. stamens of outer whorl lie opposite the petals while that of inner whorl alternate with the petals; polyandrous or basally connate; staminodia sometimes present (*Stellaria media*, *Arenaria*); stamens of outer whorl are absent in *Arenaria*; in *Stellaria media* some or all the stamens of outer whorl are absent in *Arenaria*; in *Stellaria media* some or all the stamens of outer whorl and even 23 of inner whorl are missing or present in the form of staminodes; only 8 stamens in 2 whorls in *Sagina*; 5 stamens of inner whorl are absent in *Paronychia*; anthers 2-celled, longitudinally dehiscent, introse.

9. Gynoecium: Carpels 2 (*Dianthus*, *Saponaria*), 3 (*Silene*) or 5 (*Lychnis*); syncarpous; superior but slightly inferior in *Arenaria*; unilocular; numerous ovules; free-central placentation; single basal ovule in *Paronychia*; styles and stigmas as many as the carpels.
10. Fruits and seeds: Fruit is usually a dry capsule opening by valves or teeth; rarely in achene or utricle; very rarely a berry (*Cucubalus*). Seeds with usually curved embryo around the perisperm.
11. Pollination and dispersal: Pollination is entomophilous. Seeds cannot escape from fruit until shaken by wind or animals.
12. General floral formula: Br, Brl, ⊕, ⚥, K_5, C_5, A_{5-10}, $G(\underline{2-5})$.

Description of Some Common Plant

A) *Dianthus caryophyllus.*

Habit: Annual, ornamental herb. Stem: Aerial, erect, solid, herbaceous. Leaf: simple, exstipulate, opposite decussate, sessile, united at the base, lanceolate, entire, acute, unicostate reticulate. Inflorescence: Solitary terminal or dichasial cyme. Flower: Multibracteate, bracts appear like epicalyx; pedicellate, complete, hermaphrodite, actinomorphic, hypogynous, pentamerous. Calyx: 5 sepals, gamosepalous, tubular, quincuncial or valvate. Corolla: 5 petals, polypetalous, caryophyllaceous, twisted. Androecium: 10 stamens, arranged in 2 whorls of 5 each, polyandrous, obdiplostemonous; anthers dithecous, dorsifixed, introrse. Gynoecium: Bicarpellary, syncarpous, superior; ovary unilocular at base but bilocular at top; ovules many, free-central placentation; styles 2, long; stigma simple. Fruit: Capsule. Floral formula: Br, ⊕, ⚥, K_5, C_5, A_{5+5}, $G(\underline{2})$.

B) *Stellaria media* (Chickweed).

Habit: Annual, herbaceous common weed. Leaf: simple, exstipulate, opposite decussate, sessile, ovate or oblong; entire, acute, unicostate reticulate. Inflorescence: Dichasial cyme. Flower: Bracteate, ebracteolate; pedicellate, complete, hermaphrodite, actinomorphic, hypogynous, pentamerous. Calyx: 5 sepals, free, quincuncial, green. Corolla: 5 petals, free, petals deeply bilobed appearing as if they are 10; imbricate or valvate. Androecium: Stamens 10 or reduced to 5 or 8, leaving the rest abortive in the form of staminodes; present in 2 whorls; obdiplostemonous; anthers dithecous, basifixed, introrse. Gynoecium: Tricarpellary, syncarpous, superior; unilocular; numerous ovules, free-central placentation; styles 3 but reduced; stigma 3. Fruit: Capsule. Floral formula: Br, ⊕, ⚥, K_5, C_5, $A_{10 \text{ or } 5-8}$, $G(\underline{3})$.

Taxonomy of Carnation

There are four types of pinks *viz.*, cottage, rockery, annual and cluster-head. Having been developed from *D. barbatus*, Sweet Williams is also classified as pinks, but it easily distinguishable by its sword-shaped leaves, and cluster of flowers.

Floral biology: Plants take 168 days for bud formation. The whole developmental process of floral bud can be divided into six stages. Flower bud takes an average of 16 days for its full development i.e. 6th stage which is suitable for hybridization purposes. At that time, bud reaches 2.0 cm length and 0.6 cm diameter. Anthesis time is from 10 am to 2 pm and 8 to 10 pm and is influenced by environmental factors. Cloudy weather, rains, low temperature and high relative humidity delays anthesis. Anther dehiscence commences immediately after anthesis with a peak between 9 and 10 hrs. Stigma become receptive 48 hours after anthesis (between 10-11 am) and receptivity is highest at 72 hrs after anthesis. Pollen grains are irregular in size, heavy and sticky. TTC stain is unsuitable to determine pollen viability. Pollen viability is found to be 96.26 per cent. Maximum pollen germination is seen in six per cent sucrose media. Self pollen grows slower than cross pollen. For crossing purpose, pollen supply can be increased by choosing temperature >20^0C during flower development (Kho and Baer, 1973). Rana (1999) studied some aspects of floral biology and seed set studies in *D. barbatus*. Fruit setting takes place 11 days after anthesis. On an average 30 days are required for capsule maturity. Maximum capsule set (89%) and seeds per capsule (52) were seen in artificial pollination two days after flower anthesis. Highest number of seeds per capsule (89) was obtained with hand cross pollination compared with self pollination (42 seeds).

Dyaberi *et al.* (2015) studied floral biology and seed setting in standard carnation at IIHR, Bengaluru. Each flower produces on an average 4 to 14 stamens with average of 4 to 9 pollen grains having round shape. The number of styles/ pistil varied from 2 to 4. Stigma remains receptive one day after anthesis and continues up to 3rd day of opening with maximum receptivity. Anther dehiscence occurs in bud and half opened to fully opened flowers indicating protandry and allogamy. Capsule matures in about 40 to 60 days after artificial cross pollination and highest number of seeds per capsule were recorded in genotypes IIHRCG 204 (8) and IIHRCG 210 (7). Highest pollen germination was recorded in genotype IIHRCG 202 at 1, 2 and 3 hour of incubation period. No seed set was observed in all the 12 genotypes on selfing and natural cross pollination. IIHRCG 207 was found as best parent for hybridization as having better flower quality and adequate quantity of pollen production. The genotypes IIHRCG 210 and IIHRCG 202 were suitable as good seed and pollen parents. This was attributed to compatible ploidy (either diploid or tetraploid background), male and female fertility and absence of incompatibility. Shafi Bhat *et al.* (1993) reported that best medium for pollen viability test through *in vitro* pollen germination is 4 per cent sucrose with 27.73 per cent germination. Pollen viability ascertained by staining test ranged from 89.18 per cent for fresh pollen to 31.94 per cent after 10 days of storage under laboratory conditions (20-25^{0}C).

References

Dyaberi, A., Dhananjaya, M V., Kumar, R. and Rao, T.M. 2015. *Indian Journal of Agricultural Sciences*, **85**(9): 1175-1180.

Kho, Y.O. and Baer, J. 1973. *Euphytica,* **22**: 467-470.

Shafi Bhat, M., john, A.Q. and Adbullah, M. 1991. *Progressive Horticulture*, **23**: 60-67.

Rana, N. 1999. Floral biology and seed setting studies in *Dianthus barbatus*. L. *M.Sc. (Hort.)* thesis submitted to Dr.Y.S. Parmar University of Horticulture and Forestry, Solan.

29

Casuarinaceae (Casuarina/ Beefwood Family)

It is a monotypic or unigeneric family represented by four genera containing about 91 species distributed mainly in north-east Australia and also in Fiji, New Caledonia, Malaysia and Mascarene islands. Only one species (*Casuarina equisetifolia*) occurs in India, and this is the most common species of the world.

Selected Genera

Casuarina, *Gymnostoma*, *Allocasuarina*, *Ceuthostoma*.

Economic Importance

Casuarina is important mainly as timber tree. The wood of *C. equisetifolia*, *C. cunningbamia* and *C. stricta* is used for furniture, for fuel and also for basket making. The bark of *C. suberosa* and *C. stricta* is utilized for tanning. *C. equisetifolia* is usually grown as an ornamental and also in order to reclaim sandy dunes in India and several countries.

Description of the Family

1. Habit: Xerophytic, evergreen, shrubs or trees.
2. Stem: Woody, much branched, with jointed, whorled or striate branches having prominent nodes and internodes with longitudinal ridges and grooves. Anatomically, the stem contains (i) ridges and grooves, (ii) thick cuticle, (iii) sunken stomata in grooves protected by long hairs, (iv) sclerenchyma and green palisade parenchyma in the ridges, (v) vascular bundles in two alternate rings, (vi) wood contains vessels, tracheids, fibres and parenchyma.
3. Leaf: Scale like, 4-16 on each node, whorled, minute, basally connate forming a sheath around the node, usually linear to lanceolate; leaf tips appear only as minute teeth; leaves resemble with that of *Ephedra* and *Equisetum*.

4. Inflorescence: Male flowers arranged in catkin-like erect spikes while the female flowers usually in spherical heads; in both male and female branches the internodes are short; plants are monoecious or dioecious.
5. Flower: Bracteate, bracteolate, bracteoles usually 2; incomplete, unisexual, perianth generally absent or rudimentary.
6. Male flower: Several male flowers develop at each node of the inflorescence axis; these flowers at each node remain protected by a sheath formed by the combination of bracts; each male flower is surrounded by 2 bracteoles, two small tepals or perianth lobes, and a central stamen which usually hangs out over the edge of the sheath of bracts; anthers are dithecous, basifixed, dehiscing by vertical slits.
7. Female flower: Several female flowers remain crowded in spherical heads at the ends of the branches; each female flower is surrounded by a bract, a pair of bracteoles and a centrally located gynoecium; perianth lobes are absent; the gynoecium is bicarpellary, syncarpous; ovary superior, originally bilocular, but becomes unilocular by the suppression of posterior locule; ovules 2, of which 1 usually aborts; parietal placentation; style 1, very short; stigmas 2, long linear and feathery.
8. Fruit and seeds: Fruit usually a 1 seeded, small-winged samara surrounded by 2 woody bracteoles and a bract; several such fruits remain arranged together to form cone-like dry multiple fruit. Seeds are non-endospermic and the embryo is straight.
9. Floral formula: (a) Male flower: Br, Brl, ⊠, ♂, P_2, A_1, G_0. (b) Female flower: Br, Brl, ⊠, ♀, P_0, A_0, $G_{\underline{(2)}}$.

References

Bentham, G. and Hooker, J.D. 1862-1883. *Genera Plantarum*, 3 vols., Reeve and Co., London.

Cronquist, A. 1981. An integrated system of classification of flowering plants, Columbia University, New York.

Engler, A.W. and Prantl, K. 1887-1915. *Die Naturlichen Pflanzenfamilien*, 23 vols. Leipzig.

Hickey, M. and King, C. 1988. 100 families of flowering plants, (2nd ed.), Cambridge University Press, Cambridge.

Radford, A.E. 1986. Fundamentals of plant systematic, Harper and Row Publishers, Inc. New York.

30

Combretaceae (Combretum Family)

Family Combretaceae contains about 20 genera and 600 species distributed mainly in the tropical and sub-tropical countries of the world (Rahate *et al.*, 2019). Eight genera and about 45 species of this family have been reported from India, chiefly from Assam and West Bengal. *Combretum* and *Terminalia* are two largest genera of the family. *Lumnitzera littoria* is found as a mangrove plant in Sundanban and other tidal forests of India.

Terminalia bellirica is tall tree and leaves clustered near ends of branches. Flowers are small, pale green, arise in simple spikes. Fruit is ovoid, 3 cm long, brownish & densely covered with hairs and contains 5-17 per cent tannins. Unripe fruit is used as astringent but ripe fruit is purgative. Fruit powder is used as tonic & laxative and useful in piles, leprosy, dropsy and fever. Fruit of *T. chebula* is astringent and laxative. Roasted fruits are used to treat cough, chronic ulcer, wounds & scalds, used in dyeing and tanning purposes, fruit decoction is used in bleeding and ulceration of gums

Selected Indian genera and species:

Genera	Importance
Combretum	Favourite ornamental plants which are scandant shrubs; leaves of *C. acuminatum* are anthelmintic while that of *C. roxburghii* are used in malarial fever
Quisqualis	Rangoon creeper (*Q. indica*) is a beautiful ornamental shrub
Terminalia	Favourite tree *T. arjuna*; timber of *T. alata*, *T. catappa*, *T. bialata*, *T. ivorensis* (indigo timber), *T. tomentosa*, *T. myriocarpa* are used; bark of *T. arjuna* is highly valued as a cardiac tonic; *T. bellirica* (*Behera* or *Beheda*) and *T. chebula* (*Hararh* or *Hirdha*) are the two constituents of 'Triphala', the famous Ayurvedic Indian preparation along with *P. emblica* (aonla) as its 3rd constituents; the bark and fruit of *T. alata*, *T. arjuna*, T. catappa, *T. chebula*, *T. ivorensis* and *T. citrina* are used for tanning and dyeing purposes.
Calycopteris	Leaves of *C. floribunda* are used in dysentery & malarial fever

Description of Family

1. General habit: A family of exclusively arborescent taxa, consisting of tall trees or woody twiners of lianas, such as *Combretum* and *Quisqualis*.

2. Stem: Solid with large mucilaginous sacs and abundant tannin.
3. Leaves: Simple, alternate or sometimes opposite (*Quisqualis*); exstipulate; margin entire.
4. Inflorescence: Racemose, or paniculate or simple or branched spikes (*Quisqualis*, ; rarely racemose heads (*Anogeissus*).
5. Flower: Bracteate, bisexual (rarely unisexual), actinomorphic but occasionally zygomorphic, pentamerous, epigynous; tubular floral receptacle along with the adnate tubular base of calyx form a hypanthium; rarely flowers are tri-to octomerous.
6. Calyx: Usually 5 sepals, fuse to form a tube; sometimes 4 or 8 sepals; valvate; rarely imbricate.
7. Corolla: Usually 5 or as many petals as sepals, free, located alternately with sepals, valvate, imbricate, or twisted (*Quisqualis*); petals absent in *Terminalia* (Fig. 30.2).
8. Androecium: Usually 10 or twice as many as the petals, arranged in 2 alternate whorls; stamens of the outer whorl are antisepalous; sometimes only 4 or 5 stamens; filaments inflexed in bud; anthers dithecous, versatile, dehiscing longitudinally.
9. Gynoecium: Carpels 4-5 or less, syncarpous; ovary inferior, unilocular, angular or ribbed, with as many angles or ribs as calyx lobes; ovules 2-6, anatropous, pendulous from the top of the locule; style slender and solitary, stigma pointed or capitates; ovary is surrounded by nectariferous disc.
10. Fruit and seed: Fruit leathery; one-seeded drupe, often winged. Seed non-endospermic; embryo with folded or convolute cotyledons,
11. General floral formula: Br, ⊠, ⊠, $K_{(5)}$, C_5, A_{5+5}, $G_{(4-5)}$.

References

Rahate, S., Hemke, A. and Umekar, M. 2019. *International Journal of Pharmaceutical Sciences Review and Research*, 58(2): 22-29.

31

Crucifeare (Mustard Family)

This cosmopolitan family of about 375 genera and over 4,060 species, occur chiefly in north temperate regions, especially in the Mediterranean region. About 50 genera and over 140 species have so far been reported from India. Some of the larger genera, along with their reported number of species from the world in parenthesis, include *Draba* (300), *Cardamine* (160), *Lepidium* (150), *Alyssum* (150), *Arabis* (120), *Erysimum* (100), *Sisymbrium* (90), *Heliophila* (75), *Rorippa* (70), *Iberis* (30) and *Raphanus* (8). Majority of Crucifers produce an abundance of vitamin C. Characteristic pungent odour of its members is because of sulphur compounds.

Selected Indian Genera and Species

- Vegetables: Several common vegetables of daily use belong to this family. Some of them are radish (*Raphanus sativus*), cauliflower (*B. oleracea* var. *botrytis*), Brussel's sprout (*B. oleracea* var. *gemmifera*), sengri (*B. sativus* var. *caudatus*), etc.
- Oils: Fatty oil, used for cooking, pickles, bruning, massaging, etc. is obtained from the seeds of several species of this family. Mustard oil is obtained from several species of *Brassica* such as *B. campestris* var. *sarson* and *B. campestris* var. *toria* (Indian rape). Oil is also obtained from the seeds of *B. juncea* var. *cuneifolia* (Rai), *B. alba* (white mustard). The oil cake left after the extraction of oil, is a good cattle feed.
- Ornamentals: Among the known ornamentals are candytuft, honest, sweet alyssum, wall flower, stocks, etc. are being used annual flowers.
- Medicinal: Majority of Crucifers produce an abundance of vitamin C. Characteristic pungent odour of its members is because of sulphur compounds. Cauliflower and a few other crucifers have recently attracted botanists as possible anticancer food items.

Description of Family

1. General habit: Annual, biennial or perennial herbs with a pungent watery sap; herbage commonly covered with branched or stellate unicellular hairs; rarely sub-shrubs.

2. Root: Tap root which in some genera gets swollen due to stored food and becomes fusiform (radish) or napiform (turnip).
3. Leaf: Mostly simple, alternate, often dissected, exstipulate; rarely opposite or subopposite; pubescence of simple to stellate hairs; radical leaves form a rosette whereas cauline leaves are alternate or opposite.
4. Inflorescence: Typically a raceme or corymb (*Iberus*) and nearly always without bracts or bracteoles.
5. Flower: Ebracteate, ebracteolate, complete, bisexual, actinomorphic, rarely zygomorphic, tetramerous, hypogynous.
6. Calyx: 4 sepals, polysepalous, in 2 whorls of 2 each, outer whorl is anterio-posterior.
7. Corolla: 4 petals, polypetalous, cruciform (arranged in a cross), alternate with sepals, often clawed; 2 anterior petals in *Iberis* are larger; petals reduced or scale-like Coronopsis; petals completely absent in *Lepidium*.
8. Androecium: 6 stamens arranged in 2 whorls, an outer whorl of 2 short stamens, and an inner whorl of 4 long stamens; anthers dithecous, introse, longitudinally dehiscent; stamens are only 4 in some species of *Nasturtium* while only 2 in Coronopus didymus and Lepidium and up to 16 in *Megacarpaea polyandra*.
9. Gynoecium: Bicarpellary, syncarpous, superior; unilocular but becoming bilocular due to the development of a false septum or replum; replum unites the two parietal placentae; numerous anatropous or campylotropous ovules; parietal placentation; styles 1 or obsolete; stigma capitates to bilobed; carpels 3 in *Lepidium sativum* and 4 in *Tetrapoma*.
10. Fruit and seed: Fruit siliqua (*Arabis*, *Brassica*) or silicula (*Capsella bursa-pastoris*, *Lunaria*) or occasionally a nut (*Crambe*) or rarely achene-like (*Isatis tinctoria*). Seeds are small, non-endopsermic with curved embryo and often mucilaginous testa. Fruits and seeds are characters of systematic importance in this family.
11. Pollination and dispersal: Pollination is entomophilous and dispersal of seeds may be by birds, wind, water and cattles.
12. General floral formula: Ebr, Ebrl, ⊕, ☒, K_{2+2}, C_4, A_{2+4}, $G\underline{(2)}$.

Description of Common Plant

A) *Brassica campestris* var. *sarson* Prain (Fig. 31.1)

Stem: Herbaceous, erect, branched, smooth. Leaf: Alternate, exstipulate, simple, sessile or subsessile, lower leaves lyrate, upper ones ovate to lanceolate; serrate, acute, unicostate reticulate. Inflorescence: Racemose raceme. Flower:

Ebracteate, ebracteolate, pedicellate, complete, actinomorphic, hermaphrodite, tetramerous, hypogynous, yellow. Calyx: 4 sepals, in 2 whorls of 2 each, free, slightly petaloid. Corolla: 4 petals, polypetalous, clawed, cruciform, valvate or imbricate. Androecium: 6 stamens, polyandrous, tetradynamous; in 2 whorls of outer 2 short stamens and inner 4 long stamens; dithecous, basifixed, introrse. Gynoecium: Bicarpellary, syncarpous, superior; unilocular but development of replum or false septum makes it bilocular; many ovules in each locule, parietal placentation; style reduced, stigma bilobed. Floral formula: Ebr, ⊕, ⊠, K_{2+2}, C_4, A_{2+4}, $G\underline{(2)}$.

32

Euphorbiaceae (Spurge Family)

A family of over 322 genera and 8,900 species, Euphorbiaceae is closely related to Sterculiaceae of the Malvales through the various degrees of the union of stamens and the presence of the pistillode and staminode in the male and female flowers, respectively. They are distributed mainly in tropics, but also in Mediterranean region, Australia and USA. Some of the largely represented genera along with the approximate number of their world over reported species are *Euphorbia* (2000), *Croton* (700), *Phyllanthus* (500), *Acalypha* (430), *Japtropha* (175), *Manihot* (170).

Selected Indian Genera

Acalypha, Codiaeum, Euphorbia, Hevea, Jatropha, Mallotus, Manihot, Phyllanthus, Poinsettia, Putranjiva, Ricinus.

Economic Importance

1. Fruits of *Emblica officinalis* are rich in vitamins, medicinal and also used for tanning.
2. Garden croton (*Codiaeum variegatum*) with variegated leaves, poinsettia (*Euphorbia pulcherrima* Fig. 32.1, Fig. 32.3) with beautiful red bracts, *Acalypha sanderiana, A. tricolor* and *Pedilanthus tithymaloides* (Jew's slipper), *Putranjiva roxburghii* (Child life tree) are used as ornamental purpose; *Jatropha gossypifolia* is commonly grown as hedge; *Manihot esculenta* (tapioca) yield valuable starchy food; *Manihot utillissima* (bitter cassava); *Manihot alpe* (sweet cassava); *Manihot glaziovii* yields ceara rubber; *Hevea brasiliensis* yields para rubber; fruits of *Mallotus philippinensis* yields a crimson powder used for dyeing silk, etc.
3. Oils: Castor seeds yield castor oil, croton seeds yield oil that is used as drastic purgative, Indian walnut (*Aleurites molucana*) seeds yield a drying oil, sometimes used as candles, *A. cordata* nuts yield 35-40% tung oil that is used for lacquering, varnishing, water-proofing and manufacturing oil cloth, etc.
4. Timber yielding: *Bischofia javanica* (Bhillar); *Cleistanthus* (Garari); *Flueggea microcarpa* (Dalme); Bridelia retusa (khaja).

Description of the Family

1. General habit: Herbs, shrubs and trees. They yield an acrid, milky juice and many are xerophytic.
2. Leaves: These are simple, and usually alternate. Stipules are usually present.
3. Inflorescence: This may be racemose, cymose or mixed, or a cyathium as in spurge (*Euphorbia*) and Jew's slipper.
4. Flowers: These are small, bracteate, regular and hypogynous. They are always unisexual and may be Monoecious or dioecious. Rudiments of the other sex are often present.
5. Perianth: This may be in 1 or 2 whorls, and is sometimes altogether absent. It is dissimilar in male and female flowers.
6. General floral formula (Fig. 32.2): ⊠, ♂-♀ or ⊠⊠, $P_{0\ or\ 5}$, $A_1\infty$, G_0 $1A_0\underline{G}_{(3)}$.

33

Geraniaceae (Geranium Family)

Geraniaceae is a family of about 7 genera and 830 species which are cosmopolitan in distribution and found widely in temperate and tropical regions. Only 3 genera and about 25 species have been reported from India, mainly from Himalayas. Some of the larger genera along with their approximately reported species in parenthesis include *Geranium* (430), *Pelargonium* (280), *Erodium* (80) and *Monsonia* (25). Well-known 'geranium oil' is distilled from *Pelargonium odoratissinum.*

Pelargonium (syn. Geraniums, Stork bills) is an extremely popular garden plant and has considerable economic importance in the market of ornamental plants. Its genotypes and hybrids are grown in temperate regions. Its value as a potential essential oil yielding crop is also mentioned. Geranium oil is widely used in perfumery, cosmetics, pharmaceutical, and food industries. Oil has anti-microbial, and pesticidal activities. The genus *Pelargonium* comprises more than 280 species in 13 sections of which 33 have been originated in southern parts of South Africa (Van der Walt and Vorster, 1988). It has been reported that nearly 80 per cent diversity of *Pelargoniums* is centered in Cape Province if African continent (South Africa, Zimbabwe and Mozambique). *Pelargoniums* are popular seed and vegetatively propagated bedding and pot plants with numerous phenotypes (Fish type, Ivy leaved, Regal, Angels, Fancy leaves, Scented, Oak leaf) and cultivars (Double/ Bar/ Centre/ Ring/ Vein/ Picotee pattern/ Tulip/ Carnation/ Phlox/ Birds egg flowered) available on the world market. True geranium is a bushy, aromatic plant. The stem is cylindrical, woody at the base, pubescent, green when young and turning brown with age. The leaves are highly aromatic in nature. Flowers are regular and devoid of spur like appendages.

Selected Genera and Species

- Ornamental value of Geraniaceae is because of the beautiful and attractive flowers of several of its members including *Pelargonium hortorum*, *P. zonale*, *Geranium wallichianum*, *G. pratense*, *Erodium cicutarium*, etc.

- Several species of *Erodium* and *Geranium* are weedy.

Genera	Importance
Pelargonium	Well known 'geranium oil' used in perfumery, is obtained from *P. odoratissimum*; flowers are irregular; nectar tube is present (extends from base of one sepal and adherent to side of calyx tube or pedicel); flowers are zygomorphic
Geranium	Roots & leaves of *G. robertianum* are used in treatment of diarrhea where rootstocks of *G. wallichianum* are used in eye troubles; roots & leaves of *G. nepalense* are used in treatment of kidney troubles; flowers are regular; nectar tube is absent; flowers are actinomorphic
Erodium	Several species are weedy

Description of Family

1. General habit: Mostly herbs, sometimes under shrubs; suffrutescent or arborescent; stems often fleshy; hairy.
2. Leaf: Simple and lobed or dissected, or compound; opposite or alternate; stipulate; venation mostly palmate.
3. Inflorescence: Usually cymose, sometimes umbellate (*Pelargonium*), or flowers solitary.
4. Flower: Bracteate, bracteolate, complete, hermaphrodite, actinomorphic by zygomorphic in Pelargonium, hypogynous, pentamerous; often very attractive.
5. Calyx: Usually 5 sepals, free or rarely united, imbricate, sepals 4 in *Vivania* and 8 in *Dirachma*; persistent; in zygomorphic flowers the dorsal sepal is sometimes spurred and the spur is adnate to the pedicel.
6. Corolla: 5 petals, free, alternate with sepals, imbricate or twisted; rarely petals are 2,4,8 or 0; often alternating with nectariferous glands.
7. Androecium: 5 to 15 stamens, filaments often united at the base; in 1 to 3 whorls; obdiplostemonous; staminodes often present; anthers usually versatile, 2 celled, longitudinally dehiscent; a nectar is present at the base of each stamen of inner whorl in *Geranium*.
8. Gynoecium: Pentacarpellary, syncarpous, superior ovary, pentalocular, one to two ovules in each locule; axile placentation; rarely 3 to 8 carpels; styles 3 to 5, slender and beak like; stigmas 3 to 5.
9. Fruit and seed: Fruit capsular dehiscing into 1 to 2 seeded 3 to 5 mericarps, the styles adhering to the ovarian beak. Seeds usually with curved embryos; endosperm scanty or absent.
10. Pollination and dispersal: Pollination mainly by bees which visit the flowers in search of nectar. In *Geranium*, the dispersal of seeds takes

place by the awn which suddenly curls upwards, throwing the seeds at some distance from the plant.

11. General floral formula: Br, Brl, ⊠, ⊠, K_5, C_5, $A_{(5-15)}$, $G_{\underline{(2-5)}}$.

Taxonomy of *Pelargonium*

Sections provide a convenient way of looking at the species within a genus. Sixteen sections are recognized in genus *Pelargonium*. Most of these sections contain species that are cultivated by pelargonium fanciers. Sheltema and Walt (1990) reported taxonomic revision of section *Jenkinsonia* on the basis of morphology, anatomy, and chromosome number. This section is advanced and closely related to section *Myrrhidium* and some spp. of section *Ligularia*. A reappraisal of section *Ligularia* reveal four basic chromosome numbers as X=8, 9, 10 and 11 and variation in chromosome size (Albers *et al.*, 1992).

Marais (1991) indicated that *P. aciculatem, P. confertum, P. fasciculaceum* and *P. connivers* are tuberous species belong to section *Hoarea*. Leaf structure, leaf anatomy, androecium structure, and pollen morphology support the close relationship among them.

References

Albers, E., Gibby, M., Austmann, M.1992. *Plant Systematics and evaluation*, **179**(1-2): 257-276.

Marais, E.M.1991. *South African Journal of Botany*, **57**(1): 55-66.

Sheltema, A.G. and Walt, J.J.A.1990. *South African Journal of Botany*, **56**(3):285-302.

Van der Walt, J.J.A. and Vorster, P. 1988. *Pelargoniums* of Southern Africa 3. National Botanic Gardens, Kirstenbosch, Republic of South Africa.

34

Iridaceae (Iris Family)

A family of about 66 genera and over 2,244 species, Iridaceae are distributed in both tropical and temperate regions with the chief centres of distribution in tropical America and South Africa. Chief areas of distribution in India are Kashmir, and Kumanon hills. Common species cultivated in Indian gardens include *Belamcanda chinensis*, *Crocus sativus*, *C. biflorus*, *C. speciosa*, *Gladiolus primulinus*, *G. colvillei*, *Freesia refracta*, *Iris kumaonensis*, *I. germanica* and *I. foetidissima*. Iridaceae are cultivated throughout the world for their beautiful flowers.

Gladiolus 'the queen of bulbous flowers' is one of the most important bulbous ornamentals grown commercially in all parts of the world for its attractive flowers. It is very much liked for its majestic spikes containing attractive, elegant and delicate florets which open in sequence over a long duration and hence, has good keeping quality of cut spikes. These are extensively employed as cut flower in herbaceous borders, individual beds in the gardens, for exhibition, as pot plants for various kinds of interior decoration and for making bouquets. North American Gladiolus Council, Boston established in 1910 (www.gladworld.org), and New England Gladiolus Society (www.britglad.com), came into existence during post war era (1945-55), are actively involved in popularization and improvement of gladiolus in tune with flourishing cut flower trade in western countries.

Saffron (*Crocus sativus*) is eulogized as 'Golden condiment or Golden spice'. It is a low volume, high valued & non-perishable commodity. It is commercially grown for its medicinal, perfumery, flavouring & colouring properties. Medicinal uses are stimulant, stomachic and anti-catarrhal in action. It also helps in urinary, digestive and urine troubles and used in fevers, Melancholia and enlargement of liver and spleen. It is used in colouring butter, cheese, puddings and pastry and as food additive in confectionery preparations.

Selected Indian Genera

Belamcanda, *Crocus*, *Gladiolus* (270), *Freesia*, *Iris*.

Economic Importance

1. Dried styles of *C. sativus* are used to make 'saffron', an orange-yellow dye used chiefly for flavouring and colouring dishes. Saffron is also used to give colour to the cooked rice. "Around 100,000 styles are required to produce 1 kilogram of saffron".
2. Orris root, used in perfumes and in dentifrices, is obtained from the rhizomes of *Iris germanica* var. *florentina*.
3. Roots of *Belamcanda chinensis* are an antidote to snake poison. A drug obtained from its rhizomes is used in tonsilitis.
4. Roots of *I. germanica* are diuretic and used in diseases of gall bladder.
5. Iridaceae are highly-prized decorative plants of gardens and parks. Some of these ornamental genera are *Belamcanda* (blackberry lily), *Croscomia*, *Crocus*, *Dierama*, *Freesia*, *Gladiolus*, *Iris*, *Ixia*, *Romulea*, *Sisyrinchium* (blue-eyed grass), *Tigridia* (tiger flower) and *Tritonia*.

Description of the Family

1. General habit: Perennial herbs with roots produced from corm (*Crocus*, Fig. 34.1A, B), bulb (*Gladiolus*), or rhizome (*Iris*,; *Belamcanda*); very rarely subshrubs; stems solitary or several, or plants scapose.
2. Leaf: Usually simple, exstipulate, sessile, linear to ensiform, equitant in 2 ranks; mostly basal, numerous and crowded at the base of stem; with sheathing leaf base; parallel venation.
3. Inflorescence: Racemose raceme or paniculate, or flowers solitary (*Sisyrinchium*, *Romulea*)l invariably terminal; in several genera the flowers develop on distinct aerial flowering axis or scape (*Iris*).
4. Flower: Bracteate, complete, actinomorphic or zygomorphic (*Gladiolus*, *Tritonia*), bisexual, epigynous, trimerous; showy, large and beautifully coloured; usually subtended individually or in groups by 2 spathe-like bracts.
5. Perianth: Usually 6 tepals arranged in 2 whorls of 3 each; of these the outer whorl of 3 tepals represents calyx and often petaloid, and inner whorl of 3 tepals represents corolla; tepals of both the whorls are often distinguishable from each other by colour, size or texture; free or united, but usually basally connate into a long tube (*Crocus*).
6. Androecium: Usually 3 distinct stamens situated opposite the tepals of the outer whorl; mostly epiphyllous; anthers dithecous, basifixed, extrorse, dehiscing by vertical slits.

7. Gynoecium: Tricarpellary, syncarpous; ovary inferior, trilocular, with few to many anatropous ovules in each locule, axile placentation; in *Hermodactylus* the ovary is unilocular with parietal placentation; style 1, often 3-branched, long and coloured; stigmas 3, filiform to subulate or fimbriate; ovary rarely superior (*Isophysis*); style is very short in *Eleuthrine*.
8. Fruit and seeds: Fruit a loculicidal capsule dehiscing by 3 valves; seeds with small embryo and copious endosperm, sometimes arillate.
9. Pollination: Usually entomophilous,, effected by bees and butterflies.
10. General floral formula: Br, ⊠ or ·|·, ⊠, $P_{(3+3) \text{ or } 3+3}$, A_3, $G_{(3)}$.

Taxonomy of Gladiolus

Genus *Gladiolus* belongs to the order Liliales, class Monocotyledon, sub-family Ixioideae of the family Iridaceae. Sub-family Ixioideae is further divided into Irieae, Watsonieae and Gladioleae based on form of flowers i.e. zygomorphic or actinomorphic and styles i.e. complete or divided into two and mono or bicarpillary ovary. This later sub-division comprises eight genera with *Gladiolus* L. as one of them. The genus has further been divided into four sections as:-

Section I: *Eugladiolus*- includes 100 species of Tropical and South Africa (Cape Peninsula,

Madagascar Island, Europe and western Asia).

Section II: *Habea*- includes 12 species of South Africa (Cape Peninsula) and Madagascar Island.

Section III: *Schweiggeria*- includes only two species, i.e. *G. montana* and *G. arenarius* from the Cape.

Section IV: *Homoglossum*- includes five to six species of South Africa.

Lewis *et al.* (1972) divided the genus *Gladiolus* into four sections/ groups based on number of well developed leaves (Table 34.1).

Table 34.1. Species belonging to different sections/ groups (Lewis *et al.*, 1972)

Section	Distinct characters
Plurifoliati	Having 5-8 or more well developed leaves, distichously arranged, forming a fan; contain 22 species from South Africa *viz.*, *G. appendiculatus*, *G. buckerveldii*, *G. calcaratus*, *G. cardinalis*, *G. crassifolius*, *G. cruentus*, *G. densiflorus*, *G. ecklonii*, *G. elliotii*, *G. exiguous*, *G. hollandii*, *G. invenustus*, *G. natalensis*, *G. ochroleucus*, *G. oppositiflorus*, *G. papilio*, *G. pole-evansii*, *G. saundersii*, *G. sempervirens*, *G. sericeo-villosus* & *G. varius*.
Paucifoliati	Having 2-5 well developed leaves, distichously or spirally arranged; contain 24 South African species *viz.*, *G. acuminatus*, *G. alatus*, *G. angustus*, *G. arcuatus*, *G. buckerveldii*, *G. carneus*, *G. caryophyllaceus*, *G. equitans*, *G. floribundus*, *G. gueinzii*, *G. involutus*, *G. kamies-bergensis*, *G. lapeirousiodes*, *G. leptosiphon*, *G. lewisiae*, *G. macneilii*, *G. marlothii*, *G. microcarpus*, *G. mostertiae*, *G. orchidiflorus*, *G. permeabilis*, *G. pretoriensis*, *G. rufomarginatus*, *G. salteri*, *G. scullyi*, *G. seresianus*, *G. stellatus*, *G. undulatus*, *G. uysiae*, *G. vernus*, *G. vigilans*, *G. viresens* & *G. watermeyeri*.
Unifoliati	This group has one well developed basal leaf and rest being reduced bract-like with short blades, spirally arranged on the scapes. Contain 29 South African Species *viz.*, *G. blommesteinii*, *G. brevitubus*, *G. bullatus*, *G. carinatus*, *G. citrinus*, *G. comptonii*, *G. cylinraceus*, *G. debilis*, *G. exilis*, *G. gracilis*, *G. hyalinus*, *G. inflatus*, *G. mutabilis*, *G. liliaceous*, *G. longicollis*, *G. oreocharis*, *G. ornatus*, *G. pritzelli*, *G. punctulatus*, *G. quandrangulus*, *G. recurvus*, *G. robertsoniae*, *G. rogersii*, *G. symonsii*, *G. tenellus*, *G. tristis*, *G. violaceo-lineatus* & *G. viridiflorus*.
Exfoliati	In this group leaves appear after flowering, and the scape bears 1-3 (or 4) sheathing bracts with or without small free blades. Contain 24 South African species *viz.*, *G. aurantiacus*, *G. bilineatus*, *G. brachyphyllus*, *G. brevifolius*, *G. carmineus*, *G. emiliae*, *G. engysiphon*, *G. guthriei*, *G. jonyuilliodorus*, *G. odoratus*, *G. maculatus*, *G. martleyi*, *G. monticola*, *G. nerineoides*, *G. parvulus*, *G. pillansii*, *G. stefaniae*, *G. stokoei*, *G. subcaeruleus*, *G. unguiculatus*, *G. vaginatus* & *G. woodii*.

Floral Biology of Gladiolus

Ravi kumar and Nainan (1980), Raychoudhury (1980), Singh and Singh (1985) and Shah *et al.* (1988) have studied some important aspects of floral biology in gladiolus. In *G. psittacinus*, the flower buds take 16 days to reach the full bloom stage. Wild gladiolus species flower during winter and modern cultivars of *G.* x *grandiflora* flower in summer. Flowering was earlier in cv. Spite (89 days) and late in cv. Her Majesty (99.5 days). The unfurling of petals starts early in the morning and complete opening of florets takes 22-24 hrs. Anthesis starts between 3-4 am and with peak at 8-10 am and after 11.30 am there is no anthesis. Anthers dehiscence starts at 8 am and continues up to 12 noon. Stigma become receptive one day before anthesis and remains receptive up to 2-3 days after anthesis. Stigma becomes feathery when receptive. Pollen grains are round in shape with average diameter of 106 μ, heavy, not carried away by wind and bees. Pollen viability ranges between 76 to 85 per cent. Growth of pollen tubes in *in vivo* was 8-9 times faster than in *in vitro*. Pollen stainability in acetocarmine was 98 per cent.

The structure and placement of stamens and stigma suggest that it is essentially outbred crop. Maximum pod development is observed only in hand pollinated flowers. Natural cross pollinated and bagged flowers show considerable reduction in pod formation. Self incompatibility (SI) has been found in *G. tristis* (*elemental* spp.). Floral biology along with presence of SI suggest out crossing.

References

Lewis, G.J., Obermeyer, A.A. and Barnad, T.T.1972. *Journal of South African Botany*, 10 (Suppl).

Ravi kumar, C. and Nainan, M.O. 1980. *Indian Journal of Botany*, **3**: 155-159.

Raychoudhury, N. 1980. *Lal-Baugh journal of Mysore Horticultural Society*, **25**: 36-38.

Singh, B. and Singh, M. 1985. *Progressive Horticulture*, **17**: 135-136.

Shah, A., Pant, G.G. and Lal, S.D. 1988. *Progressive Horticulture*, **20**(1-2): 74-76.

35

Labiatae or Lamiaceae (Mint Family)

A family of about 236 genera and 7,200 species, Labiatae are cosmopolitan in their distribution and their chief centre is the Mediterranean region. About 65 genera and over 400 species have been reported from India, mainly distributed in North Western parts of India and southern parts of India. Some largely represented genera along with the approximate number of their worldover reported species and common names of some of them are Salvia (500-700, sage), *Hyptis* (400), *Thymus* (300-400, thyme), *Teucrium* (300, germander), *Stachys* (300), *Scutellaria* (200, skullcap), *Coleus* (200, Jacob's coat), *Ocimum* (150, basil), *Nepeta* (150, catnip), *Lavendula* (28, lavender), *Mentha* (25, mint) and *Rosmarinus* (3, rosemary).

Salvia is a tender herbaceous annual or perennial plant and cultivated as bedding plant in many countries. Owing to its adaptability, self sowing ability and handsome appearance, it has been introduced and commanded all over the world. *Salvia* species have long history of utilization as medicinal herbs. The therapeutic value of this crop is widely recognized in traditional Chinese medicine for various ailments. Some species are strongly aromatic having horticultural, commercial and medicinal values.

Selected Indian Genera

Coleus, *Lamium*, *Lavendula*, *Leucas*, *Mentha*, *Ocimum*, *Salvia*, *Thymus*.

Economic Importance

1. Garden ornamentals: *Coleus blumei*, *C. thyrsoideus*, *Lavendula officinalis*, *Ocimum basilicum* (sweet basil), *Mentha viridis* (spearmint or garden mint or *pudina*), *Rosmarinus officinalis* (rosemary), *Salvia coccinea*, *S. officinalis* (garden sage), *S. splendens* (scarlet sage) and *Thymus vulgaris* (thyme).
2. Volatile oils:
   (i) Basil oil, obtained from *Ocimum basilicum*, *O. canum* (ram tulsi), and *O. sanctum* (holy basil or Tulsi), is used in variety of medicine and perfumery.

(ii) Lavender oil (*Lavandula latifolia, L. spica, L. vera*) is used in perfumery and cosmetics.

(iii) Mint oil (*Mentha arvensis, M. longifolia, M. spicata*) is used in nausea and vomiting.

(iv) Origanum oil (*Organum vulgare*) is used in cosmetics and soap industry.

(v) Patchouli oil (*Pogostemon benghalense, P. cablin*) is used in perfumery.

(vi) Pennyroyal oil (*Mentha pulegium*) is used in preparing synthetic menthol.

(vii) Peppermint oil (*M. piperata*) is used in pharmacy and confectionery.

(viii) Perilla oil (*Perilla frutescens*) is used in paints, varnishes and printing link.

(ix) Rosemary oil (*R. officinalis*) is used in medicines and perfumery.

(x) Sage oil (*Salvia splendens*) is used in soap industry.

(xi) Sweet majoran oil (*Majorana hortensis*) is used in perfumery and liquor industry.

3. Medicinal plants: (i) *O. sanctum* leaves are used for cough, cold and fever, (ii) *Mentha spicata* (pudina) leaves, used to preparing 'chutney', help in indigestion and rheumatism, (iii) *T. vulgaris* leaves are effective against hookworms and (iv) *M. longifolia* leaves are used in rheumatic pains and some heart diseases.
4. Camphor is obtained from oil of *Ocimum kilimandscharicum*.
5. Root tubers of some species of *Coleus* and *Ocimum* are edible.

Description of the Family

1. General habit: Usually perennial or annual terrestrial herbs with square stems, sometimes shrubs (*Orthosiphon*), rarely trees (*Hyptis, Leucoscepirium*), or vines (*Scutellaria*); plant parts usually covered with glandular hairs; usually with aromatic oils.
2. Stem: Usually quadrangular and covered with glandular hairs.
3. Leaf: Usually, simple, opposite, exstipulate; often hairy and with epidermal glands secreting volatile oils; sometimes whorled (*Dysophyla*); rarely pinnately dissected or compound.

4. Inflorescence: Often cymose, the cymes at the nodes condensed into a false whorl or *verticillaster*; rarely flowers are in simple racemes (*Scutellaria*), or solitary and axillary; very rarely in head (*Hyptis*).
5. Flower: Usually bracteate, bracteolate, complete, hermaphrodite, zygomorphic, hypogynous; rarely nearly actinomorphic (some species of *Mentha*).
6. Calyx: 5 sepals, gamosepalous or united into a funnel-shaped or campanulate tube, tube often ribbed; persistent in fruit; valvate or imbricate; usually 2-lipped (1/4 in *Ocimum* and 3/2 in *Salvia*).
7. Corolla: 5 united petals, 2 lipped (4/1 in *Ocimum* i.e. 4 in posterior lip and 1 in anterior lip, or 2/3 in *Salvia*, *Leucas*); corolla is usually differentiated into a tube and a limb; valvate (*Ocimum*), imbricate (*Salvia*) or contorted.
8. Androecium: Usually 4 stamens (*Ocimum*), didynamous, epipetalous; sometimes only 2 stamens (*Salvia*), in each of which the fertile anther lobe is attached at the anterior end of very long connective while at the posterior end at present the sterile anther lobe (Fig. 35.1); filament is usually free and rarely connate (*Coleus*); dithecous, dorsifxed, introrse, dehiscence longitudinal; usually the missing stamens are present in the form of staminodes.
9. Gynoecium: Bicarppelary, syncarpous; ovary superior, deeply 4-lobed, bilocular when young but becomes quadrilocular due to the formation of false septum; 2 ovules in each locule in bilocular condition, but in quadrilocular condition one anatropous ovule is present is each locule; axile placentation, style gynobasic; stigma bifid; 4-lobed, hypogynous, nectar-secreting disc is present.
10. Fruit and seeds: Fruit schizocarp of 1 to 4 nutlets, each nutlet containing one seed; rarely a drupe. Seeds with scanty or no endosperm and straight embryo.
11. Pollination: Pollination is entomophilous. Coloured bracts (*Salvia*), corolla and nectar is several other genera attract the insects. Usually the lower lip of the corolla forms the platform for the visiting insects. The long connective in *Salvia* also proves helpful in shedding pollens from the fertile anther lobe. Cross pollination is effected because the flowers are usually protandrous. Long-tubed flower of *Monarda* and several other genera are pollinated by butterflies and moths.
12. General floral formula: Br, Brl, ·|·, $K_{(5) \text{ or } (1/4) \text{ or } (3/2)}$, $C_{(4/1) \text{ or } 2/3}$, $A_{2 \text{ or } 4 \text{ or } 2+2}$, $G_{\underline{(2)}}$.

Description of a Common Plant

A) *Ocimum basilicum* (Sweet basil or Ban tulsi)

Habit: Cultivated aromatic herb. Stem: Herbaceous, erect, hairy, quadrangular. Leaf: Simple, opposite decussate, petiolate, exstipulate, ovate, serrate, hairy, unicostate reticulate. Inflorescence: Verticillaster. Flower: Bracteate, pedicellate, complete, hermaphrodite, hypogynous, zygomorphic, pentamerous. Calyx: 5 sepals, fused, ¼ bilabiate, valvate. Corolla: 5 united petals, 4/1 bilabiate, valvate. Androecium: 4 stamens, epipetalous, didynamous, posterior stamen absent; anthers dithecous, dorsifixed, introrse. Gynoecium: Same as in general description of the family. Fruit: Schizocarpic (carcerulus), consisting of 4 nutlets. Floral formula: Br, ·|·, ♀, $K_{(1/4)}$, $C_{(4/1)}$, A_{2+2}, $G_{\underline{(2)}}$.

Floral Biology of Salvia

Flowering of salvia is induced by cool temperature with short day length playing major role. The flower is hermaphrodite and naturally self pollinating. *Salvia splendens* cv. Feu de la Saint Jean is a quantitative long day plant. Cross pollination is mainly due to monoecy along with protandry. Generally, large flowered species are out breeders and the species with minute flowers are inbreeders.

Floral biology and pollination ecology of *S. splendens* cv. Vista Red reveal that red corolla, large tubular flowers with abundant, but dilute nectar are the flower characteristics (Sánchez *et al.*, 2002). Flower structure in salvia is heteromorphic and homostyly resulting in high cross pollination. Most of the species, possess both heterostylic I) pin type (long style and short stamen); II) thrum type (short style and long stamen) and III) homostylic (stamen and styles at same level) flowers. The species *S. coccinia, S. splendens, S. farinacea, S. pratensis, S. grahamii, S. glutinosa, S. leucantha* bear all types of flowers. The species *S. hispanica*, *S. reflexa*, *S. horminum*, *S. verbenacea*, *S. tilliifolia*, *S. aegyptica*, *S. taraxacifolia*, *S. verbenacea* bear homostylic flowers. Haque and Ghoshal (1981) studied some important aspects of floral biology in 14 species of *Salvia*. All the species are reported to be predominant out breeders and entomophilous. Ants, black bees and honey bees are pollen carriers. Cross pollination by insects is due to its showy corolla and monoecy along with protandry. It has been found that at least 186 *Salvia* species mainly from New World are pollinated by humming bird. The most important among them are *S. fulgens*, *S. atrocyanea*, *S. lanceolata*, *S. sessei*, *S. africana-lutea*, *S. oxyphora*, *S. patens*, *S. dombeyi*, *S. gravida*, *S. confertiflora*, *S. guaranitica*, *S. penstemonoides*, *S. divinorum*, *S. madrensis*, *S. disjuncta* and *S. holwayi*, etc.

Bird-pollinated *Salvia* species transfer pollen either by means of a staminal lever mechanism or by immovable stamens (Wester and Bockhoff, 2007). In out breeding species, seed setting by open pollination is low and it can be increased by hand pollination. Species with homostylic flowers show a high degree of seed setting in nature.

The anthesis occurs between June to October. The period of effective receptivity of stigma ranges from two days prior to flower opening to six hours after opening. In protandrous species like *S. verbenacea,* pollen germinability and stigma receptivity peaks on the last day (day 3) of the flower's life, which favours selfing, but if receives out crossing pollen, it produces heavy seeds (Navarro, 1997).

References

Haque, Md.S. and Ghoshal, K.K. 1981a. *Proceedings of Indian National Science Congress,* **47**(5): 716-724.

Navarro, L. 1997. *Plant Systematics and Evolution,* **207**: 111-117.

Sánchez, L.A., Picado, A., Sommeijer, M.J. and Judith, S.E. 2002. *Journal of Horticultural Science and Biotechnology,* **77**(4): 498-501.

Wester, P. and Bockhoff, R.C. 2007. *Annals of Botany,* **10**: 1-21.

36

Liliaceae (Lily Family)

A family of about 15 genera and 705 species showing cosmopolitan distribution. About 35 genera and over 195 species have been reported from India. Some of the largely represented genera along with the number of their world over reported species are *Smilax* (350), *Aloe* (300), *Asparagus* (300), *Dracaena* (150). *Ornithogalum* (150), *Haworthia* (150), *Tulipa* (100), *Fritillaria* (85), *Lilium* (80), *Colchicum* (65), *Sansevieria* (60), *Yucca* (40), *Asphodelus* (12) and *Ruscus* (7).

Lilium is a bulbous ornamental plant commonly known as lily. The flowers are associated with religious feasts and symbolism. In the language of flowers, the lily is the symbol of purity, and innocence. It is one of the sixth major genera of flower bulbs produced worldwide. *Lilium* ranks 4th among cut flower trade in international market. Certain species of *Lilium* are edible, and valued for medicinal properties also.

Tulip is one of the most important bulbous geophytes in the world and popular due to its beauty and economic value. It is the top rated flower bulb in the international cut flower market. Modern tulip cultivars, predominantly *T. gesneriana*, are grown for bulb production, cut flowers, flowering potted plants, and landscaping. The current commercial assortment still consists mainly of cultivars from *T. gesneriana*. The second group of cultivars, the Darwin hybrids consists has been obtained from interspecific crosses between cultivars *T. gesneriana* and genotypes of *T. fosteriana*. *Tulipa suaveolens*, a kind of tulip discovered wild in southern Europe, has played an important role in the evolution of the garden tulip (Bailey, 1963).

Selected Indian Genera

Aloe, *Asparagus*, *Cochicum*, *Dracaena*, *Fritillaria*, *Lilium*, *Ruscus*, *Sansevieria*, *Smilax* and *Yucca*.

Economic importance

The family is important from economic point of view in several ways, of which some are mentioned here.

1. Ornamental plants: Some of the well known garden ornamentals of Liliacae are *Agapanthus africanus* (African lily), *Asparagus plumosus*, *Dracaena*, *Fritillaria tenella*, *Gloriosa superba* (Malabar glory lily), *Hemerocallis fulva* (orange day lily), *Lilium candidum* (Lily), *L. giganteum*, *Ruscus aculeatus* (Butchers broom), *Tulipa suaveolens* syn. *T. gesneriana* (Tulip), *Yucca aloifolia* and *Y. gloriosa*.
2. Edible plants: Fleshy roots of *Asparagus officinalis* and tuberous roots of *A. racemosus* are used as vegetable.
3. Medicinal plants: (i) *Aloe barbadensis* leaves are the source of resinuous drug, used as a purgative, (ii) *Aloe vera* leaves provide mucilaginous liquid, used in piles and inflammations, (iii) Medicated oil prepared from the roots of *A. racemosus* is used for nervous and rheumatic complaints and also in skin diseases, (iv) *Colchicum luteum* roots are used in treatement of gout and rheumatism, (iv) *Fritillaria roylei* bulbs are used as expectorant and antipyretic, (vi) *Gloriosa superba* tubers prove helpful in promoting labour pains in women, (vii) *Hemerocallis fulva* flowers are blood purifier and given to women during child birth, (viii) *Iphigenta indica* corms are used in stomach pains and headache, (ix) *Paris polyphylla* rhizomes are used as anthelmintic, (x) *Scilla hyacinthiana* bulbs are used as heart stimulant, (xi) *Smilax glabra* and *S. ovatifolia* roots are used in treatment of veneral diseases, (xii) *Urginea indica* bulbs are heart stimulant and also used in skin disease and rheumatism, (xiii) Drug obtained from roots of *Veratrum viride* is used in hypertension.
4. Resins are obtained from stems of *Dracaena* and *Xanthorrhoea*.
5. Fibre plants: Excellent fibres, used for cordage, fishing nets, mattings, twines, etc., are obtained from *Phormium tenax* (New Zealand flax), *Sansevieria roxburghiana*, *Smilax glabra* and *Yucca filamentosa*.
6. Insecticides and raticides: Bulbs of *Urginea indica* are used for killing rats while that of *Veratrum album* are used as insecticides.
7. Polyploidy: Colchicine, an alkaloid obtained from *Colchicum luteum*, is used to induce polyploidy.

Description of the Family

1. General habit: Mostly perennial herbs with sympodial bulbs, rhizomes (*Paris quadrifolia*), corms (*Colchicum*), or bulbils (*Lilium*Fig. 36.2); some are woody shrubs (*Dracaena*, *Yucca*), or trees (*Xanthorrhoea*); a few are succulents (Aloe, *Gasteria*) and climbers (*Gloriosa*, *Smilax*); in some the ultimate branches are modified into phylloclades (*Ruscus*) or cladodes (*Asparagus*) and the leaves are reduced to scales.

2. Leaf: Basal or cauline, simple, exstipulate but stipulate in *Smilax* where the stipules modify into tendrils; usually alternate, less commonly whorled (*Paris*), and rarely opposite (*Gloriosa*, *Scolyopus*), sometimes fleshy or spiny margined (*Aloe*), acicular, long, fleshy and hollow in *Asphodelus* (Fig. 36.1); reduced to scales in *Ruscus* and *Asparagus*; usually parallel venation but in *Smilax* and *Paris* the venation is reticulate.
3. Inflorescence: Usually racemose raceme (*Asphodelus*), sometimes in spikes (*Aloe*), terminal panicle (*Yucca*), or monochasial cyme on *Hemerocallis*, or cymose in apparent umbels in the *Agapanthus*; flowers solitary axillary (*Gloriosa*), or solitary terminal (*Tulipa*).
4. Flower: Bracteate, usually ebracteolate except *Dianella* and *Lilium*; bisexual, actinomorphic, trimerous, hypogynous; rarely unisexual (*Smilax zeylanica*, *Ruscus*) and such species are usually dioecious; rarely tetramerous (*Paris*); slightly zygomorphic (*Lilium*, *Haworthia*).
5. Perianth: 6 tepals arranged in 2 whorls of 3 each, free or rarely united (*Aloe*); usually petaloid or sometimes sepaloid; odd tepal of the outer whorl is anterior in position; valvate or imbricate; more than 6 tepals in *Paris quadrifolia*.
6. Androecium: Usually 6 stamens arranged in 2 whorls of 3 each; rarely the stamens are 3 (*Ruscus*), 4 (*Maianthemum*) or up to 12; polyandrous; opposite the tepals; sometimes epiphyllous; filaments distinct or connate; anthers dithecous, basifixed or versatile, extrorse or introrse, dehiscing usually by vertical slits and sometimes by terminal pores; rarely synandrous (*Ruscus*).
7. Gynoecium: Tricarpellary, syncarpous, the odd carpel usually anterior; ovary superior, trilocular, with 2 rows of numerous anatropous ovules, axile placentation; rarely unilocular with parietal placentation; style usually 1; stigmas 1 or 3; rarely the ovary is inferior (*Haemodorum*); usually the nectar-secreting septal glands are present in the ovary.
8. Fruit and seeds: Fruit usually a septicidal or loculicidal capsule or a berry (*Asparagus*, *Smilax*). Seeds with curved or straight embryo and abundant endosperm.
9. Pollination: Flowers are usually insect-pollinated. Insects are attracted usually because of gaudy and scented perianth lobes and the nectar secreted in the sepal glands. In *Paris* the insects are attracted because of the foetid small and dark purple colour of the floral parts. Bright colours of tepals in *Tulipa* also attracts insects. Snails which come to eat the

fleshy tepals of *Rhodea* bring about the cross pollination.

In Yucca, the cross pollination is carried out by a special moth, *Pronuba yuccasella*. Fully expanded flowers emit perfumes and are visited by the female moth, especially during nights. She collects a lot of pollen grains from one flower and visits another flower. Life-history of this moth is intimately associated with the pollination mechanism in *Yucca*.

10. General floral formula: Br, Ebrl, $\oplus$, $\boxtimes$, P_{3+3}, A_{3+3}, $G_{\underline{(3)}}$.
11. Bulb: Storage organs of tulip bulb comprises scales only (i.e. the bulb has no leaf bases Fig. 36.3). The base of single leaf of non-flowering plant forms the outer brown skin or tunic of its bulb. Tulip bulb scales are concentric and each has sub-terminal hole through which the shoot grows. Mature bulb has 5-7 tunicated scales & attached to the base of plant and develop roots. At least one daughter bulb is initiated in the axil of each scale. An average of 7 to 8 daughter bulbs are produced per mother bulb.

Description of a Common Plant

A) *Asphodelus tenuifolius* Cav. (Fig. 36.1)

Habit: An annual weed with fibrous roots, and condensed, underground and reduced stem. Leaf: Radical, simple, extipulate, sessile with sheathing leaf bases; long acicular, acute, fleshy, hollow. Inflorescence: Racemose raceme with flowers on long scape or peduncle. Flower: Bracteate, pedicellate, bisexual, actinomorphic, trimerous. Perianth: 6 tepals arranged in 2 whorls of 3 each, free, petaloid, valvate. Androecium: 6 stamens arranged in 2 whorls of 3 each, polyandrous, epiphyllous, present opposite the tepals. Gynoecium: Tricarpellary, syncarpous, superior, trilocular, 2 ovules in each locule, axile placentation; style long, stigma trifid. Fruit: Capsule. Floral formula: Br, $\oplus$, $\boxtimes$, P_{3+3}, A_{3+3}, $G_{\underline{(3)}}$.

B) Taxonomy of *Lilium* spp.

The genus *Lilium* is classified into seven sections by Comber (1949) and later revised by Lighty (1968) and De Jong (1974). Nishikawa *et al.* (1999) performed some phylogenetic studies in the genus using sequences of internal transcribed spacer region of nuclear ribosomal DNA. The position of the species in section *Sinomartagon* was improved. The seven sections are 1. *Martagon*, 2. *Pseudolirion*, 3. *Lilium* (*Liriotypus*), 4. *Archelirion*, 5. *Sinomartagon*, 6. *Leucolirion* and 7. *Daurolirion*. Some species formerly included within this genus have now been placed in other genera *viz*., *Cardiocrinum*, *Notholirion*, *Nomocharis*, and *Fritillaria*. Major lily assortments, species and cultivars from sections are listed below.

Section I- *Martagon*- Section *Martagon* is more primitive than *Sinomartagon*. Hybrids are based on *L.* x *dalhansonii*, *L. hansonii*, *L. martagon*, *L. medeoloides* and *L. tsingtauense*. The flowers are nodding and have Turk's Cap style (with petals strongly recurved).

Section II- *Pseudolirion*- American hybrids are grouped in this section. These are mostly taller growing and originally derived from *L. bolanderi*, *L. canadense*, *L. greyi*, *L. kelleyanum*, *L. kelloggii*, *L. maritimum*, *L. michiganense*, *L. occidentale*, *L. pardalinum*, *L parryi*, *L. parvum*, *L. philadelphicum*, *L. superbum* and *L. washingtonianum*. Many are clump forming perennials with rhizomatous rootstocks.

Section III- *Lilium* (*Liriotypus*)- Candidum (Euro-Caucasian) hybrids are grouped in this section. These are mostly based on European species *viz.*, *L. candidum*, *L. chalcedonicum*, *L. kesselringianum*, *L. monadelphum*, *L. pomponium*, *L. pyrenaicum*, and *L.* x *testaceum*.

Section IV- *Archelirion*- Oriental (O-genome) hybrids are grouped in this section which are produced through crosses involving specifically *L. auratum* and *L. speciosum*, together with crossbreeds from several species native to Japan including *L. alexandrae*, *L. japonicum* (hermaphrodite flowers in July), *L. nobilissimum* and *L. rubellum*. Plants are tall having large sized (16-22 cm) bulbs, sturdy stems with broader leaves which are almost parallel to ground and dark green in colour. Early (*L. rubellum*) to late flowering (*L. nobilissimum*) habit and limited flower colour range (white or pink) is observed in these hybrids. The whole group is sometimes referred to as 'stargazers' because the flowers appear to look upwards. The flowers are large in size, highly scented, long lasting, produce good amount of pollens, resistant to botrytis susceptible to fusarium and fetch higher price e.g. Cvs. Acapulco (pink), Alma Ata, Expression, Marco-Polo (white/light pink), Montana Muscadet, Siberia, Star Fighter, etc.

Section V- *Sinomartagon*- Asiatic (A-genome) hybrids are grouped in this section. They are derived from central and East Asian species and interspecific hybrids including *L. amabile*, *L. bulbiferum* (orange, upright facing flowers), *L. callosum*, *L. cernuum* (early flowering, white or pink colour), *L. concolor*, *L. dauricum*, *L. davidii*, *L.* x *hollandicum*, *L. lancifolium* (strong stem, vigorous, hairy and bulbils forming), *L. lankongense* (flowers pink with fire spots), *L. leichtiinii* (flowers citron yellow or orange with hairy buds), *L.* x *maculatum* (Upright facing flowers), *L. pumilum* (early, dwarf and bright red coloured flowers), *L.* x *scottie*, *L. wardii* and *L. wilsonii*. Plants are medium sized having small bulbs (10-16 cm), scattered and narrow leaves growing upward, often bear more flowers and early (*L. pumilium*, *L. cernuum*) to late (*L. callosum*) in flowering. The flowers are mostly upright facing, available in wide range of colours (orange, yellow, white, pink, red, purple and salmon), less fragrant and not very strong. The flowers have dark spots with relatively short vase life and resistant to fusarium, but susceptible to botrytis and fetch lower market price e.g. Blackout, Brunello, Dreamland, Gran Paradiso, Elite (orange), Navona, Nove Cento (yellow), Nutmegger, Pollyanna (yellow), Prato, Shiraj, Solemio, Stargater, Sumplon and Vivaldi (pink), etc.

Section VI- *Leucolirion*- Longiflorum (L-genome) hybrids are grouped in this section. They are produced through crosses involving *L. formosanum*, *L. leucanthum*, *L. longiflorum*, *L. philippinensis*, *L. puerense*, *L. regale*, *L. sargentiae*, *L. sulphureum*, *L. wallichianum* and *L. wenshanense*. The hybrids are derived mainly from *L. longiflorum* (fragrant, white) and *L. formosanum* (fragrant, purple brown) species. They have year round forcing ability, outward facing flowers, trumpet in shape, pure white in colour with distinct fragrance and suitable for cut flower purpose and less often grown in gardens.

Section VII- *Daurolirion*- Comprises two species *viz.*, *L. pensylvanicum* and *L. maculatum*.

The individual *Lilium* species show certain characteristics common to genus as a whole namely adventitious root system, early bulb formation, changes in shoot structure, shoot branching, and differentiation of leaf functions, type of germination (epigeal or hypogeal), characteristics of formation of 1st leaf, degree of embryo differentiation in ripe seed and poor embryo differentiation in ripe seed. Baranova (1989) suggested that these characters would be useful for identification purpose.

C) Taxonomy of *Tulipa* spp.

Tulipa is generally distinguished from other bulbous genera in the family by having erect flowers or very rarely nodding, generally with six perianth segments (tepals), six anthers and one style with a three-lobed stigma (De Hertogh *et al.*, 1983). Genus *Tulipa* is divided into two sub-genera. I. *Leiostemones* II. *Eriostemones* (Van Raamsdonk and De Vries, 1995). Species from different sections display complimentary agronomic/ horticultural characteristics. These sub-genera are further classified into eight sub-sections (Table 36.1a and 36.1b). Garden tulips belong to sub-genus *Leiostemones*. The primary cultivars belonging to *T. gesneriana* from sub-section *Tulipa*, is a collective name given to a large number of varieties of unknown origin.

Table 36.1a. Systematics/ taxonomic classification of genus *Tulipa* (Van Raamsdonk *et al.*, 1997)

Sub-genus/section *Leiostemones*

Sub-sections				
Tulipa/ Gesneriana	*Eichleres* (Hall) Van Raamsdonk	*Tulipanum* de Reboul	*Kolpakowskianae* (Hall) Van Raamsdonk	*Clusianae* Baker
T. gesneriana L.	*T. ingens* Hoog	*T. agenensis* DC	*T. altaica* Pall. Ex Sprengel	*T. clusiana* DC
T. armena Boiss	*T. lanata* Regel	*T. systola* Stapf	*T. lehmanniana* Mercklin	*T. montana* Lindley
T. hungarica Borbas	*T. tubergeniana* Hoog	*T. kuschkensis* B. Fedtschenko	*T. tetraphylla* Regel	*T. linifolia* Regel
T. suaveolens Roth	*T. eichleri* Regel	*T. julia* C. Koch		
T. didieri Jord	*T. fosteriana* Hoog ex W. Irving	*T. aleppensis* Boiss ex Regel		
	T. greigii Regel	*T. praecox* Tenore		
	T. albertii Regel			
	T. sosnovskyi Akhverdov et Mirzojeva			
	T. praestans Hoog			
	T. kaufmanniana Regel			
	T. tschimganica Bochantzeva			
	T. dubia Vvedensky			
	T. subpraestans Vvedensky			

Sub-genus/section *Eriostemones* (Boissier)

Table 36.1b: Contd..

Sub-sections		
Australes sensu Hall	*Saxatiles* sensu Hall	*Biflores* sensu Hall
T. australis Link	*T. humilis* Herb	*T. turkestanica* Regel
T. primulina Baker	*T. pulchella* Fenzl	*T. polychroma* Stapf
T. biebersteiniana Schultes	*T. saxatilis* Sieb ex Sprengel	*T. biflora* Pallas
T. sylvestris L.	*T. bakeri* A.D. Hall	*T. sogdiana* Bunge
T. whittalii (Dykes) A.D. Hall	*T. aucheriana* Baker	*T. neustrueva* Pob
T. ophanidea Boiss Ex Heldr.		*T. tarda* Stapf
T. hageri Heldr		*T. dasystemon* Regel

On the basis of geography, specific morphological characteristics and polyploidy the nomenclature of sub-section *Biflores* is considered. Sub-genus *Eriostemones* consists of 30 species which are low growing, narrow leafed and possess hairs at inner base of perianth parts. On the basis of morphological characters evolutionary relationship and intra and interspecific variability within the species of section *Eriostemons* has been documented (Van Raamsdonk and De Vries., 1992). Based on palynological studies, Kosenko (1990) propounded the possible trends in evolution of central Asian species. Section *Eriostemones* is more homogenous in terms of pollen morphology, characterized by monosulcate pollen. Section *Leiostemones* contains species with trisulcate, inaperturate and monosulcate pollen. Boat shaped monosulcate pollen is considered the most primitive and inaperturate.

References

Baranova, M.V.1989. *Horticulture Abstract*, 1992. **62**(1): 460.

Bailey, L.H., 1963. How Plants Get Their Names, Dover Publications, Inc. New York, pp. 30-45.

Comber, H.F. 1949. *Lily Yearbook, Royal Horticultural Society, London*, **15**: 86-105.

De Hertogh A.A., Aung, L.H. and Benschop, M. 1983. 'The tulip Botany, Usage, Growth and Development'. *In*: Horticultural reviews (eds.) J. Janick (5): Westport, Conn. USA, pp. 73-90.

De Jong, P.C. 1974. *Yearbook North American Lily Society*, **27:** 23-28.

Kosenko, V.N.1990. *Horticulture Abstracts*, 1981, **61**(8): 7096.

Lighty, R.W. 1968. *Yearbook North American Lily Society*, **31**: 40-44.

Nishikawa, T., Okazaki, K., Uchino, T., Arakawa, K. and Nagamine, T. 1999. *Journal of Molecular Evolution*, **49** (2): 238-249.

Van Raamsdonk, L.W.D. and De Vries, T.1992. *Plant Systematics and Evolution*, **179**: 27-41

Van Raamsdonk, L.W.D. and De Vries, T.1995. *Plant Systematics and Evolution*, **195**: 13-44.

Van Raamsdonk, L. W. D., Eikelboom, W., De Vries, T. and Straathof, Th. P. 1997. *Acta Horticulturae*, **430**: 821-828.

37

Magnoliaceae (Magnolia Family)

Named after a French botanist Pierre Magnol, Magnoliaceae is a family of about 7 genera and 300 species, mainly distributed in the warm temperature regions of the Northern Hemisphere. The species are common in eastern Asia and eastern North America, and found also in Brazil, West Indies, and Malaysia.

Selected Indian genera and species

Genera	Species	Importance
Magnolia	*M. grandiflora* (Bari champa), *M. fuscata* (Chidi champa), *M. campbelli* (Lal champa), *M. pumila* (Jauhre champa), *M. pterocarpa*, *M. stellata*	Known for their sweet, scented, beautiful flowers; *M. acuminata* is used in making toys, tea boxes, and quality furniture.
Michelia	*M. champaca* (Champa), *M. alba*, *M. excelsa*, *M. nilogrica*, *M. fuscata*	Ornamental plant & source of volatile oil which is widely used in perfumery & cosmetics; flowers are used to cure kidney troubles & gonorrhoea. Flowers of *M. fuscata* yield essential oil which is used in hair oil
Liriodendron	*L. tulipifera* (tulip tree)	Provides tall graceful posture & bears beautiful scented flowers; yields good timber for furniture
Talauma	*T. pbellocarpa*	Yields good quality timber used in making tea boxes & furniture
Manglietia	*M. bookeri*	Provides good furniture wood
Schizandra	*S. grandiflora*	Fruits are edible
Illicium	*I. verum*	Yields an oil useful in colic disorders; *I. griffithii* fruits are used as condiment for flavouring curries
Drimys	*D. winteri*	Bark is used as an astringent & stimulant
Manglietia	*M. hookeri*	Provides good furniture wood

Description of family

1) General habit: Deciduous or evergreen, often aromatic shrubs or trees; *Liriodendron tulipifera* reaches 20 to 65 m in height; *Michelia* and *Magnolia* are medium sized trees or shrubs; *Schizandra* is a climber.

2) Leaves: Alternate, large, simple, entire, petiolate, pinnately-veined; stipulate, stipules large, protect young buds and leave a circular scar on falling; stipules absent in Wintereae; oil glands present.

3) Inflorescence: Flowers usually solitary, either axillary or terminal; flowers crowded near the tips of the branches in *Illicium*.

4) Flowers: Large, showy, bracteate (*Magnolia*, *Michelia*) or ebracteate, pedicellate, bisexual but rarely unisexual flowers while other have bisexual flowers.

5) Perianth: Except in *Illicium* and some species of *Magnolia*, *Michelia* and *Drimys*, the perianth is undifferentiated into calyx and corolla. Perianth spiral to cyclic in 3 or more series of 6 to 18 parts; polyphyllous. In *Drimys* sepals fused while petals free. Perianth lobes are arranged in whorls of threes, valvate or imbricate in each whorls; 9-12 perianth lobes in *Magnolia* while 6-15 in *Michelia*. In *Euptelea*, perianth is absent.

6) Androecium: Stamens numerous, free, spirally arranged on the basal part of floral axis; anthers long, dithecous, basifixed or adnate; introrse but extrorse in *Liriodendron*, dehiscence longitudinal, connective prolonged.

7) Gynoecium: Carpels numerous, apocarpous (free), arranged spirally on the floral axis above the stamens, superior, unilocular, one or more ovules in a ventral suture; placentation parietal; style short, one; stigma one. In *Zygogynum* and *Pachylarnax*, the carpels are rarely united.

8) Fruits and seeds: Fruit of aggregate type. It may be a follicle (*Magnolia*), samara (*Liriodendron*), or rarely berry (*Schizandra*). Seed often suspended within the follicle by a thread like funiculus, has a minute embryo in a large endosperm.

9) Pollination and dispersal: Pollination is entomophilous; dispersal takes place by those animals (e.g. birds, monkeys, etc.) which carry off the fruits.

10) General floral formula: Br, $\oplus$, $\boxtimes$ (rarely unisexual), P_{9}-∞, $A\infty$, $G\underline{\infty}$.

11) Floral diagram of *Michelia champaca* is depicted and Magnolia grandiflora.

38

Malvaceae (Mallow Family)

A family of about 244 genera and 4,225 species, Malvaceae are distributed widely in tropical and temperate regions. About 22 genera and 125 species of Malvaceae have so far been reported from India. Some of the larger genera, along with their common names and/or number of reported species in parenthesis, include *Hibiscus* (rose mallow, 300), *Sida* (200), *Abutilon* (Indian mallow, 100), *Alcea* (60), *Malva* (mallow, 40), *Lavatera* (25), *Gossypium* (cotton, 20) and *Althaea* (12). The family is recognized by *Hibiscus rosa-sinensis* (Rose of China) because of its beautiful large flowers and hundreds of its known cultivated varieties.

Hibiscus commonly called as 'Chinese hibiscus' or 'Rose of China' or 'Hawaiian hibiscus' or 'Blacking plant' or 'Shoe flower' is a flowering plant of hot and warm climate, and hence, known as 'Queen of Tropics'. Florida is the prominent hibiscus growing centre of the world. The commercial species have wide variety of uses including fibers, food, medicinal and ornamental, either as a potted plant, fence or hedge plant or as a landscape shrub. Ethno medicinal value of hibiscus is documented.

Hollyhock is an important winter flowering annual used to impart colourful backgrounds in mixed borders or to fill in vacant spaces in a newly planted shrubbery. It is very drought resistant, and does well in full sun locations that might be too hot or dry for other plants. Black hollyhock has significant place as a colouring ingredient in herbal tea, liquors, for painting Easter eggs, as a skin conditioning and dehydrating substance in cosmetic industry. Unusual height of plant and delay in flowering as compared to other winter annuals handicap its popularity (Kazemi *et al.*, 2011).

Selected genera and species

Genera	Importance
Hibiscus	*Hemps*, which are bast fibers are obtained from *H. elatus* (Cuba bast), *H. sabdariffa* (Rozella hemp), *H. cannabinus* (Deccan hemp), *Sida rhombifolia* (Queensland hemp), etc.; a fatty oil obtained from seeds of *H. cannabinus* is used in manufacture of paints, varnishes and linoleum and its refined oil is edible; among the ornamentals those include are *H. rosa-sinensis*, *H. mutabilis*, *H. schizopetalus*, *H. syriacus*, *H. sabdariffa*, etc.; a delicious chutney is prepared from sepals of *H. cannabinus* and *H. sabdariffa*; flower buds of *H. syriacus* are acidic and used as vegetable; wood of *H. elatus* is used in furniture making; roots of *H. rosa-sinensis* is used against cough; a well known drink is prepared from the calyx of *H. sabdariffa*
Hollyhock (*Althaea*)	*A. rosea* is used as ornamental; roots of *A. officinalis* is used against cough; blue dye is obtained from leaves of *A. rosea*;
Abutilon	Bags, ropes, cords, etc. are made from fibres obtained from *Abutilon asiaticum*, *A. theophrasti*, *H. falcatus*; roots of *A. indicum* are used against fever;
Malva	*M. sylvestris* (mallow) is used as ornamental & medicinal; leaves & seeds of *M. sylvestris* are used in fever; roots of *M. verticillata* are used against whooping cough;
Abelmoschus	Famous vegetable is *A. esculentus*; Fibers were obtained from *Abelmoschus moschatus*; essential oil, used in perfumery is obtained from seeds of *A. moschatus*; roots of *A. moschatus* is used in stomachache
Urena	Roots and bark of *Urena repanda* are used to cure hydrophobia, a dreaded disease caused by bite of mad dog
Pavonia	'Hina', a well-known Indian perfume, is obtained from roots of *Pavonia odorata*;
Thespesia	Toys, pencils & matchsticks are manufactured from wood of *T. populnea*

Description of family

1. General habit: Mostly annual (*Althaea*, *Malva*) or perennial herbs, sometimes growing very tall (*Lavatera*); shrubs (*Gossypium*, China rose) or trees (*Hibiscus elatus*, *Thespesia populnea*, *Kydia*); often with mucilaginous sap.
2. Root: Profusely branched tap roots.
3. Stem: Erect, branched, sturdy, young portions often covered with stellate trichomes; decumbent in *Malva rotundifolia*.
4. Leaf: Simple, often palmately lobed and palmately veined, alternate; stipulate, stipules often deciduous; petiole; ovate, cordate or reniform; margin toothed or entire; venation reticulate and multicostate divergent.
5. Inflorescence: Usually cymose or flowers solitary in the leaf axil (*H. rosa-sinensis*); raceme in *Althaea rosea*.

6. Flower: Bracteate, bracteolate, bracteoles often form calyx-like epicalyx; complete, bisexual, actinomorphic, hypogynous, pentamerous; flowers rarely unisexual and the species are dioecious, variously coloured.
7. Epicalyx: Outside the calyx, several bracteoles often form a calyx whorl called epicalyx. Epicalyx consists of 3 bracteoles (*Malvastrum tricuspidatum*, *Malva*), or 5-9 bracteoles (*Althaea rosea*) or several bracteoles (*Hibiscus radiatus*), bracteoles are free (*Malvastrum*) or fused (*Althaea*); epicalyx or involucres of bracteoles is absent in *Abutilon* (Fig. 38.2) and *Sida cordifolia*.
8. Calyx: 3 to 5 but usually 5 sepals, free or basally connate, valvate; tubular in *Hibiscus*.
9. Corolla: 5 petals, free but basally adnate to the staminal column; usually twisted but sometimes imbricate, showy and variously coloured.
10. Androecium: Numerous stamens, monadelphous; anthers 1 celled or monothecous, reniform, extrorse; dehiscence longitudinal; filaments of anthers fuse to form a staminal column around the ovary.
11. Gynoecium: A compound pistil of 1 to many carpels (often pentacarpellary, syncarpous); ovary superior, 2 to 5 or more locules, rarely 1 locule, 1 to numerous ovules in each locule; axile placentation; style 1 and branched above, or as many as the carpels; stigmas as many or twice as many as carpels, discoid or capitate.
12. Fruit and seed: Fruit is usually a capsule (*Gossypium*) or schizocarp (*Malva*), or rarely a berry (*Malvaviscus*) or samara. Seeds with curved embryo and scanty or no endosperm.
13. Pollination and dispersal: Majority of Malvaceae are insect pollinated mainly because of their large sized flowers, brightly coloured corolla and extrafloral nectarines (in some). Dispersal of seeds takes place mainly by wind (*Gossypium*), or sometimes by water (*Malva*) or by animals because of the presence of hooked spines (*Urena lobata*).
14. General floral formula: Br, ⊕,Epik$_{3\text{-}9 \text{ or } (3\text{-}9)}$, K$_5$, C$_{5 \text{ or } (5)}$, A$_{(\infty)}$, G$_{\underline{(2\text{-}5) \text{ or } (1\text{-}\infty)}}$.

Description of some common plant

A) *Abutilon indicum* Sweet.

Habit: Annual herb. Stem: Erect, herbaceous, branched, solid, hairy. Leaf: Simple, alternate, petiolate, stipulate, serrate, acute, multicostate reticulate. Inflorescence: Solitary axillary. Flower: Bracteate, ebracteolate, pedicellate, complete, hermaphrodite, actinomorphic, pentamerous, hypogynous. Calyx:

5 sepals, gamosepalous, valvate. Corolla: 5 petals, free but slightly united at the base and adnate to staminal tube, twisted. Androecium: Numerous monadelphous stamens forming staminal tube around style, monothecous, extrose. Gynoecium: Polycarpellary, syncarpous, superior, multilocular, one ovule in each locule, axile placentation, style long, stigmas as many as carpels. Fruit: Capsule. Floral formula: Br, Brl, ⊕, ☿, $K_{(5)}$, $C_{5 \text{ or } (5)}$, $A_{(\infty)}$, $G_{\underline{(\infty)}}$.

B) *Hibiscus rosa-sinensis* L. (Chinese rose, Fig. 38.2).

Habit: Typical ornamental shrub. Leaf: Simple, alternate, petiolate, stipulate, ovate, serrate, acute. Inflorescence: Solitary axillary. Flower: Ebracteate, complete, hermaphrodite, actinomorphic, large, red showy. Epicalyx: 5-8 bracteoles around the calyx form epicalyx. Calyx: 5 sepals, fused, valvate. Corolla: 5 petals, free but adnate to staminal tube, twisted, red petals with sinuous margin. Androecium: Numerous monadelphous stamens forming staminal tube around style, monothecous, extrose. Gynoecium: 5 fused carpels, superior ovary, pentalocular, one to many ovules in each locule, axile placentation; styles 5, united below and passing through the staminal tube; stigmas 5 and discoid. Fruit: Capsule. Floral formula: Ebr, ⊕, ☿, $Epik_{5-8}$, $K_{(5)}$, C_5, $A_{(\infty)}$, $G_{\underline{(5)}}$.

Taxonomy of hibiscus

Linnaeus 1st attributed the name *Hibiscus* to a malvaceous genus. Hibiscus is heterogenous consisting of at least two distinct elements, currently known as *Malvaviscus* and *Hibiscus*. The former can easily be recognized by its auriculate petals, carpel-style ratio 1: 2 and fleshy schizocarpur opposed to simple petals, carpel-style ratio 1: 1 and capsular fruits in the latter. De Candolle (1824) was 1st to make a comprehensive intrageneric classification of genus *Hibsicus* primarily based on corolla characters, number of involucels, calyx lobes, vestiture and number of seed in each locule. Hochreutiner (1900) published the last comprehensive taxonomic monograph of *Hibiscus*. No modern comprehensive monograph is available for this section. Australian taxa are listed by Wilson (1974), later on by Craven *et al.* (2003). Wilson (1993) treated the taxa that occurred in the island of Pacific Basin. Later on Wilson (1999) revised this section that occurred in Africa, Asia and Malaysia. American taxon was published by Menzel *et al.* (1983). Publication by Wilson (2006) includes census of all the presently accepted taxa that includes keys, typification, description, synonyms, geographical distribution and cytological data. Section *Furcaria* has distribution from 35^{0}N in southern United States (*H. aculeatus*) to 32^{0}S in South America (*H. diversifolius* subsp. *rivularis*) to 34^{0}S in Africa (*H. diversifolia* subsp. *diversifolia*) and to 35^{0}S in Australia and New Zealand (*H. diversifolia* subsp. *diversifolius*). Section *Furcaria* includes important fiber, food and medicinal plants such as *H. cannabinus* and *H. sabdariffa*. Many species from this section are grown as ornamentals or have ornamental potential with their large, showy, colourful flowers. *Lilibiscus* is a small section of the genus *Hibiscus* that contains about three species. These were the original species used in the development of the *H. rosa-sinensis* cultivars that exist today. Fryxell (1988) and Borssum-Waalkes (1996) provided a key for the separation of the genera in the tribe Hibisceae and genus *Hibiscus* has been described. The genus *Hibiscus* is comprised of ≈300 annual and perennial species in 11 sections. Species under different sections are mentioned in Table 38.1.

Key for identification of Hibiscus (Fryxell, 1988; Borssum-Waalkes, 1996)-

1a. Calyx lobes with prominent marginal costae and often nectarines on mid veins, when necteries absent involucel bracts bifurcate ----------- sect. *Furcaria.*

1b. Calyx lobes without marginal costae and nectarines, involucel bracts never bifurcate-------2.

2a. Stipules large, foliaceous, enclosing developing shoots, capsules 10 loculed ------sect. *Azanzae.*

2b. Stipules small, linear or lanceolate, not enclosing developing shoots, capsules 5 loculed----3.

3a. Immature capsules with 5, longitudinal stripes along the sepal line, seeds silky hairy, hairs equal or longer than the diameter of seed -------- sect. *Bombicella.*

3b. Immature capsules without stripes, seeds glabrous or pubescent, hairs shorter than the diameter of seed --------4.

4a. Calyx lobes with many, hispid nerves --------- sect. *Trionum.*

4b. Calyx lobes without prominent, hispid nerves -------- 5.

5a. Herbs, involucels very short or caducous- ------sect. *Solandra.*

5b. Sub-shrubs or trees, involucels well developed ------ 6.

6a. Leaves ovate-cordate, lobed or unlobed without foliar nectaries -------7.

6b. Leaves elliptic to lanceolate, truncate, unlobed, with foliar nectaries -------9.

7a. Corolla white turning pink, capsules globose, seeds patently hirsute half way from the hilum ------sect. *Venusti.*

7b. Corolla yellow or pale pink with a dark eye, capsules oblong, ovoid or dipressed globose, seeds glabrous or tomentose ---------- 8.

8a. Corolla yellow, capsules adpressed tomentose or glabrescent, valves glabrous within -----sect. *Trichospermum*.

8b. Corolla pink, capsules densely setose, fruits valves setose within ----- sect. *Spatula*.

9a. Calyx pubescent, stamina column included, antheriferous throughtout, involucel almost equaling the calyx ---------sect. *Hibiscus*.

9b. Calyx glabrate, stamina column exserted, antheriferous only on the upper half, involucel shorter than calyx -------sect. *Lilibiscus*.

Table 38.1. Different sections of *Hibiscus* species (Fryxell, 1988; 1993)

Section	**Species involved**
Hibiscus	*H. syriacus* L.
Azanzae DC.	*H. liliaceus* L.
Bombicella DC.	*H. hirtus* L., *H. micranthus* L., *H. talbotii* Paul.
Furcaria DC.	*H. acetosella* Welw. ex. Hiern, *H. cannabinus* L., *H. hispidissimus* Griffith., *H. radiates* Cav., *H. sabdariffa* L., *H. surattensis* L.
Lilibiscus Hochr.	*H. rosa-sinensis* L., *H.* var. *rosa-sinensis* L., *H.* var. *schizopetalus* Mast.
Solandra Hochr.	*H. lobatus* (Murr.) Kuntz.,
Spatula Hochr.	*H. platanifolius* (Willd.) Sweet.
Trichospermum Hochr.	*H. canescens* Heyne ex. Wall., *H. lunariiformis* Willd., *H. panduriformis* Burm.f.
Trionum DC	*H. trionum* L., *H. sreenarayanianus* Anil & Ravi
Venusti Ulbr.	*H. mutabilis* L.
Muenchhusia L.	*H. coccineus* L., *H. dasycalyx* L., *H. grandiflorus* L., *H. laevis* L., *H. moscheutos* L.

Species belonging to section *Muenchhusia* are commonly called as rose mallows and share common growth habit being cold hardy, herbaceous perennials, primarily wetland species and thus suitable for garden use in temperate regions worldwide. Rose mallows are popular ornamental plants mainly due to their abundant, large and colourful flowers. All species of this section have basic chromosome number of n=9 and are known to form hybrids relatively easy (Winters, 1970; Small, 2004). Kuligowska *et al*. (2016) stated that plants within *Muenchhusia* group hybridize easily while intergroup hybridization is usually hampered by hybridization barriers. According to division, *H. moscheutos* belongs to group-I while *H. coccinues* and *H. laevis* belong to group-II (Wise and Menzel, 1971; Small, 2004). Hibiscus section *Furcaria* is a circumglobal tropical and sub-tropical, natural group of plants, presently comprised of 109 recognized taxa. This section is set apart from the other sections of the large genus *Hibiscus* by each of the five calyx lobes having a prominent thickened midrib and two thickened marginal ribs (Wilson, 1994).

Sharma and Sharma (1962) observed that the heterogenous *H. rosa-sinensis* complex has been stabilized. The evolution is rather accomplished by origination of new forms through participation of nuclei with variant chromosome numbers in the formation of daughter shoots and mutations which might have been extensively observed in this species. Such evolved varieties include *H. rosa-sinensis* var. *cooperi*, *H. rosa-sinensis* var. *carneiplenus* and *H. rosa-sinensis* var. *rubroplenus* (Singh and Khoshoo, 1989). *Hibiscus* species/ cultivars have been distinguished on the basis of morphological characters. The petals, staminal column and stigma may be variously modified in double flowered varieties. Similarly, pollen grain size is considered to be characteristic of the varieties. The pollen size differs not only between varieties, but within varieties also. Besides size differences between normal and sterile grains, the normal grains also varies among themselves. Pollen morphology of 10 cultivars of Shoe flower (*H. cooperi*- Australian Single, Splendens, Rose, Albus; *H. achania malvaviscus*- Floropleno; *H. schizopetals*- Sunset and Juno) was studied by Vilasini *et al.* (1966). There was slight variation in colour and pollen grains of all varieties which had more or less similar shape. The colour of pollen grains varies from light yellow to deep yellow in most varieties and light brown in *H. achania malvaviscus*. Markose and Aravindakshan (1987a) studied pollen morphology of 34 types/varieties of *H. rosa-sinensis* and two other species, namely *H. mutabilis* and *H. schizopetalus*. Pollen grains of all species (*H. rosa-sinensis, H. mutabilis, H. syriacus*) are spherical, echinate, pantoporate and spinose and exhibited variation in colour and size. Pollen grains having straight or crooked spines are characteristics of *H. syriacus* cultivars. Pollen is sticky, spiny, light-yellow to pink in colour and cling together. Single flowered cultivars have larger pollen grains than double flowered cultivars. Pollen diameter varies from 25.91 μ (Acc. 29) to 198.58 μ (Acc. 25). *Hibiscus mutabilis* and *H. rosa-sinensis* cultivars could not be distinguished on the basis of their pollen characters (Shim *et al.*, 1988).

Floral Biology of Hollyhock

Flowering period lasts for 100-103 days with peak flowering between June-August. The life of single flower is about 3-4 days. Soundrapandian and Rao (1961) studied some aspects of anthesis and pollination in hollyhock. It possesses high degree of protandry and dehiscence of anther occurs one day earlier than the emergence of stigmatic branches. It occurs from top to bottom and entire process is completed within two hours. Flowers at the bottom of the raceme bloom first. Anthesis in hollyhock occurs between 7-9 AM. Emergence of stigmatic branches occurs one day after anthesis. Stigma remains receptive 3-5 days of flower opening. Maximum stigma receptivity occurs at 19 00 hours on 2nd day of anthesis at that time it touches their own anthers via curvature of style branching assuming autogamous fertilization. Pollen grain will remain viable for 48 hours. Although flowers exhibit protandry, self pollination and self fertilization occurs in normal course. The cut end of style very close to ovary is also observed to be receptive to pollen congenial to tube formation and successful fertilization.

References

Borssum-Waalkes, J.Van. 1996. *Blumea*, **14**: 1-251.

Craven, L.A., Wilson, F.D. and Fryxell, P.A. 2003. *Australian Systematic Botany*, **16**: 185-218.

De Candolle A.P. 1824. *Prodromus Systematis Naturalis Regni Vegetabilis*, **1**: 429-474.

Fryxell, Y. 1988. *Systematic Botany, Monograph*, **25**: 1-522.

Fryxell, P. 1993. *In*: Flora of India, **3**: 352.

Hochreutiner, B.P.G. 1900. *Annuaire Conserv Jard Botany Geneve*, **4**: 23-191.

Kazemi, M., Aran, M. and Zamani, S. 2011. *World Applied Sciences Journal*, **13**(5): 1234-1239.

Kuligowska, K., Lutken, H., Christensen, B. and Muller, R. 2016. *Breeding Science*, **66**: 300-308.

Markose, B.L. and Arvindakshan, M.1987a. *Agriculture Research Journal of Kerala*, **25**(1): 5-10.

Menzel, M.Y., Fryxell, P.A. and Wilson, F.D. 1983. *Brittonia*, **35**: 204-221.

Sharma, A.K. and Sharma, A.1962. *La Cellule*, **62**: 281-300.

Shim, K.K., Suh, B.K. and Han, B.K. 1988. *Journal of Korean Society for Horticultural Science*, **9**(1): 38-45.

Singh, F. and Khoshoo, T.N. 1989. *The Nucleus*, **32**(1-2): 62-67.

Small, R.L. 2004. *Systematic Botany*, **29**: 385-392.

Soundrapandian, G. and Rao, J.S. 1961. *Madras Agriculture Journal*, **48**: 348.

Vilasini,G., Mary, K.G. and Pillai, P.K. 1966. *Agriculture Research Journal of Kerala*, **4**: 24-27.

Wilson, F. D. 1974. *Australian Journal of Botany*, **22**: 157-182.

Wilson, F. D. 1993. *Brittonia*, **45**: 275-285.

Wilson, F. D. 1994. *Genetic Resources and Crop Evolution*, **41**: 13-25.

Wilson, F. D. 1999. *Bulletin of Natural History Museum, London, Botany*, **29**: 47-79.

Wilson, F. D. 2006. *Bonplandia*, **15**(1-2): 53-62.

Wise, D. A. and Menzel, M. Y. 1971. *Brittonia*, **23**: 425-437.

Winters, H. F. 1970. *Economic Botany*, **24**: 155-164.

39

Myrtaceae (Myrtle Family)

A family of about 150 genera and 3,800-5650 species, distributed mainly in sub-tropical and tropical regions of the world. Family is particularly abundant in Australia where some Eucalyptus trees are world's tallest angiosperms. Myrtaceae is represented in India by about 15 genera and over 170 species. Beautiful pinkish flowers of *Myrtus communis* are considered highly sacred among Greeks, Egyptians, Jews and Persians, and used in religious rites and ceremonies. Some of the larger genera of Myrtaceae along with their number of reported species include *Eugenia* (1000), *Eucalyptus* (500), *Syzygium* (500), *Myrcia* (500), *Psidium* (140), *Myrtus* (100) and *Melaleuca* (100).

Eucalyptus spp. produce important components of medicinal and pharmaceutical applications. Economic parts are leaves and terminal buds-along with trunk. Lemon scented/ Tasmanian (*E. globulus*) is high in cineol content. Deodorant is manufactured besides utilization of oil for asthma, bronchitis, perfumery, etc. Phellandrene rich type, *E. citridora* (most widely grown in hill stations of southern parts of India). The source of citronellal and menthol from leaves and terminal buds is famous for perfumery industries.

Selected genera and species

Other garden ornamentals of Myrtaceae include *Callistemon lanceolatus* (bottle brush or Laila Majnu plant), *Myrtus communis* (Myrtle), *Leptospermum laevigatum* (Australian tea tree), *Tristania conferta* (Brisbane box), etc. Valuable timber is obtained from several species of *Eucalyptus*, *Eugenia*, *Barringtonia*, and *Coreya*.

Genera	Importance
Eucalyptus	Hundreds of species of *Eucalyptus* are grown as ornamental plants; eucalyptus oil, obtained from *E. globulus*; *E. citridora* & several other species is used widely in medicine, perfumery and other industries; wood is used for paper manufacturing; valuable gum is obtained from *E. globulus* (blue gum), *E. maculata* (citron gum), and *E. rostrata* (red gum).
Melaleuca	Cajuput oil is obtained from the leaves of *M. leucadendron*;
Pimenta	Oil of bay rum is obtained from leaves and flower buds of *P. racemosa*

Genera	Importance
Syzygium	Cloves are dried flower buds of *S. aromaticum* while all spice are the unripe berries of *Pimenta dioica.*
Eugenia	Yield fruits (*E. jambolana*); spice (E. caryophylla)
Myrus	Ornamental (*M. communis*)

Description of Family

1. General habit: Trees and shrubs, varying from small creepers to giant Eucalyptus reaching to a towering height of 300-400 feet; lysigenous cavities containing ethereal oils are present in young stem, leaves, floral parts and fruits.
2. Leaves: Usually opposite, but rarely alternate, simple; exstipulate, usually entire, gland-dotted; evergreen; and coriaceous.
3. Inflorescence: Usually cymose, sometimes racemose, rarely of solitary flowers; multichasial cyme in Eucalyptus, spikes in Callistemon; solitary axillary in *P. guajava* and *Myrius communis*, racemose raceme in *Barringtonia.*
4. Flower: Bracteate, often with 2 bracteoles, bisexual, actinomorphic, penta-tetramerous, epigynous or perigynous; in several genera, the floral receptacle unites with the gynoecium at the base and often prolonged into a hypanthium.
5. Calyx: 4 or 5 sepals, free or united, valvate or quincuncial; sepals in Eucalyptus are thrown off unopened as a lid; sepals contain oil glands.
6. Corolla: 4 to 5 petals, free (*Callistemon citrinus*) or united (*Eucalyptus*), imbricate, often nearly circular; in *Eucalyptus citridora* corolla fuses with calyx to form a cup called operculum (Fig. 39.1); few *Eugenia* species lack petals.
7. Androecium: Stamens numerous; free (*Psidium*), or sometimes basally connate (*Callistemon*); usually bent inwards in bud; connectives often gland-tipped; anthers usually 2-celled, versatile, introrse and dehiscing longitudinally or sometimes apically; sometimes stamens are grouped into as many clusters as petals and are obdiplostemonous (*Syzygium*); stamens variously coloured.
8. Gynoecium: 2 to 5 or more carpels, syncarpous; ovary inferior or semi-inferior, with one to many (often 2-5) locules; 2 to many anatropous or campylotropous ovules in each locule; usually axile or rarely parietal placentation; style one, long and simple; stigma capitate. A nectar-secreting, circular disc is present on the inside of the upper part of the floral cup.

9. Fruit and seed: Fruit variable; fleshy berry (*Psidium*, *Eugenia*), capsule (*Callistemon*, *Eucalyptus*), nut (*Darwinia*) or drupe (*Pimenta*). Seeds with little or no endosperm.
10. Pollination and dispersal: Pollination entomophilous. Insects are attracted because of nectar and variously coloured stamens. Seeds are dispersed by wind, birds or other animals.
11. General floral formula: Br, Brl, ⊕, ⊠, $K_{4\text{-}5 \text{ or } (4\text{-}5)}$, $C_{4\text{-}5 \text{ or } (4\text{-}5)}$, A_{∞}, $G_{(2\text{-}5)}$.

Description of Some Common Plant

A) *Callistemon citrinus* (Curtis) Skeels.

Habit: Large tree. Leaf: Simple, alternate, exstipulate, subsessile, lanceolate, gland dotted. Inflorescence: Long pendent spike. Flower: Bracteate, subsessile, complete, hermaphrodite, actinomorphic, pentamerous, epigynous, crimson-yellow to red. Calyx: 5 sepals, fused, imbricate. Corolla: 5 petals, polypetalous, imbricate or quincuncial. Androecium: Stamens numerous, filaments long, bright red and united at the base to form a staminal sheath; dithecous, dorsifixed, introrse. Gynoecium: Tricarpellary, syncarpous, inferior, trilocular, many ovules in each locule, axile placentation. Fruit: Capsule. Floral formula: Br, ⊕, ⊠, $K_{(5)}$, C_5, A_{∞} (basally fused), $G_{(3)}$.

40

Nelumbonaceae (Lotus Family)

This family of sub-class polypetalae is represented by only single genera and three species. Members are cosmopolitan in distribution, except in very cold regions, and are favorite plants for garden pools. *Nelumbo* lotus are the 'sacred lotus' plants of the family. Indian lotus, found commonly in shallow lakes and pods, is *Nelumbo nucifera. Nelumbo*, genus of aquatic angiosperms belong to monogeneric family Nelumbonaceae is represented by two elemental species *N. nucifera* and *N. lutea*. The 1st fossil record available with Pleistocene epoch of Kashmir.The NBRI, Lucknow acts as a biodiversity conservation centre.

Among the few bulbous ornamentals which survive only in water, one of them is lotus.*N.aureavallis,* known from Eocene fossils from North Dakota, USA is an extinct species from USA. Nearly all parts are edible. Leaves are used for wrapping around other food before it is cooked. Stalks are sliced & added to salad. Flowers are cultivated for ornamental value, religious offerings, environment beautification, water purification, besides for food, Chinese traditional medicine & yield valuable perfume. Flowers recommended as cardiac tonic and used in curing diseases of liver, skin. Its dried and powdered rhizomes are used to cure piles. Due to its astringent qualities, rhizome is known for its hypoglycemic, anti-pyretic, anti-diarrheal, anti-bacterial and anti-inflammatory activity. It possesses anti-fungal activity against *Aspergillus, Trichoderma*, and *Penicilium*. Many traditional ancient medical texts report its use for skin conditions, notably ring worms & leprosy, sexually transmitted diseases such as gonorrhea and syphilis. Milky latex found in stems, leaves and flowers is used to fight bacterial infections. Lotus has been found to have 77% anti-oxidant activity. Swollen rhizome of vegetable lotus is used in culinary preparations. The roasted seeds are used as a substitute to coffee. Seeds are ground to obtain almond-flavoured flour. Seed pods (fresh or dried) are used in dry flower arrangement. *Bhasinda* is an underground stem i.e. rhizomes are edible as raw or cooked in northern parts of India.

Nutrition composition/ 100 g edible portion: Fresh rhizomes contains moisture (83%), crude protein (2.7%), fat (0.11%), reducing sugar (1.56%), sucrose (0.41%), starch (9.25%), fiber (0.80%), ash (1.10%), calcium (0.06%), thiamine (0.22 mg), riboflavin (0.06 mg), niacin (2.1 mg) and ascorbic acid (15.0 mg).

Scent of flower is due to 1-4-Dimethoxy benzene and 1-8-cineole, terpinen-4-01 and linalool. Among these 1-4-Dimethoxy benzene increases daily during anthesis and it is present at higher concentration in *N. pentapetala* or *N. lutea* than in *N. nucifera.*

Selected Indian genera and species

Genera	Importance
Nelumbo	*N. nucifera* is called as 'Lotus of India or Kamal'; fruiting torus, flowers, rhizomes, young leaves and petioles of *N. nucifera* are used as vegetable; flowers of *N. nucifera* are recommended as a cardiac tonic & also used in curing diseases of liver, skin & its dried and powdered rhizome to cure piles; flowers of *N. nucifera* yield a valuable perfume. It is frequently mentioned in ancient Sanskrit scriptures by various names including *padma* (pink lotus), *kamala* (red lotus), *pundarika* (white lotus) and *utpala* (blue lotus). Rhizomes possess anti-fungal activity against *Aspergillus niger*, *Trichoderma viridae*, *Penicilium* spp., *Candidaal bicans* and *Sacchromyces cerevisiae*. Leaves contain three alkaloids namely, nuciferine, roemerine, and nornuciferine.
Nelumbo lutea (syn. *N. pentapetala, N. luteum*)	Native to New World; commonly known as American lotus/ Yellow lotus/ Lotus lily; introduced in India from USA during 1993; yellow flower colour; thrive in 60-75 cm deep water; tolerant to cold
N. nucifera (syn. *N. speciosa, N. indica, N. nelumbo*)	Native to Old World; commonly known as Indian lotus/ Asiatic lotus/ Sacred lotus/ Chinese water lily/ Rose of India/ Sacred water lily of Hinduisum and Buddhisum. Introduced into Egypt in 523 B.C.; flower colour red, white & pink; widely cultivated for food (seeds, rhizome, petiole, leaves & flowers); flowers bowl shaped, fragrant, bisexual, protogynous
N. speciosum	Very beautiful in hot season with white and pink flowers. Require 1-1.5 m deep water large perennial aquatic herb (Fig. 40.2)

Description of Family

1. General habit: Aquatic, freshwater, perennial herbs with large rhizomes; submerged or free-floating in pods; shallow lakes, etc.It has long chain of tuberous rhizome creeping underground and recalcitrant to regenerate.
2. Root: Adventitious, branched clusters of fibrous roots develop from the root stocks or rhizomes.
3. Leaf: Simple, alternate, peltate or cordate, long petioled; submerged, floating or immersed. Leaves are waxy and bluish green, attached to long cylindrical leaf stalk and has frilled edge with leaf dia., 45-90 cm.

First leaves that appear, few in number, are flat and float on surface. They are followed by thicker, funnel-shaped leaves with slightly wavy edges. Diameter and area of leaf exceed 0.5 and 0.25 m^2, respectively in full growing period. Floating leaves and flowers have long stems, which contain air spaces to maintain the buoyancy.

4. Inflorescence: Solitary and axillary.
5. Flower: Actinomorphic, hypogynous, solitary/ axillary and pedicillate at the end of long pedicle. They are thermogenic and physiologically thermoregulatory.Flower diameter varies from 15-25 cm. Flower fully open each morning (6-7 h for 3 successive days & closes by 12 noon).
6. Calyx: 4 sepals.
7. Corolla: Flowers are termed single- when they have fewer than 25 petals (16-24); semi-double 25-50 petals and double- when they have more petals (116-160).
8. Androecium: Stamens are many (up to 230); short filaments and spirally arranged and become yellow during anthesis; anther dithecous; basifixed or adnate, introrse; dehiscence longitudinal.Flowers are self heating during anthesis and produce one million pollen grains per flower.
9. Gynoecium: Polycarpellary, very sticky and superior with one ovule in each locule; marginal or parietal placentation; numerous carpels of *Nelumbo* remain embedded separately on upper flat surface of swollen receptacle.
10. Fruit and seed: Brown, woody in texture up to 10 cm dia., containing 7-15 nut like seeds in sunken cavity. Seed had hard and impermeable seed coat and are recalcitrant in nature.
11. General floral formula: ⊕, ♀, $K_{3-\infty}$, $C_{3-\infty}$, $A_{3-\infty}$, $G_{\underline{3-\infty}}$ or $\underline{(3-\infty)}$.

Description of Some Common Plant

A) *Nelumbo nucifera* Gaertn.

Habit: Aquatic, perennial herb. Stem: Branched, rhizomatous. Leaves: Simple, large, forming a triad, of which one is present on the lower side of rhizome and scaly, another is present on upper side of rhizome and third leaf is large and reaches up to the water surface; latter is long-petioled, stipulate, peltate, upper surface waxy, multicostate reticulate. Petiole is smooth with scattered prickles. Peduncles and petioles rise high above the surface of pond water. Inflorescence: Solitary axillary. Flower: Ebracteate, pedicellate, pedicel long, large, complete, hermaphrodite, actinomorphic, hypogynous, showy. Calyx: 4 sepals, free, petaloid or greenish pink. Androecium: Numerous stamens, polyandrous, spirally arranged; filaments long, anthers basifixed, connective is prolonged as an appendage. Gynoecium: Polycarpellary, apocarpous, superior, carpels embedded in the upper part of an obconically flat but spongy receptacle; unilocular, one ovule in each locule; stigma flat. Floral formula: Ebr, $\oplus$, $\boxtimes$, K_4, C_{∞}, A_{∞}, $G\ \underline{\infty}$.

Floral biology of *Nelumbonucifera*

Blooming period is June-October. In the plains of India, lotus blooms between March- October and remains dormant in winter. Flowers open each morning for three successive days. On 1st day, the flowers are protogynous with the stigma secreting a fluid which fills the perigynous cup. Second and third-day flowers are functionally staminate, lacking the stigmatic fluid. Protogynous flower open two to several days with pollen release usually occurring from 2nd day. The fragrant flowers are self heating during anthesis and produce around one million pollen grains per flower. Bees and flies are most frequent flower visitors in wild population and contribute 88 and 50 per cent fruit set, respectively. The protogynous phenomenon account for the high degree of out crossing and there is no 'biparental inbreeding' in the population.

Kasumi and Sakuma (1998) studied some important aspects of floral biology in *N. nucifera*. Flowers fully open at 6-7 hours in morning and close by 12.00 hours. Dehiscence of anthers occurs on 2nd and 3rd day while stigma is receptive on first day. Pollen is usually viable for a few hours after shedding. Seeds mature 30 days after pollination.

Reference

Kasumi, M. and Sakuma, F. 1998. *Journal of Japanese Society of Horticultural Science*, **67**: 595-599.

41

Nyctaginaceae (Four O'clock Family)

A family of over 33 genera and 391 species, Nyctaginaceae are distributed mainly in tropics, but also in South America particularly Brazil, Asia, Ecuador, Peru, Columbia, Australia, Africa and New Zealand are the chief centers of distribution. Presently, bougainvilleas are popular ornamental plants in warm climates including India, Taiwan, Vietnam, Malaysia, Australia, Mediterranean region, the Caribbean, Mexico, South Africa, and the United States in Arizona, California, Florida, Hawaii, and southern Texas. Some of the largely represented genera along with the approximate number of their world over reported species are *Bougainvillea* (300), *Mirabilis* (60) and *Pisonia* (50). The C_3 physiology recorded directly in *Bougainvillea, Mirabilis*.

Bougainvillea is a versatile and hardy climbing shrub commonly grown in Indian gardens. It is a short day plant. C_3 and C_4 photosynthesis is reported in bougainvillea cv. Mary Palmer. It is a fast growing garden plant admired for its brightly coloured bracts, profuse flowering round the year, wide adaptability to soil and climatic conditions, and easy growing. It is, grown as hedge, pergola, standard, pot plant, and bonsai for display. In city designs, newer roads, and flyovers are introduced with bougainvilleas as dividers. Bougainvilleas also have been utilized as a kind of folk medicine developed into dietary supplements as insulin substitute and for pollution tolerance. The insulin like compound, pinitol, was first isolated from leaves of *B. spectabilis*.

Selected Indian Genera

Abronia, Allionia, Boerhaavia, Bougainvillea, Mirabilis, Pisonia.

Economic importance

1. Ornamental plants include *Bougainvillea spectabilis*, *B. glabra* and *B. peruviana* and hybrid groups *viz.*, *B.* x *buttiana* (*B. peruviana* x *B. glabra*), *B.* x *specto-peruviana* and *B.* x *specto-glabra*; *Mirabilis jalapa* (Four o'clock plant); *Boerhaavia diffusa* (hogwood) and *Abronia umbellata*.

Description of the Family

1. General habit: These are evergreen herbs, shrubs, woody climbers, or climbing vine showing anomalous secondary growth.
2. Stem: Stem turns from green to dull green brown with age and have conspicuous and curved thorns tipped with a black, waxy substance.
3. Leaves: Leaves are simple, alternate, glabrous, pubescent, opposite and exstipulate. The leaves of a pair are often unequal. Leaf blade 4 to 13 cm long, and 2-6 cm broad; globular, elliptical, obviate, ovate or cordate in shape and deep green or variegated.
4. Inflorescence: Cymose.
5. Flowers: The colourful flowers are really bracts or modified leaves, ½-2 inch long structure to which true flowers are attached at the midrib. They are regular, bisexual or sometimes unisexual (as in *Pisonia*). Flowers are nectariferous. The lower half of the flower is rather dilated, the upper half is funnel shaped and opens out as a conspicuous five red cream coloured star. Bract colour changes from their emergence to full maturity. They are hypogynous. The bracts, usually large and coloured, vary in number (3 petaloid bracts in *Bougainvillea* surrounding a group of 3 flowers; 5 sepal-like bracts in the form of an involucres in *Mirabilis*; in *Boerhaavia*, the involucral bracts are reduced to scales).
6. Perianth leaves: These are united, tubular or funnel-shaped, 5-lobed and petaloid. The lower part of the perianth is persistent in the fruit, enveloping the latter like the pericarp (and known as the anthocarp). Perianth tube opens at the tip forming a star shape. A cross-section of bougainvillea flower in depicted.
7. Androecium: There are 6-8 stamens which alternate with the perianth-lobes and dorsifixed. The number of stamens often varies greatly-there can be 1, a few or several (up to 30) by branching. The filaments often unequal. Stamen arises from the base of the flower tube. In species and most cultivars, the stigma is placed below the anthers; except in few cultivars belonging to *B.* x *buttiana* group which possess imperfect flower tubes (not star-shaped) with exerted stigma and stamens much below the level of stigma (Zadoo *et al.*, 1975).
8. Gynoecium: There is 1 carpel. The ovary is superior and 1-locular. It has a basal anatropous or campylotropous ovule. The style is long.
9. Fruit: An achene. It is single seeded and enclosed by the base of the perianth.

10. Seed: The seed has a mealy perisperm. The embryo may be straight, folded or curved.
11. General floral formula: Br, ⊕, ⊠, $P_{(5)}$, $A_{5,}$ $\underline{G}_{1}$.

Taxonomy of Bougainvillea

The domestication history of bougainvillea is 250 years old. Generic name of bougainvillea was *Bougainvillea* Commers was 1st published by de Jussieu (1789) in his work '*Genera Plantarum*'. Heimerl (1900) recognized 10 species and three of which were horticulturally important with very showy and colourful bracts. He noted great variation in colour and shape of bracts in *B. glabra*, and *B. spectabilis*. Standley (1931) described eight species under the genus *Bougainvillea*. Holttum and Standley (1947) described *B.* x *buttiana* as a separate species, but Holttum (1955) referred it as *B.* x *buttiana. Bougainvillea peruviana* is one of the putative parents of this group. Differences exist between species in the presence or absence of the star, colour and size of the star, and blooming periodicity.

Bor and Raizada (1954) attempted taxonomic classification of bougainvilleas based on hairs on the plants, tip and colour of the bracts. Nair (1965) observed differences in the size and number of brochi in pollen grains of different cultivars. Sharma and Bhattacharya (1960) exploited cytological data as an aid in interpreting the evolution and affinities of different species and varieties of bougainvillea. Swarup and Singh (1964) suggested pollen morphology to be more stable character for identification of species and cultivars. The tips of brochi are round in *B. peruviana*, acute in *B.* x *buttiana* and almost intermediate in *B. spectabilis* while in *B. glabra* there is a thick ring around the exine of pollen grain. Bougainvillea varieties grown in gardens can be classified under four botanical species. All these species differ each other in some characters like growth, habits, leaf shape, size, colour of bracts and mode of flowering. Bracts are brightly coloured, purplish red or bright red, retaining the colour when dried, mostly 2.5-4 cm long. They also suggested key to identify *Bougainvillea* species and hybrids as below.

41.1 Flower Tube Very Slender, Constrictedin the Middle

Flower tube and leaves hairless or hairy; leaves very broad, both flower tube and leaves hairless; tip of brochi round -*B. peruviana*

Flower tube sparsely hairy; hair short, re-curved; leaves with conspicuous hair on both surfaces, broad, with base at right angles to stalk; tip of brochi acute -*B.* x *buttiana*

41.2 Flower Tube Swollen in Basal Part

Flower tube and leaves more or less hairy; hair on flower tube long or short; leaves ovate to elliptic.

Hair on flower tube short (up to 0.2 mm) and curved upwards; leaves elliptic (widest in middle) with short hair, smooth to touch; thick conspicuous ring around exine of pollen grains.-*B. glabra*

Hair on flower tube long (up to 1 mm), straight and spreading: leaves ovate (widest in middle), softly hairy to touch; tip of brochi is not acute. It had iron coloured flowers instead of normal rose pink showed phytoferritin in the cambial zone. -*B. spectabilis*

Bougainvillea glabra, B. spectabilis show natural variation in shape of leaves and bracts, colour of bracts and degree of hairiness. *Bougainvillea peruviana* does not show any variability. The cultivars which do not fall within the taxonomic limits of above species can be classified into three hybrid groups *viz., B.* x *buttiana* (*B.* x *glabraperuviana*), *B.* x *spectoperuviana* and *B.* x *spectoglabra*. A summary of morphological analysis of these groups is mentioned in the Table 41.1.

Table 41.1. Morphological comparison of species and hybrid groups (Bor and Raizada, 1954; Mahadevan, 1981)

Characters	Basic species			Hybrid groups		
	B. spectabilis	***B. glabra***	***B. peruviana***	***B.* x *buttiana***	***B.* x *spectoperuviana***	***B.* x *spectoglabra***
Leaf	Broadly ovate/ovate with truncate and sub-truncate bases and acute apex	Elliptic with obtuse base and acute or acuminate apex	Broadly ovate/ ovate with truncate and sub-truncate bases and acute apex	Broadly ovate or heart shaped	Large, dark green, ovate	Small, dark green
Young shoots	Small lateral branches	Small lateral branches	Unbranched	Unbranched	Small lateral branches	Small lateral branches
Degree of hairiness	Pubescent	Glabrous (smooth, hairless)	Glabrous	Pubescent on upper & lower sides	Glabrous	-
Inflorescence cyme	On lateral branches	On lateral branches	Borne on main shoot	Borne on main shoot	On lateral branches	On lateral branches
Bract shape	Ovate, oval or Broadly ovate	Broadly ovate	Ovate	Broadly ovate	-	-
Bract colour	Purple pink or brick red	Purple or mauve or white	Rosenine purple	Crimson, magenta, burnt orange, etc	Coppery red in juvenile stage & turns to magenta or pink or cream	Mauve or purple or white
Flower tube colour	Variously coloured	Green	Creamish white	Varies with colour of subtending bract	-	-
Spine shape	Stout, large & curved	Short, slender, laterally flattened	Long, slender & straight when young & curved on maturity	Slender, laterally flattened, recurved	Straight	Numerous & curved
Blooming	Varieties bloom in response to dry hot weather, seasonal bound flowering habit	Recurrent	Bloom recurrently	-	-	-

A similar taxonomical attempt on the classification and listing of 80 cultivars has been made by Sharma (1986).

Floral Biology of Bougainvillea

The production of flowers depends on the variety and environment. Both self and cross pollination occur in bougainvillea. Relative position of stigma and anthers, their simultaneous maturity and twisting of floral tube one day after anthesis suggests predisposition for self pollination. The presence of coloured bracts and nectariferous flowers points toward adaptation for cross pollination, butterflies are the pollinators.

Two flowering peaks namely March-April and November-December are recorded (Kumar *et al.*, 1988). Anthesis takes place in the morning hours around 10 am followed by anther dehiscence and stigma receptivity. The upper part of the flower tube becomes twisted on the day following anthesis. Pollen viability ranges from 22 to 90 per cent. Pollen grains fail to germinate after self pollination, but profuse germination, pollen tube growth and good seed set takes place after cross pollination.

Pollen sterility results in no seed set. Seed set is influenced by cultivar, age of plant and environment (Muthuswamy and Lakshmanan, 1957). The plants fail to set seeds until an age of 2-4 years. A cultivar which sets seeds in one place may prove seed sterile in another place. A good seed set takes place during February-April under Coimbatore conditions. A seed set is common in Formosa, Trinidad, Lateritia, Maharaja of Mysore, Mrs. Fraser and Thomasii cultivars having persistent bracts (Pal and Swarup, 1974).

References

Bor, N.L. and Raizada, M.B. 1954. *Journal of Bombay Natural History Society,* **47**(3): 401-408.

de Jussieu, A.L.1789. *Genera Plantarum*, Paris, pp. 91.

Heimerl, A. 1900. *Classe*, **70**: 97-124.

Holttum, R.E. 1955. *Malayan Agri. Hort. Assoc. Magazine*, **12**: 25-36.

Holttum, R.E. and Standley, C.P. 1947. *Bougainvillea buttiana.* Publ. Field Mus. Chicago. Bot. Ser., **23**: 44.

Mahadevan, S. 1981. *Green News*: 3.

Muthuswamy, S. and Lakshmanan, V. 1957. *South Indian Horticulture*, **5**: 35-57.

Nair, P.K.K. 1965. *Indian Agriculture*, **9**: 53-87.

Pal, B.P. and Swarup, V. 1974. *Bougainvillea.* ICAR Publication. New Delhi.

Sharma, H.C. 1986. Horto-taxonomical studies on bougainvilleas. Ph.D. thesis. Kanpur University, Kanpur.

Sharma. A.K. and Bhattacharya, U.C. 1960. *The Nucleus*, **3**: 19-76.

Standley, P.C.1931. Nyctaginaceae and Chenopodiaceae in North western South America. Publ. Field museum, Chicago, Scr, **11**: 96-104.

Swarup, V. and Singh, B. 1964. *Indian Journal Horticulture*, **21**: 155-164.

Zadoo, S.N., Roy, R.P. and Khoshoo, T.N. 1975. *Proceedings of Indian National Science Academy*, **41B**(2): 498-502.

42

Nymphaeaceae (Water-Lily Family)

This family of sub-class polypetalae is represented by only 5 genera and over 70 species. Members are cosmopolitan in distribution, except in very cold regions, and are favorite plants for garden pools. Plants are mostly ornamental, few are medicinal or edible. Flowers of water lily are variously coloured and open just above the water surface in ponds while that of crow lily (*Nuphar*) are golden yellow and extend up and out of the water in the pond. *Nymphaea nouchali* is the Indian red water-lily. *Victoria regia* is a native of South America. Genus *Nymphaea* attained wide spread fame as ornamentals in Europe and North America in the early 19th century. Word '*Nympha*' in Greek and in Roman mythology means 'Nature-goddess'. Nymphacine is an alkaloid present in all parts except seeds and contain 47% starch and fatty oils. Flowers are used as cardiac tonic to relieve pains and cures diseases of liver and skin and rhizomes are used in tannin industry, seeds for culinary purposes and leaves as dining plates. Dried rhizome powder is used to cure piles.

Among the few bulbous ornamentals which survive only in water, one of them is water lily. Water lilies constitute an important element in tropical aquatic ornamental horticulture because of their strikingly beautiful flowers with wide spectrum of colours. Its flowers are very formal in shape and rather more attractive than lotus. Many are fragrant and there is great variation even in size of the flowers and leaves. Unripe fruits of *Nymphaea rubra* are used as vegetables and flowering stalks as vegetables and salad. Rhizomes are used in tannin industry and seeds for culinary purposes. Leaves are used as dining plates. Nymphacine alkaloid is available in all parts of the plant except the seeds which are free from this alkaloid and contain 47 per cent starch and fatty oils. Rhizomes of *N. stellata* are edible and also used to relieve cardiac pains. Rhizomes of *N. nouchali* are used to cure dysentery and diarrhoea. The vegetable '*kamal kakri*' is actually the rhizome of *N. nouchali*.

Selected Indian Genera and Species

Genera	Importance
Nymphaea	*N. nouchali* is called as 'Indian red water-lily'; *N. stellata* is known as 'Indian blue water-lily'; seeds are often eaten during times of scarcity, decoction of flowers is used against palpitation of heart; *N. alba* is known as 'white *kamal*'; *N. rubra* had red flowers; *N. coerulea* has blue flowers; *N. capensis* has sky blue flowers; *N. odorata* has pink flowers and *N. mexicana* possess bright yellow flowers; vegetable '*kamal-kakri*' is actually rhizome of *N. nouchali*; rhizomes of *N. stellata* are also edible & it is used to relieve cardiac pains; rhizomes of *N. nouchali* are used to cure dysentery and diarrhea; *N. lotus* is called as 'chota lotus', cultivated for its beautiful flowers, rhizomes are eaten and also used medicinally in dysentery (Fig. 42.2).
Euryale	*E. ferox* (syn. *Makhana* or Gorgan nut) are roasted and eaten; doctors recommend them for invalids. It is prickly aquatic herb.
Nuphar	-
Victoria	*V. regia* is called as 'Royal water-lily'; seeds of *V. regia* are also roasted and eaten aquatic plant found in Amazon river; leaves are about 3-6 feet in diameter with very beautiful flowers.

Description of Family

1. General habit: Aquatic, freshwater, perennial herbs with large rhizomes; sometimes annual (*Euryale*); submerged or free-floating in pods; shallow lakes, etc.
2. Root: Adventitious, branched clusters of fibrous roots develop from the root stocks or rhizomes.
3. Stem: Perennial, branched, rhizomatous, creeping in *Nymphaea* but somewhat erect in *Victoria* and *Brasenia*; full of aerenchyma, laticiferous vessels and intercellular spaces.
4. Leaf: Simple, alternate, peltate or cordate, long petioled; submerged, floating or emerged, large reaching up to 2 m in *Victoria*; smooth and waxy but lower portion prickly (*Euryale*); in *Nymphaea*, the mature leaves may be orbicular, sagittate or oval; *Cabomba* shows heterophilly. Three leaf types are 1) Very thin, fragile and submerged having short petiole, 2) Floating, thick in texture and 3) Aerial, leathery leaves. They emerge from rhizome in spiral form and float on water surface and have single notch from edge into centre of lily pad. Latex is present in vessels.
5. Inflorescence: Solitary and axillary.
6. Flower: Solitary, bisexual, perigynous (extra axillary), day and night blooming habit, actinomorphic; large showy, long peduncled; usually hypogynous but epigynous in *Victoria* and perigynous in *Nymphaea*; often calyx and corolla are poorly differentiated and are presented in perianth of 6 to numerous, free petals. La-ongsri *et al.* (2009) proposed

groups 1) Single flowered (< 25 petals), 2) Intermediate (25-50 petals), 3) Double flowered [> 50 petals (116-160)].

7. Calyx: 3 to many sepals, polysepalous; in *Nuphar* sepals are yellow and larger than petals; sepals 3 in *Brasenia*, 4 in *Nymphaea* and 5 in *Barclaya*. Calyx and corolla are poorly differentiated and form perianth of 6 or more free tepals.
8. Corolla: 3 to may petals, polypetalous; petals frequently grading into the stamens and therefore inner petals often form petaloid staminodia; white or variously coloured and showy; a nectar is present on abaxial side in *Nuphar*; imbricate.
9. Androecium: 3 to many distinct stamens; filaments usually extend as sterile appendage beyond the anther sacs; anther dithecous; basifixed or adnate, introrse; dehiscence longitudinal; in *Victoria* the inner stamens are sterile; in *Nelumbo* and *Nymphaea* the stamens are spirally arranged.
10. Gynoecium: Carpels 2 or 3 (*Cabomba*) to numerous (*Nelumbo*, *Nymphaea*); each carpel with 1 locule, numerous ovules, marginal or parietal placentation; ovary superior but inferior in *Victoria* and *Euryale*; stigmas mostly 1 and discoid or as many as 35 and radiate; numerous carpels of *Nymphaea* and *Nelumbo* remain embedded separately on upper flat surface of swollen receptacle.
11. Fruit and seed: Fruit is a spongy berry (*Nymphaea*) or follicle (*Cabomba*); seeds 1 to many per locule; with straight embryo and starchy endosperm; seeds sometimes arillate (*Nymphaea*).
12. Pollination and dispersal: Large, too bright, showy and often scented flowers of this family show entomophily i.e. pollinated by insects such as beetles, small flies, etc. Some *Nymphaea* species are self pollinated while Nuphar, Victoria and remaining takes place by water. However, some species of *Cabomba* and *Nuphar* are dispersed by ducks.
13. General floral formula: $\oplus$, ♀, $K_{3-\infty}$, $C_{3-\infty}$, $A_{3-\infty}$, $G_{\underline{3-\infty}}$ or $\underline{(3-\infty)}$.

Taxonomy of Water Lily

On the basis of the floral construction, family Nymphaeaceae is divided into three sub-families, i.e. Cabomboideae, Nelumboideae and Nymphaeoideae. Apocarpous condition is found in Cabomboideae, Nelumboideae while syncarpous condition is observed in Nymphaeoideae. Cabomboideae have cyclic trimerous flowers whereas, Nelumboideae have spiral pleiomerous flowers. All these three sub-families have been treated as independent families by Bessey (1915) under the names Cabombaceae and Nelumbaceae belonging to order Ranales, and Nymphaeaceae belonging to order Rhoeadales. Several taxonomists treated them on the line suggested by Bessey.

Cronquist (1981) included Nymphaeaceae under order Nymphaeales along with four more families (Nelumbonaceae, Barclayaceae, Cabombaceae and Ceratophyllaceae). However, Thorne (1983) treated Nelumbonaceae under order Nelumbonales while Nymphaeaceae and Cabomdaceae under order Nymphaeales.

Bentham and Hooker (1862-1883) treated Nymphaeaceae under order Ranales and sub-family Cabomboideae forms a link between Nymphaeaceae and Ranunculaceae. Presence of trimerous flowers and scattered vascular bundles in Cabomboideae also suggest its affinity with monocotyledons. Nymphaeaceae show affinity with Berberidaceae in possessing scattered vascular bundles and arillate seeds, and also with Papaveraceae in having latex, large and discoid stigma and type of placentation.

Genus *Nymphaea* has been divided into two main sections (Conard, 1905 and Sculthorpe, 1967) *viz*., Section I- *Apocarpiae*: It is further sub-divided into two sub-genera *viz*., *Ancephya* and *Brachyceras*. Section II- *Syncarpiae*: It is further sub-divided into three sub-genera *viz*., *Castalia*, *Lotus* and *Hydrocallis*. The species relevant to the origin of present day garden lilies are *N. odorata* Ait., *N. tuberosa* Paine, *N. alba var. rubra* Lonnroth, *N. tetragona* Georg (syn. *N. pygmaea* Aiton.), *N. lotus* Linn, *N. rubra* Roxburgh, *N. caerulea* Savgny, *N. capensis* Thunberg, *N. zanzibariensis* Casp, *N. flavovirens* Lehmann, and *N. mexicana* Zuccarini.

Morphological analysis of species and cultivars shows considerable variation in plant habit, flower size, type of flower, number of petals, number of stamens, colour of flowers, time of blooming and its duration and mode of propagation. In the elemental species, the basic colours for the flowers appear to be red, white, yellow or blue besides all shades in between exists in all the garden cultivars. This has been the result of natural hybridization or due to indiscriminate hybridization carried out by nurserymen/ amateurs.

References

Bentham, G. and Hooker, J.D. 1862-1883. *Genera Plantarum*, 3 Vols., Reeve and Co., London.

Bessey, C.E. 1915. *Annals of Moussari Botanical garden*, **2**: 109-264.

Conard, H.S. 1905. *The water lilies*: A monograph of the genus *Nymphaea*, Rubles Carnegie Inst. Washington, D.C.

Cronquist, A. 1981. An Integrated System of Classification of Flowering Plants, Columbia University Press, New York.

La-ongsri, W., Trisonthi, C. and Balslev, H. 2009. A synopsis of Thai Nymphaeaceae. *Nordic Journal of Botany*, 27(2): 97-114.

Sculthorpe, C.D. 1967. The biology of aquatic plants, Edward Arnold, London., pp. 35-47.

Thorne, R. F. 1983. *Nordic Journal of Botany*, **3**: 85-117.

43

Oleaceae (Olive Family)

A family of approximately 29 genera and 600 species, Oleaceae are cosmopolitan but occur in great diversity in temperature and tropical Asia. Some of the largely represented genera of olive family and their approximately reported species are *Jasminum* (300), *Chionanthus* (125), *Linociera* (80), *Fraxinus* (70), *Ligustrum* (40), *Syringa* (30) and *Olea* (20).

Genus *Jasminum* contains 90 species to be true in existence and use, range from landscape to hair adornments, and perfumes. It is native to Indo-Malayan region. Morphological variants from different species have been developed in different regions in India. Jasmines are one of the most important ornamental flowering plants, widely cultivated for their attractive and fragrant flowers. Jasmine has multiple uses like in the making of bouquets, garlands, decorating women's hair as *veni*, beautifying pandals and marriage halls and the like, besides preparation of perfumes and cosmetics. *Jasminum grandiflorum* is highly priced for high grade perfumes throughout the world. Jasmine oil has a significant place as an ingredient in cosmetics, confectionary, ointments, disinfectants and detergents and in pharmaceutical beverages. The word 'jasmine' has been derived from a Persian name '*Yasmun*' meaning fragrance.

Selected Indian Genera

Nyctanthus, *Jasminum*, *Fraxinus*, *Myxopyrum*, *Osmanthus* and *Olea*.

Economic Importance

Notable ornamentals are various species of *Jasminum* (jasmine), *Syringa* (lilac), *Forsythia* (golden bells), *Ligustrum* (privet), *Osmanthus* (fragrant olive), *Phillyrea*, *Chiananthus* (fringe tree) and *Nyctanthes* (Tree of sadness).

Olive (*Olea europaea*) has been a source of food since ancient times, and the oil extracted from its fruit is of high value. Flowers of several species of *Jasminum* are source of an oil used in perfumery. Valuable timber is obtained from various species of *Fraxinus* (ash timber), *Notelaea* and *olea*. An orange dye, used for colouring silk and cotton, is obtained from the flowers of *Nyctanthes*. Leaves of *N. arbortristis* (Fig. 43.1) are useful in fever and rheumatism.

Description of the Family

1. General habit: Shrubs (*Syringa*) or trees (*Olea*, *Nyctanthus*), occasionally vines (*Myxopyrum smilacifolium*).
2. Leaves: Simple (*Nyctanthus*) or pinnately compound (*Fraxinus*), usually opposite, rarely alternate (some species of *Jasminum*); often entire, exstipulate, unicostate reticulate venation.
3. Inflorescence: Axillary or terminal racemose, paniculate or thyrsiform.
4. Flower: Bracteate (*Olea*) or ebracteate (*Jasminum*), pedicellate, usually bisexual but unisexual in some species of *Olea* and *Fraxinus* (Fig. 43.1G); actinomorphic, bi-to hexamerous but usually tetramerous, hypogynous.
5. Calyx: 4 sepals, gamosepalous, valvate, rarely sepals absent (*Fraxinus excelsior,* sometimes 4-15 toothed; persistent.
6. Corolla: Usually 4 united petals, valvate or imbricate; sometimes 4-9 lobed (*Jasminum micrantha*); rarely petals absent (*Fraxinus micrantha*); sometimes deeply lobed or divided and seemingly polypetalous.
7. Androecium: Usually 2 stamens, epipetalous, alternate with the carpels; anthers dithecous, the cells usually back to back; often apiculate because of the extension of the connective, dehiscing longitudinally; stamens are 4 in *Hesperelaea* and *Tessarandra*.
8. Gynoecium: Usually bicarpellery, syncarpous, ovary superior, bilocular, with usually 2 anatropous ovules in each locule, axile placentation; style simple, stigma capitates or bilobed.
9. Fruit and seeds: Fruit drupe (*Olea*), berry (*Ligustrum*), samara (*Fraxinus*, or capsule (*Forsythia*) with 1-4 seeds. Seeds with straight embryo and little or no endosperm.
10. General floral formula: Br or Ebr, ⊕, ⊠, $K_{(4)}$, $C_{(4)}$, A_2, $G_{\underline{(2)}}$.

Taxonomy of Jasmine

Even though jasmine is under cultivation from antiquity, a lot of confusion exists in the identification of plants, because of the existence of large number of species, varieties and cultivars and their synonyms. Cytological variations also help in knowing the ploidy levels. Taxonomic descriptions of six species have made by Bor and Raizada (1946). Bailey (1949) gave the taxonomic description of 23 species of *Jasminum* as well as the keys for their identifications. The variation in morphological traits like size, and shape of leaf (simple and compound), calyx, and corolla lobes, and internodal length; shape of floral bud, petal shape, and number of whorls, and length of style, and stigma have been documented in some species. Only few varieties set seeds in nature. The spontaneous mutation, natural crossing, and simple autopolyploidy might be the possible causes for origin of variations in *J. sambac*. The important species are depicted in Figure 43.2.

Krishnaswamy and Raman (1948) gave description of four horticultural varieties of *J. sambac* namely Suji mallige, Iruvantige, Yelu-suthu mallige, and Gundu mallige. Raman (1955) grouped these varieties based on the shape of leaf and corolla lobes into two groups:

Group I- Elliptic leaves and conical buds e.g. Suji mallige, Iruvantige and forms.

Group II- Ovate leaves and globose buds e.g. Yelu-suthu mallige, and Gundu mallige.

A brief description of eight south Indian wild species of *Jasminum* has been given by Raja (1953). He considered Yelu-suthu mallige as *J. sambac* var. Trifoliatum, based on the leaves occurring in three's at the extremity of flowering branches.

Bhatnagar (1956) recorded four distinct sub-varieties in *J. sambac* on the basis of shape of flower buds, petal shape and number of whorls namely:

i) Motia Bela- with double whorl of rounded petals and globular buds.
ii) Bela- with double whorl of elongated petals and elliptic buds.
iii) Bela-Hazara Bela- with single whorl of petals and elliptic buds,
iv) Mogra- with multi-whorls of petals and round bud.

Systematics

Walpers (1852) recognized two sections based on leaflets and further classified the sections into subsections based on calyx characteristics.

Section I- Unifoliata

a. Calyx: lobes subulate and elongate
b. Calyx: sub-truncate

Section II- Trifoliata

a. Calyx: shortly dentate
b. Calyx: lacinate

Hooker (1882) studied 43 *Jasminum* species and indicated two main groups as below-

Group I- with simple leaves, calyx pubescent/glabrous, subulate/short.

Group II- with compound leaves either of trifoliate or imparipinnate.

Engler and Prantl (1897) cited about 160 species of *Jasminum* in the tropical and sub-tropical regions of Asia, Africa, Australia and over 40 in India. They considered that the simple leaf was only the transformed terminal leaflet of the imparipinnately compound leaf. They proposed four major sections as section I- *Unifoliata*, section II- *Trifoliata*, section III- *Alternifolia*, and section IV- *Pinnatofolia*. Gamble (1957) described 20 species from Madras presidency in two major groups on the basis of leaf characters as below.

Group- I: Simple leaf type

Calyx pubescent, subulate & long	*J. sambac & J. pubescens*
Calyx pubescent, linear & short	*J. arborescens*
Under calyx glabrous or nearly so, linear & long	*J. rigidum*

Group- II: Compound leaf type

Leaves trifoliate, lower leaflets wanting & calyx lobes small	*J. auriculatum*
Leaves trifoliate, lateral leaves little smaller than terminal & calyx lobes minute	*J. flexile & J. calophyllum*
Leaves imparipinnate & opposite	*J. grandiflorum*

Floral biology of Jasmine

Flowering time, duration of flowering, yield and flower longevity varies according to species and cultivars. All jasmines bloom profusely and for longer duration where summers are warm and winters are mild. Flowering is almost throughout the year in *J. grandiflorum, J. flexile, J. pubescens, J. calophyllum and J. regidum*. The peak flower production in *J. auriculatum* is in June-July and in *J. grandiflorum*, July to October while other species flower from February to October. Flower opening is not uniform in all species and it varies from species to species (Raman *et al*., 1969). Floral biology and compatibility behaviour was studied by Veluswamy *et al*. (1976). *Jasminum* species were grouped on the basis of anthesis and three patterns of anther dehiscence have been reported.

Table 43.1: Anthesis time in different species and cultivars of jasmine (Veluswamy *et al*., 1976)

Flower opening time	Species and cultivars
Early (6-7 PM)	*J. auriculatum, J. grandiflorum, J. calophyllum* and *J. paniculatum*.
Medium (7-8 PM)	*J. angustifolium, J. arborescens, J. pubescens, J. rigidum* and *J. sambac* cvs. Bold Budded, Double Mogra, Gundu mallige, Karturi malli, Madanban, Single mogra.
Late (8-9 PM)	*J. sambac* cvs. Iruvatchi, Rambanum, Suji mallige, Long flowered, Mahuria, Khoya, Juhi.
Very late (9-10 PM)	*J. communis*

Floral dimorphism expressing long and short carpel is found to exist in *J. grandiflorum* and *J. pubescens*. Two morphotypes, pin (long carpel) and thrum (short carpel) type of flowers in *J. grandiflorum* are noticed. Pin type produce profuse flowering and thrum type are shy bearers. Thrum type flowers are superior to pin type with respect to flower bud length, corolla tube length and diameter of open flower. Floral biology was studied by Srivastava and Karmakar (1986). Period of blooming varies from 3-4.5 months in different species. Maximum floral buds open during 6-7 PM. Dehiscence of anthers occurs during 4-5 PM in *J. sambac*, *J. auriculatum*, *J. grandiflorum* and dehiscence is maximum during 3-4 PM in *J. calophyllum*. An anthesis in all the species is preceded by dehiscence in 2-4 hours. Earlier flower opening under cool days compared to hot months when flower opened between 7.15 and 9.30 PM.

References

Bailey, L.H. 1949. Manual of cultivated Plants. New York: Macmillan Co.

Bhatnagar, G.S., 1956. *Science and Culture*, **21**: 613-615.

Bor, N. L. and Raizada, M. B. 1946. *Journal of Indian Botanical Society*, **46**(2): 205.

Engler, A. and Prantl, K.1897. *Die Naturlichen Pfanzen Familien*, IV. Teil, Leizpig.

Gamble, J.S. 1957. *The Flora of Presidency of Madras*, Reprint Botanical Survey of India, Calcutta.

Hooker, J.D. 1882. *Flora of British India*, III, Reeve and Co., London.

Krishnamurthy, N. and Raman, V.S. 1948. *Journal of Indian Botanical Society*, **27**: 77-83.

Raman, V.S. 1955. *Cytologia*, **68**: 252-260.

Raman, K.R., Shanmugam, A. and Ahmed shah, H. 1969. *South Indian Horticulture*, **17**: 18-27.

Raja, P.B.1953. *South Indian Horticulture*, **1**(3):189.

Srivastava, H.C. and Karmakar, P.G.1986.Germination and floral biology of jasmines. Progress in essential oil research edited by (E.J. Brunke). Walter De Gruyter & Co. Berlin-New York, pp: 485-486.

Veluswamy, P., Thangraj, T. and Muthuswami, S. 1976. *South Indian Horticulture*, **24**: 45-49.

Walpers C. G. 1852. *Annals Botanices Systematiceae*. Vol. 3, pp: 25.

44

Orchidaceae (Orchid Family)

Orchidaceae is one of the largest families of the flowering plants, represented by about 1,000 genera and 20,000 species which are cosmopolitan but primarily distributed in tropical areas. They are found in all parts of the world, even in the frozen areas of Alaska, the snow-covered areas of the Himalayas and sandy deserts of Australia and Africa. New Guinea has more orchids, native to the island, than any other part of the world. In Indian flora, this is the second largest family having 7% of the world's orchid genetic diversity contributing 1,330 species in 186 genera distributed mainly in Eastern Himalayas, Western Ghats and Khasi hills. Indian orchid wealth serves as a foundation of world's orchid industry. The North-Eastern hilly regions of India, Himalayas being the cradle, comprises eight states- Arunachal Pradesh, Assam, Manipur, Meghalaya, Mizoram, Nagaland, Sikkim and Tripura- is the abode of unique orchids. Many are now endangered due to man's invasion and "burn and grow" type of shifting cultivation. Arunachal Pradesh alone accounts for 601 species, almost 52% of the total species known in India. Arunachal's rich and colourful orchids find a place of pride. Arunachal Pradesh is called the 'Orchid Paradise of India'. The entire orchid family with about 601 species in Arunachal Pradesh is classified into six sub-families, 17 tribes, 24 sub-tribes and 111 genera. *Vanda* is found as epiphyte on mango and other trees in the plains. *Aerides* is found in southern and northern parts of India, Burma, Malaya and Indonesia, *Vanda* spp. is common in India, Sri Lanka, South East Asia and Indonesia. *Calanthe* spp. have a wide range from China, Japan, Vietnam to Australia, Southern parts of India and South Africa. The striking *Cattleya* spp. are found only in South America. *Coelogyne*, a genus of considerable horticultural value, is mostly represented in Nepal and Sikkim but several species are found in Bhutan and upper Myanmar, China, South-East Asia, the Philippines and Fiji. About 20% of recorded Indian taxa are endemic. *Dendrobium peguanum* is a rare species, *Rhizanthella gardneri* is endemic to Western Australia which grows underground and *Phaius tankervilliae* (Nun's orchid of Assam; 2n= 50) is also an endangered orchid. India has two conservation and germplasm centers at Shillong in Meghalaya and Yercaud in Tamil Nadu. Asia's largest orchidarium is situated at Tipi, Arunachal Pradesh while orchids sanctuary is situated at Sikkim.

Orchids are the most fascinating plants in the world. They belong to 2nd largest family of angiosperms of the world. It constitutes 7% of all the angiosperms and nearly 40% of monocotyledons. This is the family of greatest variety of flowers among flowering plants. Orchids occupy wide range of ecological habitats and exhibit highly specialized morphological, structural and physiological characteristics. In nature, orchid exists as a big tree & miniatures with flowers as small as minute pin head. Orchid flowers are known for their strange shape, longevity and beautiful look. Orchids are perennial plants of varying growth habits. A majority of them which grow on trees are epiphytes, some being terrestrial, growing on the forest floor or open ground. A few grow in water as semi-aquatics, and two rare genera of Australian origin are sub-terranean, at the most the tiny flower head reaching the soil surface. Some of the largely represented genera along with the number of their approximately reported species are *Dendrobium* (1,500), *Habenaria* (600), *Epidendrum* (400), *Oncidium* (350), *Odontoglossum* (200), *Cattleya* (60), *Vanda* (60), *Cypripedium* (50) and *Cymbidium* (40). International trade restrictions on orchids through CITES (Convention on International Trade in Endangered Species of Wild Fauna and Flora) to restrict the illegal collections, smuggling and export of important endangered orchids adopted in 1973. Royal Horticultural Society, London is International Authority for Registration of orchids. Germplasm is protected by using modern tools of bar coding.

Selected Indian Genera

Cattleya, *Cymbidium*, *Cypripedium*, *Dendrobium*, *Habenaria*, *Orchis*, *Paphiopedilum*, *Vanda*, *Zeuxine*, etc.

Economic Importance

1. Theophrastus (372-287 BC) and Linnaeus have mentioned the use of orchids principally as medicinal plants.
2. Many species of orchids have been used in folk medicine in India. Orchids cure variety of ailments viz. muscular pains, headache, stomachache, bone fractures, to stimulate hair growth, preventing baldness, as skin rejuvenator, curing general debility and weakness.
3. Orchids were used as anti-depressants and stimulants in Greece and Asia Minor. They are well known for their curative and aphrodisiac properties too.
4. Species of *Dendrobium* orchids are valued in Chinese medical field as source of tonic astringent, analgesic and anti-inflammatory substances.

5. *Dendrobiums* are very valuable herbs with wide medicinal and clinical application in China.
6. Pseudobulbs of *Dendrobium tokai* are used as oral contraceptives.
7. *Vanda parviflora*: Anti-viral and anti-cancerous properties.
8. *Ophioglossum vulgatum* is antidote for snake bite.
9. Drug 'chin shih hu' obtained from dried stem of *Dendrobium nobile* is being used in China as anti-pyretic and tonic.
10. Anoectochilus leaves are used as vegetable in Indonesia and Malaysia.
11. Indian *Cymbidiums*, *Dendrobium* and *Vandas* have great potential in floriculture, pharmaceutical, and tourism industry of our country.
12. Seed pods (seeds + pulp) of vanilla orchids are used as flavouring extract in culinary presentations.
13. Number of alkaloids like stilbenoids, tri-terpenoids have been isolated.
14. *Eulophia andamanensis* (Pretty Green Bay): Important medicinal indigenous orchid species registered at NBPGR, New Delhi has special characteristics viz. green flowers, long spike and duration of flowering.
15. The fruits of climbing genera *Vanilla fragrans* syn. *V. planifolia* are the source of highly fragrant extract used in perfumery and confectionary.
16. Tuberous roots of *Habenaria susannae* and *Orchis latifolia* are edible and eaten during scarcity.
17. Starchy tubers of some genera (*Cymbidium*, *Dactylorhiza*, *Eulophia*, *Orchis*, etc.) are dried to form 'Salep' of commerce, used for culinary medicinal purposes.
18. Orchids are known throughout the world for their beautiful flowers of various forms, shapes and highly attractive colours. Some orchid genera commonly grown in glasshouses are *Cattleya* (florist orchid), *Cymbidium*, *Cypripedium* (lady's slipper), *Dendrobium*, *Epidendrum* (green-fly orchid), *Habenaria* (fringe orchid), *Miltonia*, *Odontoglossum* (baby orchid), *Oncidium* (butterfly orchid), *Orchis* (showy orchid), *Paphiopedilum*, *Phalaenopsis*, *Vanda*, *Vanilla* and *Zeuxine*.
19. Important medicinal orchids are-
    - *Phaius tankervilleae*- Terrestrial species. Therapeutically important.
    - *Saccolabium papillosum*: Roots of this are used as substitute for sarsaprilla. Found in Himalayas from Kumaon eastwards to Arunachal Pradesh.

- *Bulbophyllum careyanum*: Found in base of Himalayan ranges, in tropical climate from Sikkim eastwards to Arunachal Pradesh.
- *Malaxis acuminata*- Dwells in pine forests and distributed from Himachal Pradesh to Arunachal Pradesh.
- *Dactylorhiza hatagirea*- Boiled extracts for treating fractures and fresh extracts for curing cough & stomachache. Species is native to Spiti valley.
- *Vanilla wightiana*- Used for curing cold, cough, skin burns, ulcers, viral fevers & bone fractures. Species is distributed in the southern Seshachalam hills.

Description of the Family

1. General habit: Perennial herbs; either terrestrial (*Orchis*), epiphytic (*Cattleya*, *Cypripedium*) or saprophytic (*Neottia*); orchids in temperate regions are mainly terrestrial while those in tropical regions are mainly epiphytic; sometimes climbers (*Bulbophyllum*, *Vanilla*); often with fibrous or tuberous roots or rhizomatous; epiphytic orchids usually contain fleshy pseudobulbs consisting of one or more thickened internodes of stem; aerial roots often contain velamen; mycorrhiza often present; stems leafy or scapose, may be sympodial (*Dendrobium*) or monopodial (*Vanda*); saprophytic forms usually devoid of chlorophyll (*Corallorhiza*, *Neottia*).
2. Leaf: Simple, alternate, often distichous, rarely opposite; sometimes reduced to achlorophyllous scales (e.g. saprophytic species); usually fleshy and linear to ovate in shape, often encircling the stem with a sheathing base; foliage leaves absent in some species of *Dendrophylax* and *Epipogon*; only a single leaf per shoot is present in *Pleurothallis*.
3. Inflorescence: Variable from racemose, paniculate (*Oncidium*), spicate, or flowers solitary (*Cypripedium*).
4. Flower: Usually bracteate, bisexual or rarely unisexual, zygomorphic, epigynous, showy, beautifully coloured, and extraordinarily diverse in shape, size and colour; mostly resupinate i.e. twisted 180^0 or upside down.
5. Perianth: Usually 6 tepals in 2 whorls of 3 each; the outer whorl (representing calyx) of 3 tepals is green or coloured; the inner whorl (representing corolla) of 3 tepals is petaloid, of these the posterior tepal is highly modified, often projected basally into a spur and called labellum or lip.

Labellum is strictly the uppermost petal but look as if located on the lower side of the flower in most orchids. This is caused either by the twisting of the ovary through 180^0 (as in many flowered orchids), or by the pedicel bending back over the apex of the stem (as in single-flowered orchids). This phenomenon is called 'resupination'. It enables the labellum to work as a landing place for pollinating insects.

6. Androecium: Represented by 2 fused, lateral stamens or 1 terminal stamen; anthers dithecous, introrse; pollen grains granular or bound together by viscin threads into masses called pollinia.

   Stigmas, style and stamens are adnate to form a single, highly complex structure called column or gynandrium which represents the most characteristic part of the orchid flower.

   In several orchids (*Zeuxine*) a pair of pollinia remain connected to the glandula or corpusculum with the help of a stalk called caudicle. A connection between the ovary and stamen is made by rostellum. Sometimes staminodes are also present.

7. Gynoecium: Tricarpellary, syncarpous, inferior, unilocular, many ovules in the locule; parietal placentation; rarely the ovary is trilocular and the placentation is axile (*Apostasia*); style 1 (part of the column); stigmas 3, of which 2 lateral of are often fertile and the third one is sterile forming a beak called rostellum.

8. Fruit and seeds: Fruit usually a capsule, containing a very large number of very small, non-endospermic seeds, which are distributed easily by wind.

9. Pollination: Flowers of Orchidaceae are well-adapted for insect pollination because of several characters, such as large-sized and gaudy perianth, pleasant smell, and presence of nectar secreted in nectar or sac-like rostellum. The flowers and protandrous. When an insect enters the flower, the pollinia get attached on its body by their caudicles. When this insect visits another flower, it may deposit the pollen grains on its stigmatic surface below the rostellum, thus completing the process of pollination.

10. General floral formula: Br, ·|·, ♀, P_{3+3}, $A_{(2)\ \mathrm{or}\ 1}$, $G_{(3)}$.

Floral biology

Anthesis starts in early morning (7-8 hours) and stigma remains receptive for a very long period i.e. up to 2-8 days after anthesis. Pollen germination is highest on sucrose and stigmatic extracts than inorganic salts and growth substances. Pollen of all orchids germinates better after storage at 7.2^{0}C than at 22.22^{0}C until 12 months of storage. Pollen of *Dendrobium*, *Phalaenopsis* had a relatively shorter storage life whereas that of the *D. undulatum* and *D. strebloceras* retained high viability after 12 months at 7.2^{0}C and some viability after the same period at 22.2^{0}C (Meewot and Kamemoto, 1969).

The fruit is a capsule. Seeds are non-endospermic or endosperm nucleus produced in large number are extremely light and with undifferentiated embryo. As many as 1,300 to 40,00,000 seeds per capsule are produced. The seed colour may be white cream, pale green reddish, orange or dark brown and have very diverse shapes.

References

Meewot, W. and Kamemoto, H.1969. *American Orchid Society Bulletin*, **38**: 388-393.

45

Palmae/ Arecaceae (Palm Family)

A family of 183 genera and over 2,600 species distributed widely in tropical and sub-tropical regions of the world and a few also in warm, temperate regions. About 28 genera and over 95 species have been reported from India. Chief genera along with the number of their total reported species and the common names of them are *Calamus* (300, cane palm), *Chamaedorea* (130), *Areca* (54, betel nut plam), *Coccothrinax* (50, Biscayne palm), *Copernicia* (30, wax palm), *Raphia* (30, wine palm), *Sabal* (25, cabbage palm), *Phoenix* (17, date palm), *Roystonea* (17, Royal palm), *Metroxylon* (15, sago palm), *Phytelephas* (15, ivory-nut palm), *Caryota* (12, fish-tail palm), *Elaeis* (2, oil palm) and *Cocos* (1, coconut palm).

Selected Indian Genera

Araca, Borassus, Calamus, Cocos, Nypa, Phoenix.

Economic Importance

Palm family, supposed to be second in importance to Poaceae, is used for food, oils, shelter, coconuts, copra, dates, clothing, etc. As many as 801 uses of only *Borassus flabellifer* have been mentioned in one ancient Tamil song. *Cocos nucifera* (coconut) is another such plant of this family whose almost every part of entire plant body is utilized in some ways or other, i.e. (i) root decoction is used as mouthwash, (ii) stem is used for pillars and as a fuel, (iii) leaves are utilized for making mats, fans, baskets, etc., (iv) toddy, obtained from the young inflorescences, provides an intoxicating alcoholic drink on fermentation, (v) endosperm of young fruits is edible and their milky fluid is sweet and refreshing drink, (vi) vegetable fat, obtained from dried endosperm of mature fruits is used for brushes, ropes, mats, floor coverings, etc., (viii) woody endocarp of fruits is prepared into several articles of decorations and (ix) famous coconut oil is used for anointing the body.

1. Ornamental plants: Palmae are universally known for their majestic look and handsome appearance. Some of them are *Adonidia merilli* (Manila palm), *Arenga saccharifolia* (Gomuti palm), *Caryota urens* (wine palm

or fish tail palm), *Livistonia chinensis* (Fountain palm or Chinese fan palm), *Phoenix rupicola*, *Roystonea elata* (Floridean royal palm), *R. regia* (Royal palm or bottle palm), *Sabal minor* (Bush palmetto) and *S. umbraculifera* (Cabbage palm).

2. Oils: Coconut oil is obtained from the dried ripe endosperm of *C. nucifera* while the palm oil is obtained from fleshy mesocarp of fruits of *E. guineensis* (oil palm).
3. Edible products: (i) Seeds of *Areca catechu* (betel nut palm) are sliced and chewed along with '*Paan*' by millions of people, (ii) Kernel of fruits and tender leaves of *Arenga pinnata* is edible, (iii) Young seedlings of *Borassus flabellifer* are used as vegetable and its fruits are eaten after roasting, (iv) Fruits of *Copernicia ceribera* and *Hyphaene thebaica* are edible, (v) Sago, used as an article of diet, is obtained from the pith of stem of *Metroxylon rumphii* (sago palm), (vi) Stem buds, young peduncles and immature seeds of *Nypa fruticans* are eaten raw or cooked as vegetable, (vii) Ripe fruits of *P. dactylifera* (date palm) and *P. sylvestris* (wild date palm) are eaten throughout the world for their high food value, (viii) Fruits of *Sabal palmetto* are edible, and honey is prepared from its flowers.
4. Toddy: A sap or sugary solution, obtained by cutting or tapping the stems, young peduncles or inflorescence of several palms, is used for manufacturing jaggery, and by fermentation it yields an intoxicating drink or beverage called 'toddy'. This sap is also used for manufacturing number of commercially useful products such as 'Nira', palm sugar, alcohol, vinegar, etc. Palms commonly used for this purpose are *Arenga pinnata*, *A. saccharifeira*, *B. flabellifer*, *Caryota urens*, *Metroxylon vinifera*, *Nypa fruticans*, and *P. dactylifera*. The sap of *Raphia hookeri* is the source of famous 'Bourdon wine'.
5. Hats, mats and baskets: These and many other similar articles such as umbrella handles, walking sticks, hand fans, etc. are prepared from *Borassus flabellifer*, *Nypa fruticans*, *P. dactylifera* and *P. sylvestris*.
6. Brushes: Brushes are prepared from fibres obtained from leaf stalks of *Arenga pinnata*, *Borassus flabellifer* and *Sabal palmetto*.
7. Timber: Dense stem wood of several palms (*Borassus*, *Caryota*, *Cocos*, *Hyphaene* and *Phoenix*) is used as timber.
8. Cane: Common 'cane' or 'rattan' of commerce, used for matting chairs, baskets, furniture frames, walking sticks, etc., is obtained from several species of *Calamus*, such as *C. extensus*, *C. latifolius*, *C. ovatus*, *C.*

rotundus, and *C. tenuis*. *Daemonorops adspersus*, *D. jenkinsianus* and *Korthalsia horrida* are also utilized for similar purposes.

9. Vegetable ivory: Stony endosperm of *Phytelephas macrocarpa* is used for making beads of necklaces and other similar articles as cheap substitute of ivory.
10. Wax: Wax, utilized for manufacturing candles, boot polishes, etc. is obtained from *Ceroxylon andicola*, *Copernicia cerifera*, etc.
11. Sacred writing material: Narrow strips of leaves of *Corypha umbraculifera* have been used for several sacred writings in olden days. Leaves of Borassus flabellifer were also used for several such writings by ancient Hindus.
12. Resin: Dragon's blood, a resinous exudates obtained from the fruits of *Daemonorops ruber*, is used in colouring marbels, varnishes, etc.

Description of the Family

1. General habit: Woody shrubs, trees or sometimes vines (*Calamus*, *Desmoncus*); with usually unbranched, slender to stout stem; short to over 30 m tall; and with leaves often forming a terminal cluster in the arborescent species.
2. Root: Fibrous and adventitious roots arising from the base of stem; in *Iriantea* prop roots develop.
3. Stem: Variable in different forms, such as (i) very short with leaves appear to arise from the ground (e.g. *Nypa*), (ii) thin and slender with long internodes (e.g. *Calamus*), or (iii) tall, stout, pillar like, covered by persistent leaf bases and containing a terminal cluster of leaves (e.g. *Cocos*, *Phoenix*), stem is rarely branched.
4. Leaf: Leaves are palmate or pinnate, and rarely simple; usually large with petiole base often sheathing the stem; usually in a terminal cluster; usually alternate in climbing species (*Calamus*); petiole is large, strong, smooth or spiny; ochreate stipule present in *Calamus* while leaves are ligulate in *Thrinax*; parallel venation.
5. Inflorescence: Large, much branched, paniculate or spdix (*Phoenix*), often reaching up to 1 meter, and covered by one or more boat shaped woody spathes or bracts; interfoliar i.e. arise amongst the leaves (e.g. *Borassus*) or intrafoliar i.e. arise below the leaves (e.g. *Areca*); plants monoecious or dioecious o flowers bisexual.
6. Flower: Ebracteate, sessile, actinomorphic, usually unisexual and monoecious but sometimes dioecious (*Phoenix*), rarely bisexual

(*Livistona*), trimerous, hypogynous; in *Borassus*, the male flowers are smaller than female flowers.

7. Perianth: 6 free or united tepals in 2 whorls of 3 each; tough, leathery and usually persistent; imbricate, valvate or twisted in the bud; in *Phoenix* the tepals of outer whorl are united and valvate while that of inner whorl are free and twisted; usually the tepals of outer whorl are smaller than of inner whorl.
8. Androecium: Present only in staminate or bisexual flowers; usually 6 stamens arranged in 2 whorls of 3 each (*Phoenix*); positioned against tepals; filaments short, distinct; anthers dithecous, basifixed or dorsifixed. Introrse, dehiscence by vertical slits; stamens are only 3 in *Nypa* and numerous in *Caryota* and *Phytelephas*.
9. Staminodes and pistillodes: A pistillode is usually present in the centre of each male flower. Staminodes (*Phytelephas*) or may not be present in the female flowers.
10. Gynoecium: Present only in pistillate or bisexual flowers; tricarpellary, syncarpous, or carpels are partly or completely free; ovary superior, 3 locular with a single anatropous ovule in each locule, axile placentation; in *Phoenix* the placentation is basal; sometimes the ovary is unilocular and the placentation is parietal; style extremely short or absent; stigmas 3, sessile.
11. Fruit and seeds: Fruit berry with fleshy exocarp (*Phoenix*) or drupe with fibrous epicarp (*Cocos*). Seeds with small embryo and abundant endosperm.
12. Pollination and dispersal: Pollination in palms is usually anemophilous, and the flowers are protandrous. Some palms (*Sabal*), however, are insect pollinated. Dispersal of fruit is either by animals or by water (*Cocos*).
13. General floral formula:
    (a) Male flower: Ebr, ⊕, ♂, $P_{(3)+3 \text{ or } 3+3}$, A_{3+3}, $G_{0 \text{ or pistillode}}$.
    (b) Female flower: Ebr, ⊕, ♀, $P_{(3)+3 \text{ or } 3+3}$, A_0, $G_{\underline{(3)} \text{ or } 3}$.
    (c) Bisexual flower: Ebr, ⊕, ⊠, $P_{(3)+3 \text{ or } 3+3}$, A_{3+3}, $G_{\underline{(3)} \text{ or } 3}$.

46

Primulaceae (Primula or Primrose Family)

Primrose family is represented by about 58 genera and 2,600 species which are cosmopolitan but chiefly distributed in northern hemisphere. Some of the larger genera with their approximately reported species include *Primula* (500), *Lysimachia* (200), *Dodecatheon* (50), *Dionysia* (41), *Anagallis* (30), *Cyclamen* (15), *Salomus* and *Soldanella* (11).

Selected Indian Genera

Primula, Anagallis, Androsace, Cyclamen, Primula.

Economic Importance

The family is of no specific economic importance except that of some well known decorative plants it contains. Some species of *Primula* may cause contact dermatitis. *Anagallis arvensis* is used for curing snakebite, leprosy, gout and even hydrophobia. *Cyclamen purpurascence* contains glucoside cyclamen which is poisonous. Some of the commonly cultivated ornamental plants of Primulaceae are the species of *Primula* (primrose), *Androsace* (rock jasmine), *Cyclamen* (Fig. 46.1), *Dodecatheon* (shooting star), *Douglasia*, *Lysimachia* (Loosestrile) and *Omphalogramma*.

Description of Family

1. General habit: Mostly perennial herbs, only rarely annual herbs or shrubs; plants generally perennate by means of rhizome (*Primula*) or tubers.
2. Leaves: Mostly simple, exstipulate; sometimes variously lobed or dissected; show all the three types of phyllotaxy i.e. alternate, opposite or whorled; basal or cauline; often gland-dotted or farinose.
3. Inflorescence: Flowers often borne on scapes in terminal or axillary position; solitary axillary (*Anagallis*) or umbellate (Primula) or terminal racemes (*Lysimachia vulgaris*).

4. Flower: Bracteate, ebracteolate, complete, actinomorphic; rarely zygomorphic (*Coris*), bisexual; usually pentamerous, rarely trimerous (*Pelletiera*) or 5-9-merous (*Trientalis*), hypogynous; often heterostyled.
5. Calyx: Usually 5 sepals, gamosepalous, rarely 3 to 9 sepals, valvate, persistent, foliaceous.
6. Corolla: Usually 5 petals, gamopetalous, imbricate or quincuncial; rarely 3 to 9 petals; usually rotate (*Anagallis*) to salverform (*Primula*); polypetalous in *Pelletiera*, and petals absent in *Galux*; petal lobes are reflexed sharply downward in Cyclamen (Fig. 46.2) and *Dodecatheon*.
7. Androecium: Stamens 5 or as many as corolla lobes, epipetalous, opposite the corolla lobes; sometimes 5 staminodes are also present opposite the sepals (*Samolus*, *Soldanella*);; anthers dithecous, basifixed, introrse, dehiscing longitudinally.
8. Gynoecium: Pentacarpellary, syncarpous; ovary superior, rarely half-inferior, unilocular, numerous ovules, free central placentation; style and stigma simple; heterostyly common.
9. Fruit and seeds: Fruit usually a 5-halved capsule or pyxis, variously dehiscent, generally dehiscing by teeth at tip. Seeds with small and straight embryo, in hard or fleshy endosperm.
10. Pollination and dispersal: Majority of the *Primula* species are cross pollinated. Dispersal of seeds in *Primula* is by wind but in a few species (*P. vulgaris*) seeds are dispersed by ants.
11. General floral formula: Br, ⊕, ⊠, $K_{(5)}$, $C_{(5)}$, A_5, $G_{(\underline{5})}$.

47

Ranunculaceae (Buttercup Family)

This family is represented by about 62 genera and 2,252 species, chiefly distributed in the temperate regions of the Northern Hemisphere. Eastern Asia and eastern North America are the regions of their most common occurrence. The common genera, of which over 150 or more species have so far been reported, include *Ranunculus* (400), *Aconitum* (300), *Delphinium* (250), *Clematis* (250), *Anemone* (150) and *Thalictrum* (150). Some species occur in Arctic and Alpine zones. However, they are rare in tropics. Twenty genera and over 165 species of this family have so far been reported from India. Plants of this family flower mainly during winter in our country.

Selected Indian Genera and Species

- Ornamentals: Many Ranunculaceae have showy flowers and are grown as ornamentals in gardens. Some of such ornamentals are *Aconitum napellus* (aconite), *Anemone parviflora* (wind flower), *Aquilegia coerulea* (columbine), *Caltha palustris* (Marsh marigold), *Clematis virginiana* (traveller's joy), *Delphinium ajacis* (larkspur), *Helleborus* (Christmas rose), Nigella *damascena* (love-in-a-mist), *Paeonia emodi* (Peony), *Ranunculus aquatilis, R. seleratus* (cedery peaved crow foot).

Genera	Importance
Aconitum	Used as ornamental; roots *A. napellus* are source of aconite, used as a nerve sedative, & for rheumatism; dried roots are used to cure mild heart palpitation & gastritis. Deadly poisonous alkaloid, obtained from *A. ferox*, may cause death, but is useful in snake bite; *A. heterophyllum* and *A. nepellus* of dried tuberous roots yield alkaloid 'aconitin' which is used externally in rheumatism and against fever and pains.
Anemone	*A. parviflora* is used as ornamental; *A. pulsatilla* is source of the homoeopathic drug 'pulsatilla' used to cure nervous disorders in women during menstrual periods
Clematis	*C. virginiana* is used as ornamental purpose; leaves of *C. triloba* are used to cure snake bites
Cimicfuga	*C. racemosa*- the dried rhizome contains alkaloid 'cimicifugin' which is used for the treatment of cholera.

Genera	Importance
Delphinium	*D. ajacis* is used as ornamental; seeds of *D. elatum* are insecticidal while roots of *D. denundatum* are effective in toothache. Seeds, leaves and roots of *D. zatil* are used to cure jaundice and spleen; *D. vestitum* seeds are useful in cardiac disorders
Helleborus	Flowers of *H. niger* are given in the form of syrup to mad persons; species is also useful in chronic skin infections; dried rhizome is used as purgative
Hydrastis	*H. canadensis*- Roots are effective against constipation
Nigella	*N. damascena* is used as ornamental; *N. sativa* is known as '*kala zeera*'; seeds are used as common condiment as well as in treatment of cough, asthma & fever
Paeonia	*P. emodi* is used as ornamental; tubers of *P. officinalis* are used to cure colic & uterus disorders
Ranunculus	*R. muricatus*, *R. asiaticus*, *R. repens* are used as ornamental; while juice of *R. arvensis* is used in irregular fever. *R. aquatilis* and *R. arvensis* leaves & flowers are used to cure asthma & rheumatism; *R. ficaria* is used to cure piles while *R. falcatus* is poisonous and causes skin blisters
Thalictrum	*T. alpinium* used as ornamental; roots of *T. foliolosum* are used as diuretic, purgative & tonic opthalmia and other eye diseases

Description of Family

1. General habit: Generally, annual or perennial herbs, rarely shrubs or vines (*Clematis*, *Naravelia*); perennials perennate by means of rhizome (*Cimcifuga*, *Helleborus*) or tuberous roots (*Aconitum*); some species are aquatic herbs (*Ranunculus aquatilis*).
2. Root: Tap root as well as adventitious roots; tuberous roots of *Aconitum*, *Paeonia* and *Ranunculus* store food, swell and form tuber-like structures.
3. Stem: Aerial as well as underground; herbaceous (*Ranunculus*); woody (*Paeonia*); climbing (*Clematis*); branched.
4. Leaves: Usually basal and cauline; often palmately divided with sheathing leaf bases; usually exstipulate, but stipulate in Thalictrum and Ranunculus; mostly alternate, rarely opposite (*Clematis*), simple (*Ranunculus*), pinnately compound (*Clematis*), decompounds (*Thalictrum*) or finely dissected into fine segments (*Delphinium ajacis*); aquatic species of Ranunculus show heterophylly.
5. Inflorescence: Variously formed, from solitary (Nigella) to paniculate (*Thalictrum*) or in racemes (*Delphinium*) or cymes; a single terminal flower is produced in *Anemone* and *Erantis*, but more often a cymose branching is observed.
6. Flower: Bracteate (*Clematis*) or ebracteate (*Anemone*), bracteolate (Delphinium) or ebracteolate (*Ranunculus*), usually bisexual but

unisexual in some species of *Thalictrum*, usually actinomorphic but zygomorphic in *Delphinium* and *Aconitum*, hypogynous, spurred in *Delphinium* and *Aquilegia*.

7. Calyx: 3 to many sepals, usually 5 (Ranunculus), polysepalous, valvate or imbricate, sepals are petaloid in *Aconitum* and *Delphinium* and their posterior sepal is spurred. In *Nigella* and *Anemone* there is sometimes an involucres of green leaves below the flower, usually alternate with the calyx.
8. Corolla: Usually 5 petals, polypetalous, nectariferous glands often present, sometimes petals are many or absent; posterior petals are spurred. In genera such as *Paeonia, Clematis*, *Actaea* and *Helleborus*, perianth is present; innermost perianth leaves often bear nectaries (honey-leaves) or they are reduced to small nectariferous scales.
9. Androecium: Stamens 5 to many, free, spirally arranged; anthers 2-celled, adnate, extrorse and longitudinally dehiscent; sometimes staminodes are present (*Aquilegia*); some outer stamens may be converted into honey-leaves (*Helleborus*); from 80-150 stamens in *Caltha*; stamens bright coloured in *Thalictrum*.
10. Gynoecium: Usually numerous, free carpels with either one basal ovule or several anatropous ovules showing marginal placentation; ovary superior, unilocular; style 1; stigma 1; in *Actaea* there is only 1 carpel; in *Nigella* carpels are united, ovary pentalocular or placentation axile; in *Helleborus* carpels are only basally connate.
11. Fruits and seeds: Fruit may be an etaerio of follicles (*Aconitum*), etaerio of achenes (*Ranunculus*), a berry (*Actaea*), or capsule (*Nigella*); seeds with minute embryo and oily endosperm; elaisome (an appendage which contains oily substance attractive to ants) is present on the seeds of some species of *Helleborus*.
12. Pollination and dispersal: The flowers are usually protandrous in Ranunculaceae and therefore well suited for insect pollination. Insects are also attracted because of presence of nectar, petaloid calyx, conspicuous corolla and stamens. However, some species of *Thalictrum* are wind pollinated, e.g. *T. minus*. Dispersal takes place by animals or birds (*Ranunculus*) or by wind (*Anemone, Clematis*).
13. General floral formula: Br or Ebr, ⊠ or ·|·, ⊠, K_5 or 3-∞, C_5 or P_{4-6}, $A_{5-\infty}$, G $\underline{\infty}$ or 1-∞.

Description of Some Common Plants

A) *Ranunculus sceleratus* Linn. (Celery-leaved Crow-foot.

Habit: Annual herb. Stem: Branched, erect, glabrous and solid. Leaf: Simple, alternate, petiolate, exstipulate, leaf base sheathing, divided further into obovate or cuneate segments, multicostate reticulate. Inflorescence: Cymose. Flower: Bracteate, bracteolate, pedicellate, complete, hermaphrodite, actinomorphic, pentamerous, hypogynous, spirocyclic. Calyx: 5 sepals, polysepalous, imbricate or quincuncial, petaloid, reflexed from the base. Corolla: 5 petals, polypetalous, imbricate, bright yellow, a nectar is present at the base of each petal. Androecium: Numerous stamens, polyandrous, spirally arranged; anthers dithecous, basifixed, extrose. Gynoecium: Multicarpellary, apocarpous; each carpel superior, unilocular, one ovule in each locule, basal placentation; carpels arranged spirally over an oblong receptacle; style reduced; stigma fimbriate. Fruit: Etaerio of achenes. Seed: Endospermic. Floral formula: Br, Brl, ⊕, ⊠, K_5, C_5, A_∞, G $\underline{\infty}$.

B) *Delphinium ajacis* Auct. (syn. Larkspur)

Habit: Cultivated, annual herb. Stem: Erect, branched, solid, green, hairy. Leaf: Simple or decompounds, alternate, exstipulate, deeply dissected; each lobe is linear to elliptical, smooth, with acute or acuminate apex. Inflorescence: Axillary raceme of 4 to 16 flowers. Flower: Bracteate, bracteolate, pedicellate, complete, hermaphrodite, zygomorphic, spurred posteriorly, pentamerous, hypogynous, bright blue, pink or white. Calyx: 5 sepals, polysepalous, petaloid; posterior odd sepal prolonged into a long spur; quincuncial or imbricate; blue or violet. Corolla: 4 petals, gamopetalous, valvate; two posterior-lateral petals fuse to form the spur, and this spur is situated inside the spur formed by the posterior sepal; sometimes only 2 petals. Androecium: Stamens numerous, usually 15; in 5 groups of 3 each, polyandrous; anthers adnate or basifixed, dithecous, extrorse. Gynoecium: Monocarpellary, superior, unilocular with many ovules, marginal placentation; style reduced, stigma simple. Fruit: Follicle. Floral formula: Br, Brl, ·|·, ⊠, K_5, C(4) or (2), $A_{15 \text{ or } \infty}$, G $\underline{1}$.

48

Rosaceae (Rose Family)

A family of approximately 91 genera and 4,828 species which are cosmopolitan in distribution but particularly abundant in Eastern Asia, North America and Europe. About 25 genera and 230 species of Rosaceae have so far been reported from India. Some of the largely represented genera along with their common names and approximately reported species include *Potentilla* (Cinquefoil, 500), *Prunus* (peach, plum, almond, cherry, apricot, 430), *Rosa* (Rose, 250), *Rubus* (Raspberry, blackberry, 250), *Alchemilla* (250), *Crataegus* (Hawthorn, 200), *Sorbus* (Chokeberry, 250), *Acaena* (100) and *Spirea* (100). Named after the universally famous flowering plant, rose, the Rosaceae show great diversity in types of fruit.

Roses, universally acclaimed as the 'queen of flowers', have played major role in human culture and artistry from ancient times. Among the ornamentals, roses occupy very important position both as garden plants and cut flowers for their beauty and fragrance. This unique position is attributed to great amount of variation in flowers and plant characteristics as well as their wide adaptability to varied agro-ecological conditions. Damask rose (*R. damascena*) and Edouard rose (*R. bourboniana*) are cultivated for extraction of essence, rose attar or rose oil, *otto* and other products, *gulkand*, *gulab jal* and *pankhuri* from petals. Rose oil has a significant place as an ingredient in pharmaceutical preparations *viz*., ointments, lotions, perfumes, creams and soaps and in food products including beverages, baked foods, puddings and frozen dairy desserts. Characteristics of damask rose oil as anti-HIV, anti-bacterial and anti-oxidant have been established.

Selected genera and species

Variety of delicious fruits belong to Rosaceae, and this family should be called as the 'family of edible fruits'.

Genera	Importance
Rosa	Best ornamental plant of the world; '*gulkand*' is prepared from rose petals is a good tonic, laxative as well as useful in tonsilitis; valuable perfume is obtained from *R. damascena* & other rose species.

Genera	Importance
Pyrus	Walking sticks and tool handles are prepared from wood of *P. pashia* and *P. communis*; dried bark of *P. pyrifolia* is useful in cold and cough
Prunus	Tool handles are prepared from wood of *Prunus cerasoides*;
Quillaja	Bark of soaptree (*Q. saponaria*) is powdered & used in shampoos & hair tonics
Geum	Rootstocks of *G. urbanum* are used in diarrhea & chronic dysentery
Hagenia	Female flowers of *H. abyssinica* are the source of antihelmintic drug, cusso
Potentilla	Decoction of leaves of *P. anserina* is used against anthritis and kidney stones
Rubus	Yellow Himalayan raspberry (*R. ellipticus*) possess edible fruits

Description of Family

1. General habit: Mostly trees (*Prunus*, *Pyrus*, *Eriobotrya*) and shrubs (*Spiraea*), and only some are herbs (*Fragaria indica*, *Potentilla supina*); some are prickly climbers (*Rosa moschata*, *Rubus*) or prickly shrubs (*Rosa damascena*).
2. Stem: Herbaceous or woody, usually with thorns or prickles.
3. Leaves: Mostly alternate, simple (*Prunus*, *Nuttalia*) or compound, usually stipulate, stipules paired and sometimes adnate to the petiole (*Rosa*, *Rubus*), sometimes exstipulate (*Spiraea*); compound leaves may be palmately compound (*Fragaria*) or pinnately compound (*Rosa*); extrafloral nectarines present on the back of leaves in *Prunus laurocerasus*.
4. Inflorescence: Highly variable from solitary flowers to racemose and cymose clusters; solitary (*Rosa*), racemose raceme (*Prunus mystifolia*), corymb (*Spiraea*), head or spike (*Poterium*), cymes (*Spiraea*, *Fragaria*) or thyrsoid panicles (*Eriobotrya*).
5. Flower: Mostly bracteate, bracteolate, usually actinomorphic but zygomorphic in *Hirtella* and *Parastemon*, usually bisexual but unisexual in *Poterium sanguisorba* and *Pygeum*, usually complete but apetalous in *Alchemilla*, pentamerous, usually perigynous (*Rosa*) or epigynous (*Pyrus*) and only seldom hypogynous (*Prunus*); epicalyx usually present; usually the floral receptacle along with the basal adnate portions of perianth is variously modified into hypanthium, the hypanthium is often lined with a nectariferous disc.
6. Epicalyx: Often present in several genera and alternate with calyx lobes.
7. Calyx: 5 sepals, free or basally connate, valvate or imbricate, sepals often foliaceous (*Rosa*).

8. Corolla: 5 petals, usually attached to the rim of hypanthium, petals are numerous, large and variously coloured in cultivated species of *Rosa*; petals absent in *Alchemilla* and *Poterium*.
9. Androecium: Usually numerous, distinct stamens in whorls of 5, sometimes only 5 or 10 stamens; anthers basifixed or dorsifixed, dithecous, introrse and dehisce by longitudinal slits; usually a honey-secreting disc is present in between stamens and carpels. Numerous stamens arranged in an outer whorl of 5 antipetalous pairs followed by numerous whorls of 10 stamens each in alternating positions in *Rosa*. Undermentioned are some of the commonly seen arrangements of stamens in some Rosaceous genera, though there may be some deviations from these arrangements.
10. Gynoecium: Highly variable from 1 to numerous pistils, apocarpous to syncarpous; present at the base of the floral cup, or on the sides of floral cup, or enclosed by the floral cup; pistil are either free from the floral cup or adnate to the floral cup; ovary superior to inferior with intermediate stages; styles and stigmas as many as carpel number.

    Monocarpellary (*Alchemilla*, *Prunus*) or pentacarpellary and apocarpous (*Spiraea*), or multicarpellary and apocarpous (*Fragaria*, *Rosa*, *Rubus*), or 5 carpels fused only in the basal region or carpels united completely with each other as well as with the floral cup (*Pyrus*); if the receptacle is convex or dome shaped the ovary is superior (e.g. *Fragaria*, *Potentilla, Rubus*), if receptacle is cup-shaped and the carpels are developing on its inner surface the ovary is half-inferior (e.g. *Prunus*, *Rosa*, *Spiraea*); and if the carpels are fused with one another and also with the floral cup (e.g. *Pyrus*), the ovary is inferior; placentation is marginal or axile.
11. Fruit: Various types varying from drupe (*Prunus*), pome (*Pyrus, Malus, Sorbus*), follicles (*Spiraea*, *Sorbaria*), etaerio of achenes (*Fragaria*, *Rosa*) or etaerio of drupes (*Rubus*).
12. Seed: Endosperm usually absent.
13. General floral formula: Br, ⊕ , ☿, $K_{5 \text{ or } (5)}$, C_5, A_∞, G1-∞.

Taxonomy of *Rosa* spp.

Within the genus *Rosa*, there are four sub-genera as *Eurosa, Hesperhobos, Platyrhodon* and *Hulthemia*. The sub-genus *Eurosa* contains greater than 95 per cent of *Rosa* species and is further sub-divided into 11 sections (Table 48.2). Out of these, *Indicae* and *Synstylae* are distinguished by their exerted styles. These two sections are diploid (2n=14) and contain species of importance to garden rose development. The short styled sections include *Pimpinellifoliae* (diploid and tetraploid) which contains *R. spinosissima* and *R. foetida* an important garden rose ancestor and the *Gallicanae* (tetraploid) contains the important ancestral species. *Rosa gallica* and the ancient hybrids which originated from it are the Damasks, the Centifolias, the Mosses and the Albas. The largest section *Cinnamomeae* contains 48 species of which *R. rugosa* and *R. cinnamomea*, have played minor role in the development of modern roses.

Table 48.2: Different sections of *Rosa* spp. (Cairns, 2000)

Section	**Characters**
Banksianae	White and yellow flowered roses from China
Bracteatae	Three spp, two from China and one from India
Caninae	Pink and white flowered species from Asia, Europe & North America
Carolinae	White, pink and bright pink flower species all from North America
Chinensis	White, pink, yellow, red and mixed coloured roses from China & Burma
Gallicanae	Pink to crimson and striped flowered roses from Western Asia & Europe
Gymnocarpae	One species in western North America (*R. gymnocarpa*), others in east Asia
Laevigatae	Single white flowered spp. from China
Pimpinellifoliae	White, pink, bright yellow, mauve and striped roses from Asia & Europe
Rosa (syn. Sec. *Cinnamomeae*)	Whtie, pink, lilac mulberry & red roses from everywhere, but north Africa
Synstylae	White, pink and crimson flowered roses from all areas

Pollen types are also considered characteristics of the species. Morphometric analysis of pollen exine patterns in 125 spp. and rose cultivars revealed that pattern of ridging and perforations were specific to botanical sections or cultivar groups and appeared to be transmitted effectively from wild species to cultivars (Veda and Tomita, 1989). No gene flow to wild *Rosa* species from the cultivated rose had occurred. An advantage of this method is that a phenotypic marker or trait is not required to measure gene flow (Nakamura *et al.*, 2011).

Floral biology of rose

Klimenko (1972) studied floral biology of floribunda roses at different phases of flower development. The flowers are protogynous. Mathew (1985) observed that pollen grains germinate on stigma 24 hours after anthesis. Maximum stigma receptivity is one or two days after anthesis and period of receptivity is longer in winter than spring. The hip formation is faster and better in spring than in winter. Pollination done at 12 O' Clock in the day gives highest hip set. Self pollination gave higher percentage of hips set than those observed in cross pollination.

References

Cairns, T. 2000. *Modern Roses XI, The World Encyclopedia of Roses*, Academic Press, San Diego.

Klimenko, Z.K.1972. *Horticulture Abstracts*, **42**(2): 4434.

Mathew, R. 1985. Studies on pollination on seed set and germination in rose. *M.Sc. thesis*, IARI, New Delhi.

49

Rubiaceae (Madder Family)

A family of about 611 genera and 13,150 species, Rubiaceae are chiefly distributed in tropics, but some are distributed in temperate or even arctic regions. About 75 genera and 275 species have been reported from India, chiefly in tropical and sub-tropical eastern Himalayas. Some of the largely represented genera with their approximately reported species (Hickey and King, 1988) include *Psychotria* (700), *Galium* (400), *Ixora* (400), *Gardenia* (250), *Mussaenda* (200), *Asperula* (200), *Cephaelis* (180), *Rondeletia* (120), *Coprosoma* (90), *Guettarda* (80), *Rubia* (60), *Coffea* (40) and *Cinchona* (40). Two most important plants of the family are *Coffea arabica* (coffee-yielding plant) and species of *Cinchona* (quinine-yielding plant).

Economic Importance of Selected Genera and Species

Some of the common ornamental plants grown in gardens and roadsides include *Anthocephalus cadamba* (Kadam), *Cephalanthus occidentalis*, *Galium verum* (Ladies Bedstraw), *Gardenia jasminoides* (Jasmine), *Hamelia patens*, *Hamiltonia suaveolens* (Padera), *Ixora arborea*, *I. coccinia*, *Morinda tinctoria*, *M. frondosa*, *M. luteola*, *Pentas lanceolatea* and *Rondeletia*.

Quinine, the famous drug, used throughout the world in treatment of malaria, comes from the bark of *Cinchona calisaya*, *C. ledgeriana* and *C. officinalis*. Other medicinal plants *Cephaelis ipecachuanha* (Ipecac) roots are used against amoebic dysentery and pyorrhoea. *Gardenia gummifera* provides gum used as a carminative stimulant. *Paederia foetida* is used in indigestion and stomach troubles. Roots of *Rubia tinctoria* (Madder) contain dye 'alizarin' and 'purpurin', while that of several species of *Morinda* contain red (*M. tinctoria*, *M. bracteata*) and yellow (*M. citrodora*) dye. Red dye is also obtained from root bark of *Oldenlandia umbellata*, *Rubia cordifolia* and *Morinda tinctoria*. Useful timber is obtained from *Adina cordifolia*, *Anthocephalus cadamba*, *Ixora ferrea*, *Mitragyna parviflora* and *Randia spinosa*.

Description of Family

1. General habit: Many trees (*Cinchona officinalis*) or shrubs (*Mussaenda luteola*, *Hamelia patens*); however, those in north temperate regions are mostly herbs (e.g. species of *Galium*, *Oldenlandia*); sometimes hook climber (*Uncaria*) or vines.
2. Leaves: Simple, entire, opposite decussate or whorled; stipulate, stipules usually interpetiolar and rarely intrapetiolar, sometimes foliaceous and become as large as the leaf blades so that the leaves appear whorled, or sometimes reduced to glandular setae (*Pentas*).
3. Inflorescence: Basically a dichasial cyme, dichasia sometimes grouped into globose heads (*Adina*, *Nauclea*); rarely flowers solitary (*Gardenia*, *Randia*).
4. Flower: Bracteate, sometimes bracteolate, bisexual, usually actinomorphic, rarely slightly zygomorphic, penta-or tetramerous; usually epigynous, only rarely perigynous (*Synaptantha*) or very rarely hypogynous (*Gaertnera*).
5. Calyx: 4 or 5 petals, polysepalous, valvate, rarely greatly reduced sepals (*Morinda*); in *Mussaenda* one of the sepals of some flowers becomes enlarged and bright coloured.
6. Corolla: 4 or 5 petals, gamopetalous; usually salverform, rotate or infundibular; valvate (*Ixora*, *Mussaenda*), twisted (*Gardenia*) or imbricate (*Rondeletia*).
7. Androecium: 4 or 5 stamens, epipetalous, inserted in the corolla tube or at its mouth, alternate with the corolla lobes; anthers 2-celled, basifixed, introrse, dehiscing longitudinally.
8. Gynoecium: Usually bicarpellary, syncarpous, inferior ovary, bilocular, with 1 to many anatropous ovules in each locule; axile placentation; style simple or bifid; stigma capitate; epigynous disc often present; 5 carpels in *Gardenia* an *Hamiltonia*; ovary half inferior in *Synaptantha* or even rarely superior in *Gaerinera*.
9. Fruit and seeds: Fruit a capsule (*Cinchona*, *Oldenlandia*), berry (*Mussaenda*), schizocarp (*Galium*), or even drupe. Seeds with a small straight or curved embryo in endosperm; sometimes winged.
10. General floral formula: Br, ⊕, ⊠, $K_{4 \text{ or } 5}$, $C_{(4) \text{ or } (5)}$, $A_{4 \text{ or } 5}$, $G_{(2)}$.

Description of a Common Plant

A) *Mussaenda luteola* Dilile:

Habit: An ornamental, perennial shrub. Leaf: Opposite decussate, simple; stipulate, stipules interpetiolar; sub-sessile, acute; unicostate reticulate. Inflorescence: Dichasial cyme. Flower: Bracteate, pedicellate, complete, hermaphrodite, younger flowers actinomorphic, older flowers zygomorphic, pentamerous, epigynous. Calyx: 5 sepals, free but sometimes fused, valvate; in zygomorphic flowers one of the sepals is modified into a large, yellow, leaf-like structure. Corolla: 5 petals, fused, valvate, hypocrateriform, throat and mouth of corolla hairy. Androecium: 5 stamens, polyandrous, epipetalous, alternipetalous; dithecous, basi-or dorsifixed, introrse. Gynoecium: Bicarpellary, syncarpous, inferior, bilocular, with many ovules on T-shaped placentae, axile placentation; style long, stigma bilobed. Fruit: Berry. Floral formula: Br, ⊕ or ·|·, ⊠, K_5, $C_{(5)}$, A_5, $G_{(2)}$.

50

Scrophulariaceae (Snapdragon Family)

Represented by about 65 genera and 1,730 species, Scrophulariaceae are cosmopolitan in their distribution. About 60 genera and 360 species have been reported from India, chiefly from the Himalayas. Some of the genera along with their approximately reported species from world over are *Pedicularis* (500), *Calceolaria* (300-400), *Scrophularia* (300), *Verbascum* (300), *Veronica* (300), *Penstemon* (280, *Linaria* (150), *Castilleja* (150), *Selago* (150), *Sutera* (130), *Mimulus* (100), *Antirrhinum* (40) and *Digitalis* (20). The common names of the family (snapdragons, foxglove or figwort family) are because of its genera *Antirrhinum* (snapdragons), *Digitalis* (foxglove) and *Scrophularia* (figwort).

Selected Indian Genera

Antirrhinum, Digitalis, Linaria, Lindenbergia, Mazus, Verbascum, Veronica.

Economic Importance

1. Ornamental plants: Include *Angelonia angustifolia*, *Antirrhinum majus* (snapdragon), *Calceolaria scabiosaefolia* (slipperwort), *Castilleja* (Indian paintbrush), *Chelone glabra* (turtlehead), *Collinsia grandiflora*, *Digitalis purpurea* (foxglove), *Linaria vulgaris* (toadflax), *Mimulus luteus* (monkey flower), *Nemesia versicolor* (Nemesia), *Paulownia tomentosa*, *Pedicularis canadensis* (housewort), *Russelia juncea* (coral blow), *Torenia asiatica* and *Veronica bungifolia* (speedwell).
2. Medicinal plants: Include *Bacopa monnieri* (used in epilepsy and insanity), (ii) *Digitalis purpurea* (used in congested heart failure), (iii) *Herpestis cuneifolia* (used in rheumatism), (iv) *Picrorhiza kurroa* (used as a laxative and cathartic), (v) *Scoparia dulcis* (used in fever, cough and bronchitis), and (vi) *Verbascum sinensis* (used in dysentery).

Description of the Family

1. General habit: Mostly herbs or undershrubs and only a few are big shrubs or trees (*Paulownia*); several are climbers (*Maurandia*), and some are saprophytic or parasitic (*Hyobanche*); *Lathrea* is a chlorophyll-less root parasite while *Euphrasia* and *Pedicularis* are semi-parasitic.

2. Leaf: Simple or rarely pinnately compound, extipulate; alternate (*Verbascum*), opposite (*Mimulus*) or whorled (*Russelia*); Ambulia shows heterophylly; both stem and leaves are generally hairy.
3. Inflorescence: Variable from racemose (*Mazus*) or cymose (*Verbascum*), or flowers solitary axillary (*Linaria*, *Lindenbergia*), spike (*Digitalis*), panicle (*Scrophularia himalayensis*).
4. Flower: Bracteate, bracteolate, bisexual, zygomorphic but sometimes nearly actinomorphic (*Verbascum*); upper leaves and bracts are sometimes brightly coloured (*Castilleja*); hypogynous; usually with an annual disc.
5. Calyx: Usually 5 united sepals, imbricate, quincuncial or valvate; sepals are 4 in *Veronica* and *Scoparia*; sometimes 2 anterior sepals are united (*Calceolaria*).
6. Corolla: Usually 5 united petals, often bilipped (2/3); sometimes only 4 petals are present, and the one placed anteriorly is smaller than the others (Veronica, Fig. 50.1A), frequently the anterior is spurred (*Linaria*) or with a gibbous (*Antirrhinum*, Fig. 50.1B) to saccate base; in *Verbascum* corolla is regular with all the 5 equal petals.
7. Androecium: 4 stamens (*Mazus*), sometimes only 2 stamens (*Veronica*); rarely all the 5 stamens are fertile (*Verbascum*); didynamous, epipetalous; sometimes the posterior stamen is represented by a staminode (*Antirrhinum*, *Scrophularia*, *Penstemon*); fertile stamens are alternate with corolla lobes; anthers dithecous, occasionally connivent, dehiscence mostly longitudinal and rarely poricidal.
8. Gynoecium: Bicarpellary, syncarpous, superior, bilocular, numerous anatropous or amphitropous ovules in each locule, axile placentation; carpels usually anterior-posteriorly placed; style simple, stigma bilobed; nectar-secreting disc present below this ovary.
9. Fruit and seeds: Fruit usually a capsule, rarely a berry (*Leucocarpus*) or drupe surrounded by persistent calyx. Seeds numerous, small, endospermic with straight or slightly curved embryo.
10. Pollination and dispersal: Flowers usually adapted to insect-pollination. Insects are attracted because of coloured corolla and nectar-secreting hypogynous disc. Flowers which have long-tubed corolla (e.g. *Scrophularia*) are pollinated by wasps while those having short-tubed cocolla (e.g. *Veronica*, *Verbascum*) are pollinated by bees. Dispersal of seeds usually take place by water, birds or animals.
11. General floral formula: Br, Brl, ·|·, ⊠, $K_{(5)}$, $C_{(5)}$, $A_{4 \text{ or } 2}$, $G_{\underline{(2)}}$.

Description of a Common Plant

A) *Mazus japonicus* (Thunb) Kuntze [syn. *M. rugosus* Lour.]

Habit: Small annual herb. Leaf: Radical, simple, exstipulate, sessile but lamina narrowed basally into the petiole, spathulate, dentate. Inflorescence: Racemose raceme raised on a scape. Flower: Braceate, pedicellate, complete, hermaphrodite, zygomorphic, bilipped, pentamerous, hypogynous. Calyx: 5 sepals, poly-or gamosepalous, valvate or quincuncial. Corolla: 5 petals, united, bilabiate (2/3) personate, posterior lip bilobed while anterior lip trilobed. Androecium: 4 stamens, polyandrous, epipetalous, didynamous, filaments curved apically; anthers dithecous, basifixed, introrse. Gynoecium: Same as in family description. Fruit: Capsule. Floral formula: Br, Ebrl, ·|·, ⊠, K_5, $C_{(2/3)}$, A_{2+2}, $G_{\underline{(2)}}$.

51

Solanaceae (Nightshade Family)

A family of about 90 genera and 3,000 species distributed in both tropical and temperate regions. Central and South America are the chief centres of distribution where over 40 genera are found. About 15 genera and over 90 species have been reported from India. Some of the larger genera along with their approximately world over reported species (Hickey and King, 1988) are *Solanum* (1700), *Cestrum* (150), *Physalis* (100), *Lycium* (80-90), *Nicotiana* (66), *Capsicum* (50), *Petunia* (40), *Hyoscyamus* (20) and *Datura* (10). *Solanum tuberosum* (potato) and *Nicotiana tabacum* (tobacco) are the two most utilized plants of Solanaceae.

Petunias, popular bedding plants, with variety of forms, sizes and shapes are employed for landscape planning, garden designs and container gardening. The numerous phenotypes *viz*., Grandifloras, Multifloras, Millifloras, Waves, Hedgifloras, etc. are complex hybrids. The outstanding marketable series of *P*. x *hybrida* have been developed for upright, spreading and semi-trailing to trailing habits.

Selected Indian Genera

Atropa, Capsicum, Cestrum, Datura, Nicotiana, Petunia, Physalis, Solanum, Withania.

Economic Importance

1) Ornamental plants: Common plants of ornamental value of Solanaceae are *Brunfelsia calycina* (fragrant flowers), *Cestrum diurnum* (Day jasmine), *C. nocturnum* (Night jasmine), *Hyoscyamus niger* (Henbane), *Nicotiana alata* (white-flowered), *Petunia hybrida* (Petunia pink flowered), *Salpiglossis sinuata*, *Schizanthus pinnatus*, *Solanum dulcamara* (violet coloured), *S. jaminoides* (white and blue coloured), etc.
2) Food plants: (i) Tubers of *S. tuberosum* (potato) are used throughout the world as a common vegetable and also used for the production of starch, dextrin, several alcohols and some other industrial products, (ii) Fruits of *Solanum lycopersicum* are the tomato, used as a delicious vegetable

and eaten raw, (iii) Fruits of *Solanum melongena*, (egg plant) are eaten as vegetable, (iv) *Physalis peruviana* (raspberry) yield the delicious edible fruits.

3) Medicinal plants: (i) *Atropa belladona* (Belladona) roots yield a powerful alkaloid 'atropine' used in belladona plasters, tinctures, etc. for relieving pain, and also for dilating pupils of eyes for eye-testing, (ii) *Datura stramonium* leaves and flowers are the source of the drug 'stramonium', used in asthama, and its seeds are deadly poisonous, (iii) *Hyoscyamus niger* leaves yield the alkaloid hyoscyamine used in treating asthama and whooping cough, (iv) *Mandragora autumnalis* roots are used as sedative and hypnotic, (v) *Nicotiana tabacum* leaves are used as sedative, anti-spasmodic and also in iradicating animal pests, such as lices, (vi) *Solanum surattense* roots are used in asthma and leaves in rheumatism, (vii) *Withania coagulans* fruits are used in curing asthma and liver troubles, (viii) *Withania somnifera* roots are used in curing cough and rheumatism.

Description of the Family

1. General habit: Mostly annual (*Physalis minima*), biennial or perennial herbs (*Withania somnifera*), sometimes shrubs (*Brunfelsia*) or tree (*Solanum giganteum*, *S. verbascifolium*); rarely vines (*Lycium sinensis*) or lianas.
2. Stem: Herbaceous or woody; erect or twining, or creeping; sometimes modified into tubers (*Solanum tuberosum*); often with bicollateral vascular bundles.
3. Leaves: Usually simple, alternate, exstipulate; entire; sometimes variously dissected or lobed; opposite or sub-opposite in upper part of the plant; pinnately compound in tomato and potato; unicostate reticulate.
4. Inflorescence: Generally cymose, or sometimes solitary flowers; sometimes extra-axillary scorpioid cyme called 'rhipidium' as in *Solanum nigrum*; or umbellate cyme (*W. somnifera*).
5. Flower: Bracteate (*Petunia*) or ebracteate (*Withania*), pedicellate, bisexual; actinomorphic or weakly zygomorphic due to oblique position of the ovary; penamerous, hypogynous.
6. Calyx: 5 sepals, gamosepalous, valvate, persistent; rarely the sepals are 4 or 6 often enlarging to envelope the fruit (*Physalis*, *Withania*).
7. Corolla: 5 petals, gamopetalous, rotate to tubular (*Solanum*, *Cestrum*) or bell shaped (*Atropa*) or infundibuliform (*Petunia*); usually alternate

with sepals; rarely bilipped and zygomorphic (*Schizanthus*); usually valvate, sometimes convolute (*Datura*).

8. Androecium: 5 stamens, epipetalous, alternipetalous; filaments usually of unequal length; stamens are only 2 in *Schizanthus* and 4 and didynamous on *Salpiglosis*; anthers dithecous; usually connivent i.e. forming a cone by coming close together but not actually fusing (e.g. *Solanum*); anthers opening either lengthwise or by terminal pores.
9. Gynoecium: Bicarpellary, syncarpous, carpels obliquely placed in the flower; ovary superior, bilocular, numerous anatropous or slightly amphitropous ovules in each locule; axile placentation; often the placentae are swollen; sometimes carpels divided by a false septum; style simple, with entire or 2-lobed stigma; sometimes the carpels are more than 2 (*Capsicum*) and also the number of locules is 3 to 5 (*Nicandra*); a hypogynous disc usually present.
10. Fruit and seeds: Fruit usually a berry (*Physalis*, *Withania*) or septicidal capsule (*Datura*). Seeds with copious endosperm and straight or curved embryo.
11. Pollination and dispersal: It is entomophilous. Bright and showy petals, presence of nectar secreted by nectariferous disc and strongly scented flowers of several members of this family are the chief attractions for insects. Flowers are protogynous and both self pollination and cross pollination occur. Only self pollination occurs in cleistogamous flowers of *Salpiglossis*. Dispersal of seeds takes place mainly by birds and animals and sometimes by water (*Atropa*).
12. General floral formula: Br, or Ebr, ⊕, ♀, $K_{(5)}$, $C_{(5)}$, $A_{(5)}$, $G_{\underline{(2)}}$.

Description of a Common Plant

A) *Solanum nigrum* (Black nightshade)

Habit: A small annual herb. Leaf: Simple, alternate but opposite in the floral region; petiolate, exstipulate, ovate, entire or slightly lobed, acute, unicostate reticulate. Inflorescence: Extra-axillary scorpioid cyme called 'rhipidium'. Flower: Ebracteate, pedicellate, hermaphrodite, actinomorphic, pentamerous, hypogynous, white. Calyx: 5 sepals, fused, valvate, persistent. Corolla: 5 petals, gamopetalous, valvate. Androecium: 5 stamens, free, epipetalous, filaments short; anthers conniving and forming an envelope around the style, dithecous, basifixed with apical pores. Gynoecium: Bicarpellary, syncarpous, superior, bilocular, may ovules in each locule, axile placentation, septum oblique and placentae are highly swollen; style long and hairy at the base; stigma bifid. Fruit: Berry. Floral formula: Ebr, ⊕, ⊠, $K_{(5)}$, $C_{(5)}$, $A_{(5)}$, $G_{\underline{(2)}}$.

Taxonomy of Petunia

The early literature on taxonomy of petunia is quite confusing. The first *Petunia* species was described in 1793 as *Nicotiana axillaris*. In 1803, the genus *Petunia* was described in Argentina as taxon to *Petunia nyctaginiflora*. Since, *Nicotiana axillaris* and *P. nyctaginiflora* are very closely related *P. nyctaginiflora* was transferred to *Nicotiana* as *N. nyctaginiflora*. By 1825, the taxon was commonly known as *P. nyctaginiflora* Juss. In 1888, the name of this taxon was changed to *P. axillaris*. *Petunia axillaris* has didynamous stamen and short wide corolla tube. In 1930, it was described as *P. parodii*. In 1979, the status of this taxon was changed to a sub sp, *P. axillaris* sub sp. *parodii* (Steere) Cabrera. Wijands *et al.* (1986) proposed that *P. nyctaginiflora* be conserved as the lecotype.

The first monograph on petunia was published in which two sub-genera *Pseudonicotiana* (corolla is hypocrateriform and filaments are attached to the middle of the tube) and *Eupetunia* (corolla is infundibuliform and filaments are attached the below middle of the tube) were recognized by Fries (1911). He placed *P. axillaris* in *Pseudonicotiana* and *P. integrifolia* in *Eupetunia*. He also separated *P. integrifolia* into three distinct species as *P. violacea*, *P. inflata* Fries and *P. occidentalis* Fries based on comparison of morphological characters. According to him, one of the main features of distinguishing these two taxa is the pedicel condition in the fruiting stage. In *P. integrifolia*, pedicles are deflexed whereas in *P. inflata* they are inflexed. *Petunia violacea* has largest flowers and pendent pedicels while *P. occidentalis* has smallest flowers and erect pedicels. The flower size and position of the pedicel of *P. inflata* is intermediate. In addition, Fries recognized diminutive subsp. of *P. violacea* (*P. violacea* sub sp. *depauperata* Fries) with very small flowers and leaves.

Smith and Downs (1966) recognized *P. integrifolia* (*P. violacea*), *P. inflata* (*P. integrifolia* var. *integrifolia*) and *P. occidentalis* as the same taxon (*P. integrifolia* var. *depauperata* (Fries) Smith and Downs. In earlier days they recognized four *P. integrifolia* taxa as distinct species *viz*., *P. reitzii, P. saxicola, P. littoralis* L.B. Smith and Downs and *P. scheideana*.

An extensive divergence in petunia, floral pollination syndromes indicate that insect visitation has certainly huge impact on the evolutionary history of *Petunia* species. Wijsman and De Jong (1985) separated *Petunia* into two genera based upon chromosome number, flower morphology and breeding behaviour. The type species (*P. parviflora* Juss) and all those species with chromosome number 2n=18 remained in *Petunia* and species with 2n=14 in *Stimoryne* Rafin. In 1836, *P. integrifolia* was described as *Stimoryne purpurea*.

Dichotomous key for identification of Petunia (Griesbach, 2007)

1. Pedicel deflexed in fruiting stage
2. Stigma above anthers of longest stamen
3. Corolla tube long and narrow; main stem erect, sub-glabrous; leaves oblong- lanceolate, hirsute, base petiolate, apex acute-

   P. mantiqueirensis

3. Corolla tube short and campanulate; main stem erect, hirsute; leaves oblong-elliptical, sub-glabrous, base attenuate, apex obtuse-

   P. bonjardinensis

2. Stigma at same level as anthers of longest stamen
3. Main stem erect, glabrous; leaves oblong-lanceolate, sub-glabrous, base petiolate, apex acute; corolla infundibuliform, tube paler than limb-

   P. scheideana

3. Main stem erect, hirsute; leaves oblong-lanceolate, sub-glabrous, base petiolate, apex acute; corolla infundibuliform, tube paler than limb

   - *P. guarapuavensis*

2. Stigma between anthers of longest stamen and shortest stamen
3. Main stem prostrate
4. Leaves prostrate, broad spatulate, sub-glabrous, apex rotundate, base attenuate; stem hirsute, branched at lower unflowered internodes; corolla narrow campanulate, tube paler than limb

   -*P. altiplana*

4. Leaves erect, linear to oblong, apex obtuse, base attenuate; stem unbranched at lower unflowered internodes
5. Stem and leaves glabrous; corolla tube broad infundibuliform, tube darker than limb-

   P. littoralis

5. Stem and leaves hirsute
6. Corolla limb ≈30 mm wide, tube broad infundibuliform, tube darker than limb-

   P. i. ssp. depauperata

6. Corolla limb 50 mm wide, tube narrow infundibuliform with cylindrical base, tube same intensity as limb-

P. riograndensis

3. Main stem erect
4. Corolla tube short, broad infundibuliform
5. Anther lobes with channel; stem hirsute; leaves oblong, sub-glabrous, apex acute, base attenuate-cuneate; tube paler than limb-

P. interior

5. Anther lobes without channel; stem hirsute; leaves oblong, hirsute, apex acute, base attenuate
6. Corolla tube not intruded, tube darker than limb, limb mouth round-

P. i. ssp. integrifolia

6. Corolla tube intruded, tube lighter than limb, limb, mouth reniform-

P. bajeensis

4. Corolla tube long, narrow infundibuliform with cylindrical base, same color intensity as limb
5. Leaves linear-lanceolate, hirsute, apex attenuate, base petiolate; stem hirsute; corolla limb purple-

P. reitzii

5. Leaves elliptical, sub-glabrous, apex sub-acute, base petiolate; stem sub-glabrous; corolla limb red-

P. saxicola

1. Pedicel not deflexed in fruiting state
2. Corolla hypocrateriform, tube cylindrical
3. Stamen and stigma above corolla tube; corolla limb red-

P. exserta

3. Stamen and stigma within corolla tube
4. Corolla limb white
5. Corolla tube ≈40 mm long, limb ≈25 mm wide-

P. a. ssp. axillaris

5. Corolla tube ≈60 mm long, limb ≈45 mm wide-

P. a. ssp. parodii

4. Corolla limb purple- *P. secreta*

2. Corolla infundibuliform; limb colored
3. Corolla tube base cylindrical, limb ≈20 mm wide- *P. occidentalis*
3. Limb ≈35 mm wide- *P. i. ssp. inflata*

References

Fries, R.E. 1911. *Kungl. Svenska Vetenskapsakademiens Handlingar* **46**: 1-72.

Griesbach, R. J. 2007. *Petunia x hybrida.* In: Flower Breeding and Genetics, Issues, Challenges and opportunities for 21st century (ed. Anderson N.O.). Springer press, pp. 301-336.

Hickey, M. and King, C. 1988. 100 families of flowering plants (2nd ed.). Cambridge University Press, Cambridge.

Smith, L.B. and Downs, R.M. 1966. *Herbario Barbosa Rodrigues*, pp. 261-291.

Wijands, D.O., Bos, J.J., Wijsman, H.J.W., Chneider, F.S., Brickell, C.D., and Zimmer, K. 1986. *Taxon*, **35**: 748-749.

Wijsman, H.J.W. and De Jong, J.H. 1985. *Acta Botanica Neerlandica*, **34**: 337-349.

52

Verbenaceae (Verbena Family)

A family of about 86 genera and 3,000 species distributed chiefly in tropics and sub-tropics. About 22 genera and over 30 species have been reported from India. Some of the genera along with their number of approximately reported species and common names of them are *Clerodendron* (400, glory-flower), *Verbena* (250, vervain), *Vitex* (250, chaste tree), *Lippia* (220, Frogfruit), *Lantana* (150), *Callicarpa* (140, beauty-berry), *Stachytarpheta* (100) and *Tectona* (3, teak).

Verbena is popular ornamental plant suitable for flower gardens, hanging baskets, and balcony boxes. They are excellent both as ground cover for rockeries, and as bedding types. These plants abundantly flower in the seeding year and, thus, are grown as annual plants.

Selected Indian Genera

Clerodendron, Duranta, Lantana, Tectona, Verbena, Vitex.

Economic Importance

1. Timber yielding plants: Teak, one of the most famous timber-yielding plants of the world, belongs to Verbenaceae. It is used for high quality furnitures, heavy constructions, ship building, musical instruments and several other similar purposes. Some other timber-yielding plants are *Gmelina arborea*, *Premna benghalensis* and *Vitex altissima*.
2. Garden ornamentals include *Aloysia iriphylla*, *Caryopteris incana*, *Clerodendron bungei*, *C. fragrans*, *C. interne*, *C. siphonanthus*, *C. splendens*, *C. thomsoniae*, *C. trichotomum*, *Duranta repens* syn. *D. plumieri* (golden dewdrop), *Holmskioldias anguinea*, *Lantana camara*, *Petrea volubilis* (Queen's wreath) and several species of *Verbena* (Vervain) such as *V. bipinnatifida* and *V. hybrida*.
3. Verbena oil, used in perfumery, is obtained from the leaves of *Aloysia triphylla*.

Description of Family

1. General habit: Herbs, shrubs (*Lantana*) or trees (*Tectona*); often lianas (*Clerodendron*, *Vitex*), sometimes of a xerophytic nature (several species of *Verbena*); a few are mangrove shrubs (*Avicennia*).
2. Stem: Herbaceous or woody, twining or sprawling in climbers; often quadrangular.
3. Leaf: Usually simple, sometimes pinnately (*Peronema*) or palmately (*Vitex*) compound; usually opposite, rarely alternate or whorled; exstipulate; highly reduced in xerophytic species (*Verbena*).
4. Inflorescence: Racemose raceme (*Duranta*) or umbel (*Lantana*) or a spike, often with an involucre of coloured bracts; or consisting of dichotomous cymes, or panicled cymes (*Tectona*).
5. Flower: Bracteate, often bracteolate (*Duranta*, *Lantana*), complete, hermaphrodite, zygomorphic (rarely actinomorphic, e.g. *Physopsis*), pentamerous, hypogynous.
6. Calyx: Usually 5 united sepals, persistent; usually lobed or toothed; valvate; rarely the sepals are 4, 6 or 8; sometimes coloured.
7. Corolla: Usually 5 united petals, or as many petals as sepals, often bilipped with a narrow tube; rarely campanulate, typically lobed with the lobes generally unequal, usually imbricate.
8. Androecium: Usually 4 stamens, didynamous, epipetalous, alternate with the corolla lobes; anthers dithecous, introrse, longitudinally dehiscent; posterior stamen is generally missing or present in the form of staminode; only 2 fertile stamens and 3 staminodes are present in *Oxera*; in *Tectona* all the 5 stamens are fertile.
9. Gynoecium: Usually bicarpellary, syncarpous; ovary superior; originally bilocular but divided into 4 locules by the formation of false septum in each locule; 2 ovules in each carpel i.e. 1 ovule in each locule after septation; sometimes 2 ovules in each locule (Fig. 52.1A); axile placentation; ovules anatropous to orthotropous; style terminal and only rarely sunk in between the lobes of the ovary; stigma lobes usually as many as carpels; rarely carpels are 4 (Fig. 52.3 *Duranta*) or 5 (*Genunsia*); ovary is unilocular with free-central placentation in *Avicennia*; in *Clerodendron* and *Verbena* the ovary is unilocular with parietal placentation when young but on maturity the placentae unite at the centre and the placentation becomes axile.

10. Fruit and seeds: Fruit is generally a drupe, and rarely a capsule, berry, or schizocarp (nutlets). Seeds non-endospermic with straight embryo.
11. Pollination and dispersal: Pollination is mainly by butterflies and bees, and seeds are dispersed mainly by birds or animals.
12. General floral formula: Br, Brl, ·|·, ♀, $K_{(5)}$, $C_{(5)}$, A_{2+2}, $G_{\underline{(2)}}$.

Description of a Common Plant

A) *Lantana camera* Linn.

Habit: A small shrub. Leaf: Simple, exstipulate, opposite decussate, petiolate, ovate, crenate or serrate, acute, unicostate reticulate. Inflorescence: Umbel or a compound spike with flat top. Flower: Bracteate, bracteolate, hermaphrodite, zygomorphic, pentamerous, hypogynous. Calyx: 5 sepals, untied , valvate. Corolla: 5 petals, united, quincuncial, corolla bilipped (4/1) in which 1 anterior petal is large and pointed; white, yellow, red or pale-purple. Androecium: 4 stamens, didynamous, epipetalous, posterior stamen is generally missing; dithecous, basifixed, introrse. Gynoecium: Bicarpellary, syncarpous, superior, bilocular, 1 ovule in each locule; axile placentation; style simple, stigma knob like. Fruit: Drupe. Floral formula: Br, Brl, ·|·, ♀, $K_{(5)}$, $C_{(4/1)}$, A_{2+2}, $G_{\underline{(2)}}$.

B) Taxonomy of *Verbena* spp.

The genus can be divided into diploid North American and a polyploid South American lineage, both with a base chromosome number n=7. The European species is derived from the North American lineage. It seems that verbena as well as mock vervains (*Glandularia*) evolved from the assemblage provisionally treated under the genus name *Junellia*; both other genera were included in the Verbenaceae until the 1990s. Intergeneric chloroplast gene transfer by an undetermined mechanism has occurred at least twice from the *Vervains* to *Glandularia*, between the ancestors of the present day South American lineage and between *V. orcuttiana* or *V. hastata* and *G. bipinnatifida*. In addition, several species of *Verbena* are of natural hybrid origin; the well known garden vervain has entirely muddy history. The relationships of this close knit group are hard to dissolve the standard methods of computational phylogenetics.

The genus *Verbena* is divided in to two sections *viz.*, *Glandularia* and *Verbenacea* (Table 52.1). Nobiles is one of the five groups of section *Verbenacea*. Species belonging to both the sections differ morphologically, anatomically and cytologically and are cross incompatible. Species belonging to section *Glandularia*, have same chromosome number and dissimilarity in their karyotypes speaks of genomic differentiation during speciation.

Table 52.1: Section-wise characters of genus *Verbena* (Arora, 1972)

Section	Characters	Species
Glandularia	X=5, large size & same chromosome number, but dissimilar in karyotype indicates genomic differentiation during speciation.	*V. hybrida* Voss., *V. tenuisecta* Briq., *V. peruviana* (L.) Britton., *V. platensis* Spreng, *V. incisa* Hook, *V. phlogiflora* Cham. and *V. aubletia* Jacq.
Verbenacea	X=7, small chromosome size and symmetrical in karyotype.	*V. hastata* L., *V. urticifolia* L.

Thirteen *Verbena* species were analyzed from taxonomic point of view including ornamental *V. bonariensis* (containing griselinoside), *V. peruviana* and *V. canadensis* (this too contain no iroid glucosides) *Verbena hybrida* and *V. bonariensis* show only sexual reproduction and mostly out crossing (Damtoft *et al.*, 1989). Bailey (1901) considered that garden verbena has been derived from crosses between the following four species.

1. *V. peruviana* (L.) Britton. (syn. *V. chamaedrifolia* Juss.)
2. *V. phlogiflora* Cham. (syn. *V. tweediana* Niven.)
3. *V. incisa* Hook.
4. *V. platensis* Spreng. (syn. *V. teucrioides* Gill & Hook)

Of the four species, *V. incisa* is probably a variety of *V. phlogiflora*, but others are regarded as distinct. The modern cultivated forms have characters from all wild types e.g. the colour of flowers may be bright scarlet, scarlet-magenta or very pale; the inflorescence may be extremely flat corymb, very long spike or intermediate; the eye varies in size and may or may not have a coloured inner rim.

Bailey (1902) considers that variations of the modern garden strains appeared to have arisen as mutations after the original crosses were made e.g. 1. Pure white flowers 2. Yellow flowers (said to have appeared in 1896) 3. Deep bluish purple flowers (first appeared in 1854), 4. Compact habit (arose in 1870) 5. Striped or flaked flowers (known already in 1849).

Yuan and Olmstead (2008) considered that the phylogenetic framework of *Verbena* complex results from *Junellia* (a paraphyletic and most ancestral genera), both *Glandularia* (polyploid) and *Verbena* (diploid) (monophyletic and have been derived from within *Junellia*). The later is more ancient from South American verbena group to the common ancestor of major *Glandularia* linkage and commonly assumed introgressive hybridization. The *Verbena* complex include *Verbena* (x=7), *Glandularia* (x=5) and *Junellia* (x=10) with each genus containing 40-50 species (Sanders, 2001). The relationships among these genera are complex are they were often combined in the genus *Verbena*.

References

Arora, O.P. 1972. Cytogenetical studies of ornamental verbenas. *Ph. D.* thesis. Kanpur University, Kanpur.

Bailey, N. L. 1901. Cyclopedia American of Horticulture, New York.

Bailey, L.R. 1902. Cyclopedia of American of Horticulture, New York

Damtoft, S., Jensen, R. and Nielsen, S.J. 1980. *Taxon*, **28**(5/6): 525-528.

Sanders, R.W.2001. The genera of Verbenaceae in the south-eastern United States. Harvard Pap. Bot. **5**: 303-358.

Yuan, Y.W. and Olmstead, R.G. 2008. *Molecular Phylogenetics and Evolution*, **48**: 23-33.

53

Violaceae (Violet Family)

A family of about 23 genera and 800 species, Violaceae are distributed in both tropical and temperate regions. Only three genera (*Hybanthus*, *Vahila* and *Viola*) have been reported from India. Famous pansies, violas and violets belong to this family. Some of the larger genera with their approximate number of species in parenthesis include *Viola* (500), *Rinorea* (340) and *Hybanthus* (150). Majority of Indian Violaceae are found in Himalayas and hills of southern and western India.

Garden pansy is an annual or perennial herb planted in flower beds worldwide owing to its pretty shape and contrasting colour markings, looking like a butterfly. It is called as 'King of annuals'. In horticultural industry, there are variety of forms, sizes, and shapes of pansies which are employed for landscape planting, garden design, and container gardening. With the advent of F_1 hybrids, pansies are known as high quality bedding plants. Dried flowers of *Viola odorata* Vern. are used against cough, influenza, and lung troubles while oil obtained from flowers is used in perfumes and flavourings.

Selected Indian Genera and Species

- Except for garden favourites, such as pansies, violas and violets, the family is of little economic importance. Over 100 species of *Viola*, are being grown as ornamental plants.

Genera	Importance
Viola	*V. tricolor* and *V. odorata* are cultivated; dried flowers of *V. odorata* are used against cough, influenza and lung troubles; oil obtained from flowers of *V. odorata* are used in perfumes and flavourings; its leaves yield an essential oil used in scenting soaps
Hybanthus	Roots of *H. enneaspermus* are used against urinary tract infections and gonorrhea, and also to cure bowel troubles of children

Description of Family

1. General habit: Usually perennial or annual herbs (*Viola*, *Hybanthus*) or shrubs, rarely trees or woody climbers (*Agatea*).
2. Root: Usually tap roots.

3. Stem: Usually short, solid, erect, usually unbranched.
4. Leaves: Alternate or rarely opposite, simple; stipulate, stipules often large and leafy (some *Viola* species) and small.
5. Inflorescence: Flowers solitary or arranged in racemose racemes or panicles.
6. Flower: Bracteate, invariably bracteolate, bisexual, zygomorphic (*V. odorata* Fig. 53.1) to actinomorphic (*Rinorea*); sometimes cleistogamous and hidden underground; pentamerous, hypogynous.
7. Calyx: 5 sepals, persistent, polysepalous or slightly connate, uniformly sized, imbricate or quincuncial.
8. Corolla: 5 petals, mostly unequal, polypetalous, anterior petal larger and often spurred to hold the nectar; imbricate; equal sized petals in *Rinorea* and other arborescent genera.
9. Androecium: 5 stamens, usually alternate with petals, free or connate, often forming a cylinder round the ovary; abaxial stamens often spurred at the base; filaments very short, connective usually prolonged at the apex, dithecous, introrse, 2 anterior stamens are spurred in *Viola* (Fig. 53.1).
10. Gynoecium: Tricarpellary, syncarpous, superior, unilocular, ovules numerous, parietal placentation; style simple, stigma capitates, truncate or lobed.
11. Fruits and seeds: Fruit a loculicidal capsule or rarely a berry. Seeds endospermic; smooth, winged or with tomentum; with straight embryo; elaiosome is present in some species of *Viola*.
12. Pollination and dispersal: Pollination entomophilous in most of the Violaceae. However, cleistogamous flower (*V. bicolor*) are self pollinated. Dispersal of seeds is usually by animals, birds, or by water.
13. General floral formula: Br, Brl, $\oplus$ or $\cdot|\cdot$, $\boxtimes$, K_5, C_3, A_5, $G(\underline{3})$.

Description of Common Plant

A) Taxonomy of *Viola tricolor*:

Numerous taxonomic studies on species complexes have been published over 20th century based largely on chromosome numbers (Clausen, 1927; 1929) and morphology (Backer, 1953). Recent molecular phylogenetic analysis by Ballard (1996) and Ballard *et al.* (1999) has clarified the composition and relationships of main groups. Chromosome studies of the genus growing in Europe and North America are helpful in working out its taxonomy and distributional ecology. The geographical distribution of the spices coincides with that of section itself. Clausen (1967) suggested that in genus *Viola*, six chromosome series was prevalent in section *Chamaemelanium*, 12 in sect. *Plagiostigma*, 10 in sect. *Rostellatae* and variable numbers are seen in sect. *Melanium* though 6 and 10 are more prevalent.

Section *Melanium* includes so called pansies, is a morphologically well defined group of 80 to 100 species. It is composed of three groups: *Elongatae, Delphiniopsis* and *Xylinosium*. Most spp. are herbaceous, acaulescent with frontly flattened flowers, yellow throat, big divided leaf like stipules as well as entire stipules and a well developed spur of variable length among species. The side petals are upturned (downturn in other sections) and the bottom petal is enlarged, serve as landing platform for insect pollinators.

The woody stems, linear leaves and long spur flowers have been traditionally interpreted as primitive characters by taxonomists (Beattie, 1974). Woody sections *Xylinosium* and *Delphiniopsis* might have arisen from within *Melanium* (Ballard, 1996). Both these sections have frontally flattened flowers with yellow corolla throat and share cytological features of *Melanium* section.

The floral structures are essential for diagnoses within the genus *Viola*. Shape of pistil especially of style and stigma prove crucial in very early classification. Violets of sect. *Melanium* have characteristic stigma morphology. The stigma is cup shaped with an opening (hole) on the top and typical lip below the hole. On its outer surface stigma is covered with papillae and hairs differing in size, abundance and distribution depending on the species. Structure of violet flower is an evolutionary trait. Species of sect. *Melanium* develop exclusively chasmogamous flowers with no cleistogamy which is a characteristic feature for sect. *Viola*. Pollen heteromorphism is another diagnostic character for species within sect. *Melanium* and is independent of polyploidy in sect. *Melanium*, unlike in sect. *Viola*, having selective advantage.

The zinc violets (sect. *Melanium*) are classed at the rank of sub-species as *V. lutea* ssp. *calaminaria* for the yellow zinc violet and ssp. *westfalica* for its blue counterpart. Although the violets examined (*V. lutea*, *V. lutea* ssp. *calaminaria*, *V. lutea* ssp. *westfalica*) are closely related, but there is no evidence that *V. lutea* ssp. *westfalica* is a descendent of *V. tricolor* (Kuta *et al*., 2012). Further, they opined that both zinc violets are evolved from *V. lutea* by hybridization or mutations (structural chromosomes or gene mutations. Hildebrandt *et al.* (2006) opined that blue zinc violets are aroused by autopolyploidization from *V. tricolor*.

Mukherjee and Khoshoo (1969) traced origin and evolution of garden pansy chronologically from its ancestors, *V. tricolor* L. (n=13) and *V. lutea* Hudson (n=24). They stated that these two species hybridized each other when grown together. The resulted seedling progenies were variable in growth, shape, size and colour of flowers. Many of the cultivars of both these types were lost during war and after that mostly hybrid types were available. Loss of suppressors of flower colour and size was due to reshuffling of chromosomes, and as a result of weak genetic differentiations between these two parents. The chromosome number in garden pansy is stabilized at n=24 (8x), although the range in variation is from 2n=44 to 55. This could also happen through repeated backcrossings to *V. lutea*. Garden pansy (*V.* x *wittrockiana*) is therefore, a modified *V. lutea.* The plants with 2n=48 have superior selective value because of vigour and larger flowers with greater colour diversity. The nature of ploidy in the garden pansy has been reported to be segmental alloploidy. Hybridization is common throughout the genus *Viola* and leads to fertile derivatives in spite of ploidy differences. Many species from sect. *Melanium* are believed to be originated from hybridization events. Hybridization combined with allopolyploidy and aneuploidy account for high diversity in chromosome number.

Floral Biology of *Viola tricolor*

In different species flowering starts in January-March and ends in May. Peak flowering is in April. At the start of anthesis, apices of posterior petals separate from those of laterals by 0.5 to 1 mm, so that interior of flower is clearly visible. The anthesis has two distinct phases. The 1st is characterized by retention of pollen within the androecium (which lasts for 1-6 days) and 2nd is characterized by gradual displacement of pollen from anthers to the petals (which lasts from 4-6 days till end of anthesis). Pollination during the first phase is effected only by few insect species causing inbreeding and second phase by wide variety of insects causing outbreeding (Beattie, 1969). Valentine (1962) reported that all the above mentioned three species are self compatible, but insect exclusion experiments indicated that self pollination is uncommon in *V. riviniana* Rchb. Floral biology and pollination mechanisms in two *Viola* species *viz.*, *V. cerasifolia* and *V. subdimidiata* was studied by Freitas and Sazima (2003).

In *Viola*, cv. Swiss Giants Extra Finest Mixed flowers open mainly in the morning, a few flowers open at 13 hours. The peak period of anthesis is between 8-10 AM. The cloudy weather delays the anthesis. Red spot develops 24 hours before dehiscence on the posterior anther at distal end and the suture opens up zipper like up to the base. Anther dehiscence occurs first in the posterior anther followed by lateral and anterior anthers. Pollen grains after liberation get deposited into hairy groove of the lower petal and are retained there for 7-8 days. Generally dehiscence of posterior anther (in 47% flowers) occurs prior to anthesis day. On the day of anthesis, posterior anther dehisces completely and the dehiscence in rest of the anthers is completed within 72 hours after anthesis. Protandrous type dichogamy is observed in pansy. Stigma remains receptive for 8 days after anthesis with a peak on 2nd day of anthesis. The period of effective receptivity remains up to four days after opening. After 6th day, the receptivity drops quickly. Stigma receptivity could be observed from thick mucilage exudation from the style region. The style pivots upon a distinct area of flexure near the base and returns to resting position after deflection by visiting insects. Deflection may be sideways or upwards. Pollen is most fertile (97.58%) immediately after anther dehiscence and remains viable up to 8 days after start of anthesis. It also remains viable even shade from the stamens. Pollen grains germinate only in 5 per cent sucrose solution. Pansy pollen can be stored up to 90 days under controlled conditions (4°C) without appreciable loss of viability (John *et al.*, 1988).

References

Baker, M.S. 1953. *Madrono*, **12**: 8-18,

Ballard, H. E. 1996. Phylogenetic relationships and infrageneric groups in *Viola* (Violaceae) based on morphology, chromosome numbers, natural hybridization and Internal Transcribed Spacer (ITS) sequences. Ph.D. Thesis, University of Wisconsin, Maddison.

Ballard, H. E., Sytsma, K. J. and Kowal, R. R. 1999. *Systematic Botany*, **23**(4): 439-458.

Beattie, A. J. 1969. *New Phytologist*, **68**(4): 1187-1201.

Beattie, A.J. 1974. *Ann. Mo. Bot. Gard.*, **61**: 781-793.

Clausen, J. 1927. *Annals of Botany*, **41**(4): 677-714.

Clausen, J. 1929. *Annals of Botany*, **41**(4): 741-764.

Clausen, J. 1967. *Taxon*, **16**: 271-279.

Freitas, L. and Sazima, M. 2003. *Annals of Botany*, **91**(3): 311-317.

Hildebrandt, U., Hoef-Emden, K., Backhausen, S., Bothe, H., Bozek, M., Siuta, A. and Kuta, E. 2006. *Plant Systematics and Evolution*, **257**: 205-222.

Jhon, A.Q., Saini, S.S. and Sharma, P.P. 1988. *Haryana Journal of Horticulture Science*, **17**(1-2): 1-8.

Kuta, E., Bohdanowicz, J., Slomka, A., Pilarska, A. and Bothe, H. 2012. *Plant Systematics and Evolution*, **298**(2): 445-455.

Mukherjee, I. and Khoshoo, T.N. 1969. *Nucleus*, **12**: 178-186.

Valentine, D.H. 1962. *Preslia*, **34**: 190.

54

Zingiberaceae (Ginger Family)

A family of 56 genera and over 1,300 species, Zingiberaceae are distributed in tropical regions, chiefly in Indomalaysia. About 17 genera and over 115 species have been reported from India, chiefly in Western Ghats and Eastern Himalayas. Some largely represented genera along with number of their world over reported species are *Alpinia* (250), *Amomum* (150), *Costus* (150), *Zingiber* (90), *Kaempferia* (70), *Curcuma* (60), *Hedychium* (50), *Globba* (50) and *Elettaria* (7).

Selected Indian Genera

Alpinia, *Amomum*, *Costus*, *Curcuma*, *Elettaria*, *Hedychium*, *Zingiber*.

Economic Importance

1. Spices and condiments: Seeds of *Amomum cardamom* (*Badi elayachi*) and *Elettaria cardamomum* (true cardamom), roots of *Alpinia officinarum* (gangal root) and rhizomes of *A. gangala* (Siamese ginger), *Curcuma longa* (turmeric) and *Zingiber officinalis* (ginger) are used as spices, condiments and flavouring agents. Ginger oil, obtained from rhizomes of *Z. officinalis*, is used in perfumery and for medicinal purposes. Turmeric is also used for medicinal purposes.
2. *Abir* Abir, the famous scented powder, is prepared from *Hedychium spicatum* (*kafur-kachri*).
3. *Zedoary*: The product 'Zedoary', used as a tonic and in perfumery, is obtained from the tubers of *Curcuma zedoaria*.
4. Cultivated ornamentals: Some of the ornamental genera cultivated in green houses and gardens for decorative purposes are *Alpinia* (shell ginger), *Brachychilum*, *Cardamom*, *Cautleya*, *Costus*, *Globba*, *Hedychium*, *Kaempferia*, and *Roscoea*.

Description of the Family

1. General habit: Perenial, often aromatic herbs, usually with sympodial tuberous (*Curcuma*) or horizontal (*Zingiber*, Fig. 54.3) rhizomes; rarely

with fibrous roots (*Cautleya*); aerial stem, if present, is short; sometimes an apparent stem is formed by the rolled-up leaf sheaths.

2. Leaf: Alternate, sessile or petiolate, distichous or spirally arranged, with sheathing leaf bases; ligulate, ligule present at the junction of blade with petiole or sheath; leaf blade large, linear to elliptical, with closely parallel-pinnate venation.
3. Inflorescence: Variable from a compact spike (*Alpinia*, *Zingiber*), raceme (*Curcuma*), panicle (*Globba*), or flowers solitary; cymose in some species of *Globba*; each flower or group of flowers subtended by a large bract; in several genera the inflorescence develops on leafless scapes.
4. Flower: Bracteate, pedicellate or sessile, bisexual, zygomorphic, only rarely actinomorphic, trimerous, epigynous, large, brightly coloured, usually aromatic or fragrant; perianth differentiated into calyx and corolla.
5. Calyx: 3 united sepals forming a tubular or spathiform, green or herbaceous calyx; the odd sepal is anterior; calyx tube is very short in *Globba* and splitted in *Roscoea.*
6. Corolla: 3 unequal-lobed petals, free or more or less united into a tubular corolla with the posterior lobe usually the largest; delicate and usually showy.
7. Androecium: Basically 6 stamens in 2 whorls of 3 each; 2 lateral stamens of the outer whorl modify into staminodes while the 3rd anterior stamen of this whorl is usually absent or highly suppressed; 1 median or posterior stamen of the inner whorl is fertile and 2 lateral stamens of this whorl unite to form a labellum; fertile stamen is 2-celled, usually epipetalous, with its filaments usually slender and deeply grooved.
8. Gynoecium: Tricarpellary, syncarpous; ovary inferior, usually trilocular with numerous anatrapous or semianatropous ovules on axile placentation, or unilocular with 3 parietal placentae (*Globba*); style usually filiform and more or less enveloped in the groove of filament of fertile stamen; stigma protruding beyond the anther lobes; usually epigynous nectar-secreting gland present.
9. Fruit and seeds: Fruit usually a 3-valved, loculicidal capsule, or rarely fleshy, indehiscent, berry-like. Seeds often arillate, and each with straight embryo and abundant endosperm.
10. General floral formula: Br, ·|·, ⚥, K_3, $C_{(3) \text{ or } 3}$, A_1, $\overline{G}_{(3)}$.